Heterochrony in Evolution

A Multidisciplinary Approach

TOPICS IN GEOBIOLOGY
Series Editors: F. G. Stehli, DOSECC, Inc., Gainesville, Florida
D. S. Jones, University of Florida, Gainesville, Florida

Volume 1 SKELETAL GROWTH OF AQUATIC ORGANISMS
Biological Records of Environmental Change
Edited by Donald C. Rhoads and Richard A. Lutz

Volume 2 ANIMAL–SEDIMENT RELATIONS
The Biogenic Alteration of Sediments
Edited by Peter L. McCall and Michael J. S. Tevesz

Volume 3 BIOTIC INTERACTIONS IN RECENT AND FOSSIL BENTHIC COMMUNITIES
Edited by Michael J. S. Tevesz and Peter L. McCall

Volume 4 THE GREAT AMERICAN BIOTIC INTERCHANGE
Edited by Francis G. Stehli and S. David Webb

Volume 5 MAGNETITE BIOMINERALIZATION AND MAGNETORECEPTION IN ORGANISMS
A New Biomagnetism
Edited by Joseph L. Kirschvink, Douglas S. Jones, and Bruce J. MacFadden

Volume 6 *NAUTILUS*
The Biology and Paleobiology of a Living Fossil
Edited by W. Bruce Saunders and Neil H. Landman

Volume 7 HETEROCHRONY IN EVOLUTION
A Multidisciplinary Approach
Edited by Michael L. McKinney

Heterochrony in Evolution

A Multidisciplinary Approach

Edited by
Michael L. McKinney
University of Tennessee
Knoxville, Tennessee

Plenum Press • New York and London

Library of Congress Cataloging in Publication Data

Heterochrony in evolution: a multidisciplinary approach / edited by Michael L. McKinney.
p. cm.—(Topics in geobiology; v. 7)
Bibliography: p.
Includes index.
ISBN 0-306-42947-0
1. Heterochrony (Biology). I. McKinney, Michael L. II. Series.
QH395.H47 1988
575—dc19 88-22573
CIP

A Division of Plenum Publishing Corporation
233 Spring Street, New York, N.Y. 10013

Printed in the United States of America

I dedicate this book to the memory of my father, Dewey Amos McKinney, with love and respect

—"The Perfessor"

Contributors

Victor Ambros Biological Laboratories, Harvard University, Cambridge, Massachusetts 02138

William L. Fink Museum of Zoology and Department of Biology, University of Michigan, Ann Arbor, Michigan 48019

David W. Foster Department of Geology, Museum of Invertebrate Paleontology, and Paleontological Institute, University of Kansas, Lawrence, Kansas 66045

Dana Geary Department of Geology and Geophysics, University of Wisconsin, Madison, Wisconsin 53706

Stephen Jay Gould Museum of Comparative Zoology, Harvard University, Cambridge, Massachusetts 02138

Edward O. Guerrant, Jr. Department of Botany and Plant Pathology, Oregon State University, Corvallis, Oregon 97331; and Department of Biology, Lewis and Clark College, Portland, Oregon 97219

John C. Hafner Moore Laboratory of Zoology and Department of Biology, Occidental College, Los Angeles, California 90041

Mark S. Hafner Museum of Natural Science and Department of Zoology and Physiology, Louisiana State University, Baton Rouge, Louisiana 70803

Douglas S. Jones Florida State Museum, University of Florida, Gainesville, Florida 32611

Roger L. Kaesler Department of Geology, Museum of Invertebrate Paleontology, and Paleontological Institute, University of Kansas, Lawrence, Kansas 66045

Neil H. Landman Department of Invertebrates, American Museum of Natural History, New York, New York 10024

David R. Lindberg Museum of Paleontology, University of California, Berkeley, California 94720

Michael L. McKinney Department of Geological Sciences, and Graduate Program in Ecology, University of Tennessee, Knoxville, Tennessee 37996-1410

Kenneth J. McNamara Western Australian Museum, Perth, Western Australia 6000, Australia

John M. Pandolfi Department of Geology, University of California, Davis, California 95616. *Present address:* Australian Institute of Marine Science, Townsville M.C., Queensland 4810, Australia

Brian T. Shea Departments of Anthropology and Cell Biology and Anatomy, Northwestern University, Chicago, Illinois 60611

Brian N. Tissot Department of Zoology, Oregon State University, Corvallis, Oregon 97331

Preface

> . . . an adult poet is simply an individual in a state of arrested development—in brief, a sort of moron. Just as all of us, *in utero*, pass through a stage in which we are tadpoles, . . . so all of us pass through a state, in our nonage, when we are poets. A youth of seventeen who is not a poet is simply a donkey: his development has been arrested even anterior to that of the tadpole. But a man of fifty who still writes poetry is either an unfortunate who has never developed, intellectually, beyond his teens, or a conscious buffoon who pretends to be something he isn't—something far younger and juicier than he actually is.
> —H. L. Mencken, *High and Ghostly Matters, Prejudices: Fourth Series* (1924)

> Where would evolution be,
> Without this thing, heterochrony?
> —M. L. McKinney (1987)

One of the joys of working in a renascent field is that it is actually possible to keep up with the literature. So it is with mixed emotions that we heterochronists (even larval forms like myself) view the recent "veritable explosion of interest in heterochrony" (in Gould's words in this volume). On the positive side, it is obviously necessary and desirable to extend and expand the inquiry; but one regrets that already we are beginning to talk past, lose track of, and even ignore each other as we carve out individual interests.

This book is an effort to offset some of these negative aspects. I have tried to bring together experts from a number of disparate fields covering anthropology, botany, developmental biology, paleozoology, and zoology. The authors were asked to contribute chapters that represent syntheses of their particular areas. However, my goal was not only cross-fertilization, but also education, to produce a book that is intelligible and stimulating to interested parties of all backgrounds, not a collection of specialized contributions. [A more general book on heterochrony (by McKinney and McNamara), also designed to that end, is currently being written and will also be published by Plenum.]

This book is organized into three main parts. Preceding the first part is a preview and synthesis by Gould of the chapters in the book. As one of the major analysts of heterochrony and evolutionary theory in general, he offers much insight (with characteristic flair). Reading this chapter first (and perhaps last) should provide an excellent perspective on the rest.

The first part contains five contributions summarizing the pragmatic issue of analyzing heterochrony. Mine is a general treatment relating allometry and size to heterochrony. The two that follow it tackle the description of often complex shape changes caused by heterochrony. Tissot focuses on the more traditional multivariate method while Foster and Kaesler present some newly developed, and sometimes controversial, approaches. In the two final chapters of the section, Fink and Jones discuss what are probably the two major obstructions to heterochronic analysis. Phylogenetic relationships (Fink) are crucial to identifying the precise heterochronic processes involved in evolutionary divergence but are often difficult to reconstruct (though some may balk at his focus on cladistic methods to resolve this). Neither can we identify the precise heterochronic process without a knowledge of ontogenetic age. Jones discusses ways of obtaining that data.

The second part of this book contains case studies documenting heterochronic changes in a number of major groups, selected to cover a wide range of lifestyles. The studies by Guerrant and Pandolfi are most welcome attempts to apply heterochronic views to plants and colonial marine animals (respectively). Nektonic (ammonites) and benthic (gastropods) invertebrates are covered by Landman, Geary, and Lindberg. Finally, rodents and primates represent the mammals in chapters by Hafner and Hafner and by Shea. Both of these are particularly enlightening regarding heterochronic implications for taxonomy and ecology. Shea's impressive summary includes a number of other important issues as well, such as the size–time problem and the need to revise our traditional thinking on heterochrony in human evolution.

The last part contains chapters summarizing knowledge and ideas on the causes and importance of heterochrony in evolution. Ambros discusses the important work that he (and others) have done on the genetic basis of heterochrony. Until this link between genes and morphology is mapped out by work such as this we are not unlike the chemists of the 19th century, classifying observable patterns, but not knowing the underlying atomic interactions. The black box enclosing morphogenesis will, it is hoped, soon be removed. McNamara's contribution presents the results of his extensive compilation of heterochrony from the fossil record (including many of his own studies, which are clearly some of the best ever done for their thoroughness and insight). It is about time that someone tried to collate and make sense of the heterochronic morass in the paleontological literature, for there is much valuable information there, but much misinformation, too. But whatever is presently obscure about the fine details of heterochrony, one thing is clear from his chapter: the process itself is rampant as a mode of phylogenetic change. The next step is to see what this prevalence does to alter our perspective on the evolutionary process. In the final chapter I have tried to analyze how our growing knowledge of heterochrony may affect the Darwinian (or neo-Darwinian) paradigm. At the very least it will supplement it with an increased understanding of the influence (often subtle) of development as an intermediary between genes and selection. However, there is a possibility that it could substantially contradict some of the orthodox views. Perhaps rapid evolution and internal directionality are important after all. My conclusions are probably vague, but I feel the data are also vague; we can only search for more and, importantly, continue to digest and synthesize what we have.

I (and Doug Jones, my co-convener) would like to thank the Paleontological Society for sanctioning a symposium on heterochrony at the 1986 Geological Society of America national meeting. The seeds of this book were planted there. I also thank the many reviewers; some are anonymous, but most are acknowledged in the individual chapters. Finally, I am most grateful for the work, cooperation, and enthusiasm of the contributors.

M. L. McKinney

Knoxville

Contents

Chapter 8 • Heterochrony in Colonial Marine Animals

John M. Pandolfi

Chapter 9 • Heterochrony in Ammonites

Neil H. Landman

Chapter 10 • Heterochrony in Gastropods: A Paleontological View

Dana Geary

Chapter 11 • Heterochrony in Gastropods: A Neontological View

David R. Lindberg

Chapter 12 • Heterochrony in Rodents

John C. Hafner and Mark S. Hafner

Chapter 13 • Heterochrony in Primates

Brian T. Shea

III. Cause, Abundance, and Implications of Heterochrony

Chapter 14 • Genetic Basis for Heterochronic Variation

Victor Ambros

Chapter 15 • The Abundance of Heterochrony in the Fossil Record

Kenneth J. McNamara

Chapter 16 • Heterochrony in Evolution: An Overview

Michael L. McKinney

Chapter 1

The Uses of Heterochrony

STEPHEN JAY GOULD

1. The History of Heterochrony

Ernst Haeckel, following the scattershot theory for an enduring recognition by posterity, coined terms as a government might issue promises—continuously and with some repetition. Most, like the vows of statesmen, were soon forgotten, but some have endured in the sense that Haeckel intended, notably ecology, ontogeny, phylogeny, and Monera. (Haeckel almost pulled off the neat trick of naming our ancestral genus before its discovery—for he predicted, by applying his biogenetic law to modern children, the characters of *Pithecanthropus* 30 years before DuBois found an appropriate fossil for the name on Java. But the species has since been promoted to *Homo erectus*, and Haeckel's genus rests on the ash heap of synonymy.) In some cases, however, Haeckel would only be able to view the survival of his names as, at best, a pyrrhic victory, but more as a bitter joke unconsciously played against his scattershot approach—for the names live, but with meanings contrary to Haeckel's original intent. Heterochrony, the theme of this book, is such a Haeckelian term converted into its opposite in current usage. The odyssey of its transformation is as quirky a tale as the phylogeny of many lineages; the reasons behind the shift in meaning also tell an interesting story crucial to the major concerns of this book.

Haeckel coined the term heterochrony (1875, 1905) as a prominent exception to the rule of evolution by change in developmental timing as he conceived it—universal recapitulation as expressed in the biogenetic law. Ordinary development, or *palingenesis* in Haeckel's phrase, produced an even and integrated acceleration of the entire adult ancestor into earlier and earlier ontogenetic stages of descendants. But Haeckel also defined a class of exceptions to this harmonious

STEPHEN JAY GOULD • Museum of Comparative Zoology, Harvard University, Cambridge, Massachusetts 02138.

acceleration: *coenogenesis*, or *Fälschungsgeschichte* (falsified history). Among the categories of coenogenesis, Haeckel recognized the differences among parts that we would now call mosaic evolution. True palingenesis pushed the *entire* ancestral adult back into earlier ontogenetic stages. But individual organs might become displaced relative to each other in either time or location, thereby dissociating the ancestral adult into pieces with their own ontogenetic histories.

Haeckel referred to displacement in location as *heterotopy*. Our reproductive organs, for example, now develop in mesoderm. This situation must record a heterotopy, or movement in location from one germ layer to another, because the first metazoans had not yet developed mesoderm, but must have possessed reproductive tissue (arising ancestrally in ectoderm or endoderm).

Haeckel defined displacements in time as *heterochrony*. The embryonic heart of vertebrates, for example, develops far earlier in ontogeny than its advanced phylogenetic status would warrant. The heart has, by heterochrony, been accelerated more intensely than other organs.

Note the two outstanding differences between Haeckel's and modern usage. First, we regard heterochrony as the definition of evolutionary change in developmental timing; Haeckel coined the term for exceptions to rules of change in developmental timing. Second, Haeckel devised heterochrony to describe the pathway of development in one organ relative to the pathways of *other organs in the same animal;* we use the term for the course of a trait relative to the ontogeny of the *same trait in an ancestor* (or related form).

The first alteration arises from a change in concept that marked a major advance in our understanding of evolution. Haeckel knew that his idealized passage of coherent ancestral adults into earlier stages of descendant ontogenies did not occur in practice, and that heterochronies dismembered the old coherence into separated pieces, but he did view dissociation as an exception—he called it a falsification—masking the proper path of evolution. But other recapitulationists, particularly E. D. Cope in America (Cope, 1887), interpreted dissociability and different trajectories of parts as an evolutionary norm, and therefore viewed heterochrony not as exceptional, but as part and parcel of the biogenetic law *redefined for organs* rather than whole organisms. Cope argued, for example, that each organ will be accelerated at a different rate, but acceleration remains the primary pathway of change in developmental timing. Thus, we have a nearly universal recapitulation, but of each part separately. No whole ancestral adult may be identified in any ontogenetic stage of a modern organism.

The second alteration was, I suspect, more an error on de Beer's (1930) part than a change in concept. This was, however, a felicitous error—and it has, in any case, become so canonical that we must live with it. Haeckel restricted his definition to single organisms—heterochrony is displacement in time of one feature relative to another in the same creature. De Beer generalized the concept by altering its meaning in a major way, while retaining the essential idea of change in timing. A heterochrony became any change in timing of a feature relative *to the same feature* in an ancestor. This new definition certainly jibes with the etymology of "other time," but it is not Haeckel's concept. Heterochrony became, for de Beer, the definition of change in developmental timing itself; while, for Haeckel, heterochrony was a major exception to the kind of change in developmental timing that he viewed as normal in evolution. How ironic, but sensible.

A term for an exception has become—first by a major change in concept, then by an error in application—the definition of the phenomenon itself.

The story of this quirky functional shift (the essence of evolution, *nicht wahr?!*) provides more than mere antiquarian interest for us today. As I argued at length in Gould (1977), Haeckel may have spoken a brave line about causality and mechanism, but he cared precious little for these subjects in his work and definitions. Haeckel's interest lay in the recovery of history—in tracing the pathways of phylogeny. He did not push the biogenetic law as an aid in understanding evolutionary mechanisms, but as a guide to tracing phylogenies by locating ancestral states in descendant ontogenies. This interest set his definitions and concerns. His *summum bonum* would be a complete ancestor harmoniously pushed back into the early ontogeny of descendants—this, and only this, would be palingenesis, or true repetition. Anything that dismembered the old ancestor stymied the recovery of phylogeny, and therefore counted as coenogenesis, or falsified history. No matter that heterochronies might preserve the causal principle of universal recapitulation by accelerating all characters (though at differing rates); they complicate phyletic inference by dissociation and therefore must rank as exceptions to the primary generality. Today we are more interested in studying ontogeny as a source of insight into mechanisms of evolutionary change. For us, therefore, the tie of heterochrony to dissociation—either in Cope's revision of Haeckel or in de Beer's meaning—embodies the important principle that all parts can be accelerated or retarded both relative to other parts and to ancestral conditions. We have an entire panoply of causal possibilities, and heterochrony is a fine and appropriate term for the general principle of evolution by change in the timing of development.

2. Problems

This book records a veritable explosion of interest in heterochrony during the past decade. Having never been addicted to modesty, I will take some credit for focusing this concern in a long book (Gould, 1977), although in fact I wrote *Ontogeny and Phylogeny* as a personal antiquarian indulgence (Gould, 1977, p. 2), and so it would have remained, if remarkable advances in the genetics and formalisms of development had not made ontogeny tractable, and if a loosening of Darwinian orthodoxies had not fostered a renewed interest in the shaping and constraining powers of inherited form and developmental pathways against the optimizing power of externally driven adaptation. I felt no presentiment of these fertilizing changes when I began to write; *Ontogeny and Phylogeny* was lucky to be in the right place at the right time.

If heterochrony shall be the empirical focus of the organismic biologist's approach to development as an evolutionary force, then we must ask if this notion of change in developmental timing can shoulder such a burden. Or are its ambiguities too great, its promises too empirically intractable, or its occurrences simply too infrequent? In this chapter I shall focus almost exclusively on results presented in this book to argue that problems can be overcome and promises fulfilled. Much fruitful work has already been accomplished, and the conceptual tools for continued progress are in place.

I see three major problems with studies of heterochrony as now usually practiced by organismic biologists:

1. Inadequate criteria, particularly a lack of information about actual timing, combined with a recognition that size cannot serve as a surrogate for time. No other issue or procedure is so widely recognized and discussed in this volume, forming a major theme in the chapters of McKinney, Tissot, McNamara, Foster and Kaesler, Jone, Geary, and Shea. The attention to this concern demonstrates a proper recognition that heterochrony is a causal claim about timing, not merely a description relative to any old available standard. I particularly appreciated McKinney's frank admission (Chapter 2) that his linkage of prolongations (leading to neoteny or hypermorphosis) to stable environments—perhaps the most elegant of paleontological studies in heterochrony during recent years (McKinney, 1986)—might be compromised if larger size does not, in general, betoken longer times.

2. Too distant and complex a pathway from underlying causes to morphological expression. Most organismic biologists, paleontologists in particular and *faute de mieux*, infer changes in developmental timing from their morphological expressions in phenotypes. If each phenotypic change could be tied, directly and without ambiguity, to an underlying alteration of developmental timing, this path of indirect inference to cause would suffice. But the watchword of complexity—forming both the joy and frustration of our discipline—is the possibility of curious and multifarious pathways between cause and expression, thus precluding any direct inference of one from the other.

I made a primary, heuristic separation in *Ontogeny and Phylogeny* between evolutionary change by alteration in timing of features already present in ancestors—heterochrony as used in this book—and change by genesis of novelty. I continually argued that a focus on changes in timing for features already present can provide only a *minimum estimate* for evolution by alteration of developmental rates, for what we perceive as novelty can also arise as an expression of underlying changes in timing. (We might, for example, view an extra digit as a novelty, but its developmental basis might be a simple prolongation of rapid embryonic rates in cell proliferation at the limb bud, leading to a larger primordium that, given inherited rules of differentiation, provides "leftover" material for an extra digit.) Heterochronies, in other words, do not exhaust the domain of evolution by changes in developmental timing; they are only the subset of such changes that produce phenotypic expressions isomorphic with the underlying alteration of timing.

In an important article, Alberch (1985) makes the crucial distinction between causal and temporal sequences in ontogeny—illustrating with examples of coat color in zebras and morphogenesis of digits in amphibians that a continuous change in timing of underlying regulatory mechanisms may yield phenotypic results that we interpret as novel, or at least not heterochronic in the sense of representing an ancestral morphology simply shifted in time or rate. Alberch concludes (1985, p. 46): "A paradox emerges, since the resultant morphology in the derived species is not represented in the primitive (ancestral) ontogenetic sequence, in spite of the fact that it has been produced by a regulation in the timing and developmental rates of the ancestral ontogeny."

Geary provides an excellent example in this book. Her well-documented evolutionary sequence of *Melanopsis impressa* via *M. fossilis* to *M. vindobonensis* represents a clear peramorphocline in the underlying pathway of allometric transformation, as the early stages become both more intense and shorter in duration. Morphology, however, does not display a sequence of classical recapitulation, but rather the transformation of a gently expanding shell with a basically triangular outline into an apically pointed and subsequently shouldered descendant (the point and the shoulder being the morphological expressions of peramorphic intensification of the first two allometric phases).

3. Pitfalls of the enumerative tradition. Natural history must always face the special problem that its generalities rest upon relative frequencies, not eithers and ors ruled by laws of nature. Theory will not tell us whether paedo- or peramorphosis is more common in nature, or whether correlations can be established between style of heterochrony and type of evolutionary change. Theory may suggest and predict, but the tests must rest upon relative frequencies established by observation.

Questions of relative frequency in natural history pose two special problems. First, nature provides such a multiplicity of cases; how can one establish a proper relative frequency? Any partisan can present an impressive list of cases, but what if they represent 1% of a totality, or what if they all record a rare and special ecological circumstance? Still, if one could take a random sample among equal cases, one might answer such questions by appropriately exhaustive compilation. The second problem with enumeration is that one cannot make such a list, given the uniquenesses and homologically based iterations of our evolutionary tree. A species is not a coin flip or the cast of a die. A brachiopod expert might be overwhelmed with the dominant frequency of some heterochronic process, but the conclusion might not apply to bryozoans, or even to brachiopods of another time and place. How can we establish relative frequencies among such disparate and unique objects?

3. Solutions

The major strength of this book lies in the thought and documentation provided for tackling the three problems just outlined.

1. Better definitions and formalisms. I know of no other field so previously plagued by a frightful jargon built on frustrating inconsistency (Gould, 1977, pp. 209–211). Any progress required a refiner's fire leading to a simple set of consistent and operational definitions based on causal processes, not convenient descriptions. This we have achieved in the current definitions of heterochrony, with clear criteria of size, age, and developmental stage [see de Beer (1930) and my simplifications in Gould (1977, pp. 221–228)].

But the implementation of these definitions requires a formalism. My clock models (Gould, 1977) were only semiquantitative, and too cumbersome for general use. The proper formalization and conversion to operational utility by Alberch *et al.* (1979) has been adopted by nearly all authors in this volume, and represents a welcome consensus that finally provides the necessary common basis for advance. Further modeling, as in Slatkin (1987) on the genetics of heterochrony, will provide additional, important tools.

But formalisms are only frameworks. We may recognize the need for standardization by age, for example, but can proceed no further without the goods. Jones, in his review of sclerochronology, shows that marks of time often pervade morphology and that even paleontologists, precluded in principle from assessing age directly, need not despair.

All paleontologists interested in this subject eventually become stymied if they cannot assess absolute ages. To cite a personal example, I was able to show (Gould, 1972) that a celebrated and counterintuitive reinterpretation of the *Gryphaea* sequence (Burnaby, 1965) had been falsely based on an inproper standardization at constant size. Burnaby compared descendants to ancestors of the same size and reached the astonishing conclusion that coiling had decreased in a lineage long famed for its trend to increased coiling. But size had increased in this sequence as well, and Burnaby therefore compared adult ancestors with more juvenile descendants. Since coiling increases in ontogeny, Burnaby matched a fully coiled ancestor with a descendant that would continue to grow and coil—hence the false appearance of phylogenetic decrease in coiling. I was able to show that descendants, at their larger sizes, maintained the same degree of final coiling as ancestors—and that the allometric shifts noted by Burnaby had the effect of preserving a constant amount of coiling. But here I was stymied, and so remain. Having developed a theory for the ecological basis of differing styles of heterochrony (Gould, 1977), I longed to know whether the larger size of descendants represented a hypermorphic extension of time to maturation or merely a faster growth rate to achieve an adult shell of larger size in the same amount of time. The assessment of heterochronic style depends crucially on the answer to this question. *Gryphaea* shows prominent growth lines with potential interpretations in terms of astronomical periodicities—so my question may well be tractable. But no one has yet made such a study, and this issue remains unresolved.

Several cases in this book await the same resolution. I have already discussed McKinney's important work on style of heterochrony in echinoids. Shea's studies of primates are pervaded with a need for data about age before proper assessments of heterochrony can be made. He points out that size cannot be used as a surrogate, for large does not mean old (gorillas reach their maximal primate size in the same time required by other species to attain much smaller bulk). To cite just one example, the Old World talapoin and South American marmosets are the smallest of monkeys, but would be misinterpreted if their common diminution were attributed to the same heterochronic process. Firm data are still lacking, but ecology and morphology suggest that talapoins grow slowly to reach their small size at average ages for larger bodied relatives, whereas marmosets may be progenetically truncated in an ecological context of benefit for enhanced reproductive turnover.

2. Penetrating to underlying bases. It was not so long ago that D'Arcy Thompson, our great student of organic integrity mathematically expressed, complained of the subtle and perverse influence that methodological limitations can place upon concepts of reality:

> For the morphologist, when comparing one organism with another, describes the differences between them point by point and "character" by "character." If he is from time to time constrained to admit the existence of "correlation" characters . . . yet all the while he recognizes this fact of correlation somewhat vaguely, as a phenomenon due to causes which, except in rare instances, he can hardly hope to trace; and he falls readily

> into the habit of thinking and talking of evolution as though it had proceeded on the lines of his own descriptions, point by point and character by character. (Thompson, 1942, p. 1036)

Heterochrony is a global phenomenon, not a catalogue of disembodied characters. A first step toward penetration from externalities of phenotypic results to underlying processes involves the replacement of univariate measures with an understanding of covariance via multivariate analysis. In studies of ontogeny, factor and principal components analyses possess the happy property of almost always producing a first axis that can be interpreted as a general size factor with major allometric components built in [see Jolicoeur (1963) and later work, summarized in the chapter by Tissot]. This axis provides a multivariate measure of an important criterion for standardization in heterochronic study, and also permits the direct assessment of components in shape orthogonal to, and therefore mathematically independent of, this criterion. [There is no mystery, or any "deep" principle, embodied in this common interpretation of the first axis; it represents no inevitable rule and cannot be assumed as a correct reading *a priori*. The first axis usually represents size for the banal reason that most parts of a creature get bigger as its body grows. A technique that passes its first axis through the center of a cluster of vectors representing a positive manifold of correlation coefficients must capture some sort of average increase in size. The allometric effects are "built in" (so long as their expressions are linear and therefore captured in the correlation matrix) because they represent the morphological pattern of size increase.] In this volume, Tissot summarizes heterochronic work in multivariate morphometrics, while Foster and Kaesler discuss the burgeoning field of shape analysis and its various techniques.

This emphasis on multivariate work in no way compromises the tradition of bivariate allometry, which was, after all, developed by Huxley in order to penetrate the veil of static morphology and recover underlying aspects of growth by circumventing time (so often unavailable and so complexly nonlinear when available) and plotting one aspect of form against another. McKinney discusses the allometric tradition and presents methods for dealing more directly with the generally circumvented, yet so important, measure of time itself. Geary, as previously discussed, presents a fine example of why underlying growth patterns, recoverable in bivariate analysis, rather than morphological results must be assessed in order to identify evolutionary changes in developmental timing.

In a second and more direct step, we are finally getting some exciting, immediate data on the underlying genetics of evolutionary changes in timing. Goldschmidt (1923, 1938) demonstrated the operation of "rate genes" and developed, as we all know, some controversial theories about their role in macroevolution. In this volume, Ambros summarizes his important work on heterochronic mutations in *Caenorhabditis elegans*. This research is particularly important to students of heterochrony for three reasons: (1) it represents the first attempt to provide direct genetic information in the framework of traditional concepts and terms of heterochrony; (2) it advances past earlier studies in its ability, thanks to general progress in molecular biology, to move beyond formalisms of pedigrees to actual mechanics of gene action in development; (3) one could not hope for a more elegant experimental subject than *C. elegans*, for we know the lineage and ontogenetic history of every cell in this small nematode and therefore have an optimal map for the plotting of heterochronic changes.

Ambros has been able to show that single-gene mutations can cause heterochronic changes with phenotypic effects known in the natural variation of this species. He has identified heterochronic mutations that both accelerate and slow down aspects of development (precocious and retarded mutants in his terminology). Finally, in a stunning affirmation for those of us interested in separating the various possible causes of heterochronic expression, he has been able to show that these mutations produce their heterochronic effects not by altering the rate of development, but by changing the time of onset [pre- or postdisplacement in the terminology of Alberch *et al.* (1979)].

3. The enumerative tradition. The work of McNamara should immediately remove one basic fear that the enumerative tradition is destined either to disperse or succor: we are not investigating a minor gloss upon evolution; heterochrony is both a frequent and an important phenomenon in nature.

As to what enumeration can teach us about modes of heterochrony, McNamara also makes the valuable point that hopes of recovering true natural frequencies from percentages of cases reported in literature may be hopelessly compromised by shifting preferences and fashions (see also Gould, 1977). When recapitulation was king in Haeckel's day, cases of acceleration greatly outnumbered examples of retardation, while all forms of paedomorphosis, if considered interesting enough to report at all, were deemed degenerative. Following Garstang's (1922) reforms and Bolk's (1926) hypothesis of human neoteny, paedomorphosis gained the upper hand, but surely not because nature had changed.

In the light of these limits, I believe that enumerative arguments have resolving power in two types of circumstances, both fortunately frequent in occurrence. First, when patterns arise incidentally from work done for other reasons and not at all directed toward the study of heterochrony (I realize, of course, that this claim relies in part upon the slippery testimony of negative evidence, and that correlations may exist between the actual object of research and the uninvestigated heterochrony). McNamara's exhaustive compilation, for example, revealed a great increase in the frequency of peramorphosis compared with paedomorphosis in later Paleozoic vs. Cambrian trilobites. I do not think that trilobitologists have different aims or tendencies when working in various parts of the Paleozoic—so this pattern may reside in nature. This claim also, and intriguingly, matches various macroevolutionary notions about the link of paedomorphosis to evolutionary novelty, and the high frequency of novelty in the early history of groups. Similarly, McNamara found a high frequency of progenesis in echinoderm groups that add plates during ontogeny, but a predominance of neoteny and acceleration in groups that grow primarily by allometric modification of plates present from an early age. Pandolfi documents a greater frequency of astogenetic vs. ontogenetic heterochrony in colonial organisms (nearly a threefold difference), but here I have no confidence in the natural basis of this result (and neither does Pandolfi), because workers have focused more attention on patterns of colony growth than the ontogeny of "persons" (to use this vernacular term in a technical sense common in the 19th century literature, and still used by students of siphonophores).

Second, when a structural argument can be advanced for viewing the documented frequency as an expectation: Landman, for example, reasserts the old claim that recapitulation does, overall, show a dominant relative frequency in ammonites [a view that I also accepted, albeit with reluctance given my own prejudices, in Gould (1977)]. (This case may be a type example for shifting social preferences. Ammonitology was a bastion of documentation during the heyday of recapitulation, and therefore became the great hunting ground for a claim about frequent paedomorphosis when fashions shifted. Now that both pera- and paedomorphosis have equal status as heterochronic expectations, perhaps we can get a more balanced view of the true natural frequencies.) Landman presents a structural argument that would predict a dominance for peramorphosis. Ancestral ammonites had simple sutures; sutural complexity increased in ontogeny. Since so many lineages exhibit trends to increasing sutural elaboration, the obvious channel for this presumably adaptive result is terminal addition to the inherited ontogeny of increasing sutural complexity. Similarly, McNamara's claim about different relative frequencies of heterochronic modes in echinoderm classes (see above) has a structural basis in varying modes of growth.

I need hardly add, for it is the basis of all comparative biology, that none of these arguments have any force unless they are placed into the context of a well-established genealogical nexus, as Fink argues so well. (Note how many statements about genealogy are folded into my description of Landman's claim about ammonites in the last paragraph.)

4. The Importance of Heterochrony

Why do we care? What is so special about heterochrony? What can it teach us about evolution? Surely, we cannot simply rely upon the established fact that heterochrony is common ("So is birdshit," as the anonymous private retorted to the boast of his Air Force drill instructor, "I'm airborne"). Heterochrony must also be both distinctive and informative. I believe that heterochrony has captured the interest of evolutionary biologists in the last decade because it speaks so crucially to two of the major issues most responsible for recent ferment in evolutionary theory: rates of change, and the dialectic of "inside" vs. "outside" influences (structural and developmental "constraints" vs. environmental selection) upon evolutionary pattern.

I agree with McKinney that the issue of rate has either been exaggerated or not treated in the most fruitful way in traditional literature about heterochrony. After all, heterochronic processes are not strictly linked to special rates; they may participate in all magnitudes from the most rapid to the quotidian. Progenesis, the implicated phenomenon in most hypotheses about rapid, phylum-level transitions [Hardy's (1954) "escape from specialization," for example], has also produced an array of small-scale, highly specialized, local, and restricted adaptations to dead-end environments (Gould, 1977). Moreover, arguments about putative phylum-level transitions have too often been promulgated in the speculative mode that once gave heterochrony such a bad name among thoughtful scientists concerned with testability, and led to Kleinenberg's (1886) celebrated quip that Haeckel's gastraea was but a "*mageres Thiergespenst*," or slim animal-specter.

More fruitful for insights about rate have been various attempts (apparently successful, judging from numerous consonant examples in this volume) to link styles of heterochrony with particular ecological circumstances that imply differing rates of change and evolutionary significances. This work has been doubly fruitful because it has also established linkages between immediate ecological reasons and retrospective macroevolutionary meanings—a tie that had not been previously made and that (in its absence) had left the literature of heterochrony with its traditional taint of too much untestable speculation about unobservable transitions.

As the major theme of my book (Gould, 1977), I tried to illustrate the promise of ecological linkages by arguing that paedomorphosis was merely a result with two basic causes of opposite significance—progenesis (truncated ontogeny) and neoteny (delayed somatic development). I linked progenesis to unstable *r*-selective environments that often exert strong selection for rapid maturation and cycling of generations, and neoteny to more stable ecologies that favor fine tuning of morphological adaptations. Although the framework of *r*–*K* selection theory has not survived unscathed (Stearns, 1980), this basic linkage of heterochronic mode to environmental situation seems well affirmed now. (Progenesis, while arising for immediate adaptive reasons often lodged in accelerated maturation itself, gains its macroevolutionary power by presenting to future selective contexts a rapidly generated, unusual mosaic of juvenilized and adult characters in an organism freed from rigid morphological monitoring because selection has worked primarily upon timing of maturation.)

Among examples in this volume, I have already discussed McKinney's work on heterochronic mode and environmental stability in echinoids, and McNamara's documentation of enhanced progenesis during an early period of great evolutionary flexibility in trilobites. Lindberg notes that brooding species in broadcasting clades are usually progenetic. Guerrant detects a high frequency of progenesis in two situations that favor rapid maturation in angiosperms. The cleistogamous flowers of dimorphic species [Darwin (1877) devoted a whole book to this subject] guarantee a seed set by pollinating themselves within the bud, while larger, chasmogamous flowers assure outcrossing; the cleistogamous flowers are often progenetic dwarfs. Similarly, flowers of autogamous species are often progenetic and smaller than those of outbreeding ancestral taxa. Guerrant writes that "the ability to flower and mature seeds earlier in the spring than their outbreeding ancestors is considered to have been of primary selective importance for the evolution of autogamy in species of at least three genera of annual wildflowers."

Hafner and Hafner present a particularly complete and interesting hypothesis in this mode by arguing that kangaroo mice are *r*-selected and progenetic, while kangaroo rats are neotenic with long lives and gestation periods and small litters. They also regard pocket gophers as hypermorphic, noting that neonatals look like mature pocket mice. The Hafners place all these regularities into sensible ecological contexts and use their findings to suggest important revisions of phylogeny—an interesting gloss upon Fink's reminder that all heterochronic work is severely compromised without well-established genealogical patterns; the path that links heterochrony and genealogy can be a two-way street of mutual implication. The Hafners write: "If we are correct, the kangaroo mouse (or 'dwarf kangaroo rat' as it was originally called . . .) is not merely a scaled-down version of the neotenic kangaroo rat, but is a separate, unique progenetic form."

On the subject of constraint and adaptation, heterochrony is perhaps our major source of morphological data for questions of internal capacities and channeling that must complement any study of selective pressures by "outside" ecologies: which organisms do change, and in what ways can they?

If I may offer a personal testimony to the role of this theme in a broadened view of evolutionary processes and patterns, I began my career as a committed adaptationist [see especially Gould (1970)]. I always had a strong fascination for themes of morphological integration, but I yearned to recast what I originally viewed as outworn vestiges of vitalism into a strictly selectionist framework. [My first major article (Gould, 1966) argued that essentially all allometries are adaptations because changes in parameters of power functions may have as strongly selectionist a basis as any aspect of static morphology—an extreme position that even so committed a Darwinian as Julian Huxley (1932) did not support for this subject.] I began *Ontogeny and Phylogeny* largely to show that all heterochronies could be interpreted as adaptations, once the proper ecological correlations were established.

In writing this book, I spent months reading the major works of the greatest morphologists in the formalist tradition—particularly von Baer, Geoffroy Saint-Hilaire, and Richard Owen, not to mention Bateson and D'Arcy Thompson in our century. I emerged from this superb tutorial with a growing appreciation for limits upon the shaping power of immediate selection imposed by what these great morphologists called "laws of form," and we would more likely label constraints of inherited genetic and developmental programs. These formal causes of morphology are as much a determinant of evolutionary pathways as any efficient cause of shaping by natural selection. I did work out correlations of heterochronic modes with ecological contexts, but I also came to understand how coherences of form, expressed in heterochrony, delineated the channels along which organisms push back upon forces of selection to produce evolutionary change.

Heterochrony is perhaps our best empirical mode for the study of developmental constraint, especially since we now have direct genetic evidence (Ambros), sophisticated techniques for identification and statistical analysis (Tissot, and Foster and Kaesler), and a rich compendium of examples from the natural history of all major groups (chapters in part III). For example, Ambros demonstrates a separate genetic control over gonadal and nongonadal development in *C. elegans*. The dissociability implied by these separate controls permits the freest possible range for heterochrony. (Other groups do not show such mosaicism and cannot therefore express much heterochrony. Hormonal control of sexual maturation produces a maximal dissociation of somatic and gonadal differentiation in urodeles, thus potentiating the greatest expression of heterochrony among vertebrates. The high frequency of neoteny in salamanders may not be resolved by the simple goods and bads of immediate selective contexts, but requires this structural capacity for dissociation. Neoteny might suit a frog's tadpole just as well, but the option is apparently not available.)

Ease of dissociability can also be inferred from morphological results. In trying to explain the high frequency of heterochrony in cephalopods, Landman notes that growth and development are easily dissociated in *Nautilus*, where individuals raised in aquaria under unusually favorable conditions may reach maturity much earlier than in nature, thereby producing abnormally small adult shells.

Possible pathways of future change may also be "prefigured" in the available variation of ancestral stocks (as other potential pathways, perhaps more adapted, are foreclosed by a lack of variation). In documenting the paedomorphic origin of *Melanopsis pygmaea* from *M. sturii*, Geary shows that intermediate variants are present in populations of ancestors, and that the main determinants of paedomorphosis are the most variable characters in samples of the ancestral species.

It is no accident that the fortunes of heterochrony flowered when scientists took the internal constraints of form seriously and hoped to induce the regularities of organic transformation thereby, or that interest waned during the heyday of strict selectionism, or that our fascination has been kindled again as we struggle anew to comprehend the interaction of selection and internal channels of possibility. Louis Bolk did not originate the theory of human neoteny. The first important observation was made by none other than Etienne Geoffroy Saint-Hilaire, who observed a juvenile orangutan in the Paris zoo and described its uncannily human characters of both morphology and behavior in 1836:

> In the head of the young orang, we find the childlike and gracious features of man . . . We find the same correspondence of habits, the same gentleness and sympathetic affection . . . On the contrary, if we consider the skull of the adult, we find truly frightening features of a revolting bestiality. (Geoffroy Saint-Hilaire, 1836, pp. 94–95).

Geoffroy searched all his life for unifying principles in the generation of animal form. He thought that he had found his single channel in the vertebra as a common element, and he wrongly homologized the external skeleton of arthropods with the vertebrate backbone, even accepting the peculiar corollary that insects literally dwelled within their own vertebrae. His vertebral theory was a grand error (though homeoboxes may yet restore some notion of underlying unity across the gap of protostomes and deuterostomes). But his search for principles of structure, and his good sense about heterochrony as a primary hunting ground, may continue to guide our own explorations.

References

Alberch, P., 1985, Problems with the interpretation of developmental sequences, *Syst. Zool.* **34**:46–58.

Alberch, P., Gould, S. J., Oster, G. F., and Wake, D. B., 1979, Size and shape in ontogeny and phylogeny, *Paleobiology* **5**:296–317.

Bolk, L., 1926, *Das Problem der Menschwerdung*, Gustav Fischer, Jena.

Burnaby, T. P., 1965, Reversed coiling trend in *Gryphaea arcuata*, *Geol. J.* **4**:257–278.

Cope, E. D., 1887, The theory of evolution, in: *The Origin of the Fittest*, Macmillan, New York [reprinted from *Proc. Acad. Nat. Sci. Phila.* (1876)].

Darwin, C., 1877, *The Different Forms of Flowers on Plants of the Same Species*, J. Murray, London.

De Beer, G. R., 1930, *Embryology and Evolution*, Clarendon Press, Oxford.

Garstang, W., 1922, The theory of recapitulation: A critical restatement of the biogenetic law, *J. Linn. Soc. Zool.* **35**:81–101.

Geoffroy Saint-Hilaire, E., 1836, Etudes sur l'orang-outang de la ménagerie, *C. R. Acad. Sci.* **3**:1–8.

Goldschmidt, R., 1923, Einige Materialen zur Theorie der abgestimmten Reaktionsgeschwindigkeiten, *Arch. Entwicklungsmech.* **98**:292–313.

Goldschmidt, R., 1938, *Physiological Genetics*, McGraw-Hill, New York.
Gould, S. J., 1966, Allometry and size in ontogeny and phylogeny, *Biol. Rev.* **41**:587–640.
Gould, S. J., 1970, Evolutionary paleontology and the science of form, *Earth Sci. Rev.* **6**:77–119.
Gould, S. J., 1972, Allometric fallacies and the evolution of *Gryphaea:* A new interpretation based on White's criterion of geometric similarity, in: *Evolutionary Biology*, Vol. 6 (T. Dobzhansky *et al.*, eds.), pp. 91–118.
Gould, S. J., 1977, *Ontogeny and Phylogeny*, Harvard University Press, Cambridge.
Haeckel, E., 1875, Die Gastrula and die Eifurchung der Thiere, *Jena Z. Naturwiss.* **9**:402–508.
Haeckel, E., 1905, *The Evolution of Man*, Watts & Co., London [translated from *Anthropogenie*, 5th ed.].
Hardy, A. C., 1954, Escape from specialization, in: *Evolution as a Process* (J. Huxley, A. C. Hardy, and E. B. Ford, eds.), pp. 146–171, Allen and Unwin, London.
Huxley, J., 1932, *Problems of Relative Growth*, MacVeagh, London.
Jolicoeur, P., 1963, The multivariate generalization of the allometry equation, *Biometrics* **19**:497–499.
Kleinenberg, N., 1886, Die Entstehung des Annelids aus der Larve von *Lopadorhyncus*. Nebst Bemerkungen über die Entwicklung anderer Polychaeten, *Z. Wiss. Zool.* **44**:1–227.
McKinney, M. L., 1986, Ecological causation of heterochrony: A test and implications for evolutionary theory, *Paleobiology* **12**:282–289.
Slatkin, M., 1987, Quantitative genetics of heterochrony, *Evolution* **41**:799–811.
Stearns, S. C., 1980, A new view of life-history evolution, *Oikos* **35**:266–281.
Thompson, D'A. W., 1942, *Growth and Form*, 2nd ed., Macmillan, New York.

I

Analysis of Heterochrony

Chapter 2

Classifying Heterochrony

Allometry, Size, and Time

MICHAEL L. McKINNEY

1. Introduction

Heterochrony is evolution via change in timing (and/or rate) of development. However, this oft-repeated definition threatens to dull by repetition the important fact that virtually *all* evolution involves such changes somewhere in the chain of developmental events. Whether size, shape, or behavior, phylogenetic change almost invariably springs from a change of rate or timing in the ontogeny of descendant individuals. [Since development is a series of highly contingent, interwoven processes, it is much simpler to change the rate or timing of ontogenetic processes rather than accommodate the exponentially cascading effects from changing the processes themselves. This still leads to qualitatively different individuals, including new tissues (Raff and Kaufman, 1983).] Thus, if one is to analyze evolution, a working knowledge of how to analyze heterochrony is critical.

In fact, heterochrony can be analyzed in a number of ways. Much depends on the quality and kind of data available and the goals of the investigator. However, at the core of all such analyses is the comparison of ontogenies among related

MICHAEL L. McKINNEY • Department of Geological Sciences, and Graduate Program in Ecology, University of Tennessee, Knoxville, Tennessee 37996-1410.

lineages. This usually involves morphological comparisons of shape and size change. In this chapter I will show that any approach using size and shape information alone is inadequate to determine heterochronic processes; we must also have ontogenetic age data. This casts suspicion on many previous heterochronic studies, including my own (McKinney, 1984, 1986). However, since size and shape changes are ubiquitous and of interest in their own right, a solely descriptive allometric terminology is proposed. Next, an integration of allometry with time is discussed, with the presentation of a heterochronic classification scheme based on size and time alone; shape can be omitted because it is really only a derivative of size (ratios thereof), allowing us to simplify things greatly. The result clarifies the hierarchical nature of change: that parts of an individual can undergo heterochronies that differ from one another or even the rest of the body. Finally, I try to relate allometry and heterochrony to a biologically meaningful (albeit simple) model of development.

2. Allometry in the Heterochronic Trinity

2.1. Introduction

Both allometry and heterochrony have been major themes in evolutionary thought for many years. Yet they have been developed more or less independently [see Raff and Kaufman (1983) for review]. In spite of an intuitive awareness by many evolutionists that they are related, this relationship has remained largely unarticulated. Therefore, to be explicit at the outset: heterochrony, as it is now generally accepted, denotes change in ontogenetic timing, which can be classified through observations of ontogenetic timing, size, and shape [see Alberch *et al.* (1979) for the landmark paper, which had many roots in Gould (1977)]. Thus, allometry (and even visual comparisons), which deals with size and shape, includes only two of the three heterochronic trinity of time, size, and shape. Therefore, while these two most tangible aspects of heterochrony are the most readily accessible to the neontologist and paleontologist, they alone are not enough to provide a full understanding of the heterochronic changes in an evolutionary sequence. We must also have knowledge of ontogenetic age. (Indeed, if one accepts my arguments below that shape is really only a derivative of size, then the heterochronic trinity becomes a duality; in that sense, allometry has only one of two factors, being based solely on comparisons of size change in two or three dimensions.)

Nevertheless, allometry is an integral part of heterochronic analysis and does provide useful ontogenetic information even without age data (indeed, in some cases size-based comparisons are more informative, as discussed below). This, plus the vast abundance of allometric studies in the literature, leads me to give a brief review of allometry itself before considering its relation to heterochrony. Bivariate allometry is the focus here because its simplicity allows the relationship to be more easily visualized; however, most of the basic conclusions apply to multivariate (Tissot, this volume) and visual (McNamara, 1986) comparisons.

2.2. Bivariate Allometry

The importance of size and shape differences among organisms is self-evident and has been seriously pursued since at least Galileo's time [see Gould (1966) for thorough review]. While Thompson (1917) is usually associated with the earliest rigorous attempt to analyze size and shape change, it was Huxley (1932) who succeeded in producing the more descriptively useful bivariate allometric expression $y = bx^K$, where x and y typically represent dimensional morphological measurements. It is rarely noted that this equation is the solution to the differential equation

$$\frac{dy}{dt}\frac{1}{y} = k\frac{dx}{dt}\frac{1}{x}$$

so that

$$k = \frac{dy/y\ dt}{dx/x\ dt}$$

In other words, k is the ratio of the specific growth rates of the traits. [For an interpretation of b see White and Gould (1965).] Biologically, this can be related to the mechanics of growth and cell division (Laird, 1965; Katz, 1980); it is generally descriptive because these processes are multiplicative and therefore follow regular rules of proportional change.

Computation of the parameters b and k is usually accomplished by logarithmically transforming both sides of the equation:

$$\log y = \log b + k \log x$$

This rectilinear equation is much easier to fit a curve to and is generally quite accurate as an approximation of the parameters. The preferred method of regression is usually not least-squares, because it assumes x to be truly independent. In most cases, x is no more "independent" than y (e.g., it is no more accurately measurable), so that reduced major axis or major axis regression is the preferred method (e.g., Imbrie, 1956; Davis, 1986). Once computed, the slope yields k, the growth ratio. Where both variables are linear, areal, or volumetric metrics and $k=1$, isometry obtains; where k is greater or less than 1, positive or negative allometry results, respectively. In the latter cases, the trait y increases at a greater or lesser (respectively) pace relative to x. If x and y are of different dimensions, the line of isometry will not be 1; e.g., if x is body weight (proportional to volume) and y is arm length, isometry will equal 0.33.

While plots of the raw x and y values on log-log paper are valid, it may be more telling to plot the log-transformed values of x and y on standard arithmetic scales, since the points are often not as crowded together and patterns are clearer (e.g., McKinney and Schoch, 1985).

In most heterochronic studies, one wishes to choose some measure of body size as the variable x, usually body or skull length, or perhaps body weight. The choice of x is obviously crucial in its effects on scaling estimates and it is sometimes troublesome. However, because most body measurements are highly intercorrelated, the relative scaling will usually be very similar. Thus, as long as one is consistent in the use of that x, satisfactory results should follow.

Problematica

This brief description of bivariate allometric techniques would be incomplete without noting some of the complexities often encountered in such analyses. For one, the growth ratio k is not always constant (Gould, 1966; McKinney, 1984). In such cases the plot of log-transformed x and y values will be curvilinear. In McKinney (1984) I suggested fitting the points with a polynomial curve of the appropriate order to achieve a significant fit (via an F-test). The primary use of this curve is to allow estimation of k at each point of the curve. That is, the first derivative of the polynomial will yield an estimate of the slope k at any given log x. We can then say more than only that the growth ratio was changing; we can estimate how much it was changing.

A more major problem in allometric studies (especially where heterochrony is of interest) can lie in the data base itself. The investigation should compare size and shape within and among ontogenetic series to be meaningful. While a living growth series is desirable, paleontologists especially must make do with fossilized stages. Alberch (1985a) has pointed out that some ontogenetic series do *not* have comparable stages among related groups. In these cases, much of the traditional approach outlined here is useless. Thus, we must not only compare an entire ontogenetic series in these studies, but we must use series that have comparable stages of development conserved. However, this does not seem to be a common problem, at least judging from paleontological examples.

A much more common problem is omission of juvenile stages, as occurs in comparing only adults. Use of such static allometries results in loss of information and may render the regressions misleading. However, the extent and significance of the loss are uncertain and may vary, depending on the relationship between static adult and ontogenetic allometry (Cheverud, 1982). Where possible, analysis of genetic and phenotypic covariance patterns is useful (Lande, 1979; Atchley *et al.*, 1981) and can be related to bivariate allometry [see discussion in Shea (1985)].

There is also the crucial problem of phylogenetic relationships among the ontogenies being compared. This is discussed by Fink (this volume), but is worth reiterating here. Most heterochronic studies so far have used traditional methods; however, there is a growing number using cladistics (e.g., Emerson, 1986).

3. When Age Data Are Absent: Allometry Is Not Heterochrony

3.1. Introduction

The methods discussed above are of great value in comparing ontogenies. When this is done, regular, usually simple, allometric differences are seen among the taxa. But what do they mean developmentally? As seen below, such size and

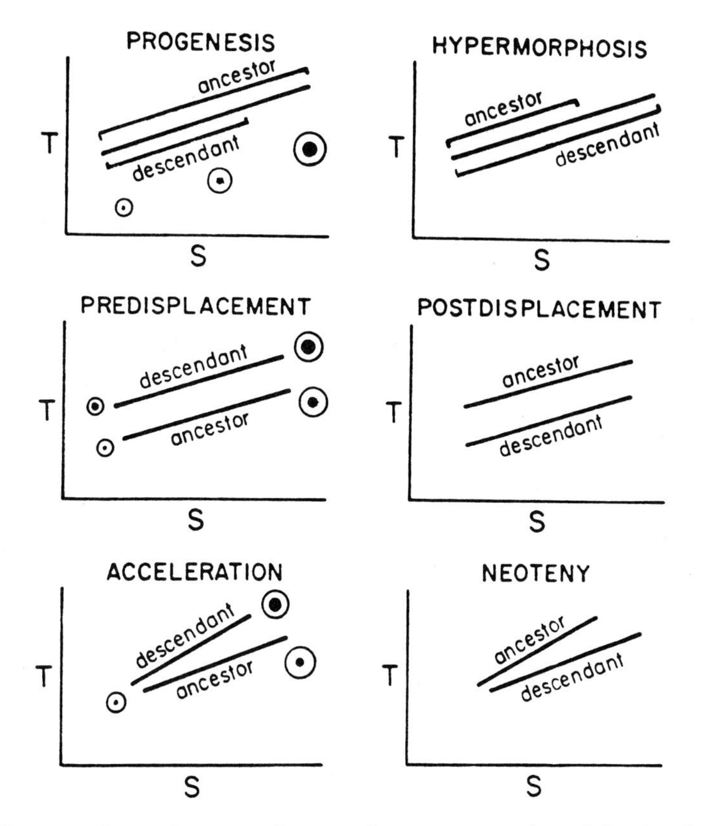

Figure 1. "Allometric" heterochrony, as shown in the ontogenetic plots of closely related species. *S* denotes body size (length, weight, etc.), and *T*, trait measurement; in these plots *S* and *T* are assumed to be logarithmically transformed. Sketches illustrate changes in hypothetical body size (light circle) and trait (dark circle). All changes are viewed relative to body size, e.g., progenesis = maturation at smaller size, not necessarily age. See Table I and text.

shape change can be attributed to timing or rate change in a "global" (i.e., whole body) or more restricted growth field.

How then to determine the type of heterochrony? The best heterochronic scheme we now have is that of Alberch *et al.* (1979), which delimits six major kinds of heterochrony (I omit proportional giantism and dwarfism, which are rare because organisms almost always change proportions due to the differential needs of size change). These six types are caused by changes in only three timing parameters: onset timing (early growth onset causes "predisplacement," late onset causes "postdisplacement"), offset timing (early growth offset causes "progenesis," late offset causes "hypermorphosis"), and growth rate (increased rate causes "acceleration," while decreased rate causes "neoteny"). McNamara (1986) has explained this in more detail.

To apply allometric data to this scheme, in McKinney (1986) I originally used the logic shown in Fig. 1 which uses the common assumption that "size equals time." More precisely, this holds that both (or more) of the groups being compared increase in size at the same rate, that they increase in size as the same function of time (age). It thus "cheats" by equating two of the three factors, reducing a three-dimensional problem to a two-dimensional one. Where this assumption holds true, the scheme in Fig. 1 works (even if the lines curve, as even log-transformed data may). However, this equating of size and age is hazardous: it is palpably not true in some cases, as elegantly shown by Shea (1983; also see Emerson, 1986). Even worse, it may *commonly* be false if studies of mollusks (Jones, this volume), primates (Shea, this volume), and carnivores (below) are an indication.

But does this nonequation of age with size always or even often invalidate the purely allometric approach? In this section, by modeling permutations of varying growth rates of body size, I will show that size–age disparity can indeed invalidate the size-based approach. (In later sections, the mechanics of such variation are more fully explored.)

In Fig. 2, the topmost graph shows that increased (or decreased) size along the same allometric trajectory may be obtained by delaying offset time (or early offset if decreased size), causing an extension (or truncation) of the growth period before maturation stops it (even with indeterminate growth, slowing down of growth usually occurs with maturation). Thus, we see the traditional hypermorphosis and progenesis. However, we also see that larger (or smaller) size along the same trajectory can occur via an increase in growth rate. That is, where the larger (or smaller) species is not the same age at a given size, growth rate differences may explain the relationship. In short, the larger (or smaller) species may have grown faster (or slower) during the same period, not at the same rate for longer (or shorter) periods. Shea (1983) documents this in primates and calls these rate processes "rate hypermorphosis" and "rate hypomorphosis." I would prefer "rate progenesis" for the latter, to be consistent with the Alberch *et al.* (1979) scheme. On the basis of allometric plots alone we cannot distinguish the two kinds of hypermorphosis or progenesis.

In the middle graph of Fig. 2, consider the three basic permutations of the size–age relation, where one species is younger than, older than, or the same age as the other. In the case of A, where species Y is actually older than X, it may not be truly "predisplaced." It has more of the trait at equal size because it is actually older and the trait may have been growing longer. In the lower graph of Fig. 2, we see that changes in allometric slope can also result from a number of size–age variations. In the cases of time-homologous points D and E, species Y is not truly "accelerated" relative to X, because the trait did not increase faster with time, only with size.

Thus, we see that allometric plots alone cannot with certainty distinguish any of the three major heterochronic processes of offset timing, onset timing, or rate change, because they cannot divorce size from age. The heart of the problem is that we are observing trait change as a function of size instead of time when in fact *size itself is also a variable function of time*. These same problems apply to multivariate (Tissot, this volume) and qualitative (McNamara, 1986) methods of heterochronic analysis as well, so long as they rely on size as the basis of comparison. Importantly, this raises doubts about the precise heterochronic assignments in the many studies that omit age data, including my own noted above, and even McNamara's excellent compilation (this volume, but see his discussion).

3.2. An Allometric Lexicon

Rather than discard the allometric terms of Fig. 1 completely, I would suggest that we retain them for convenient reference. Size-based information is still in

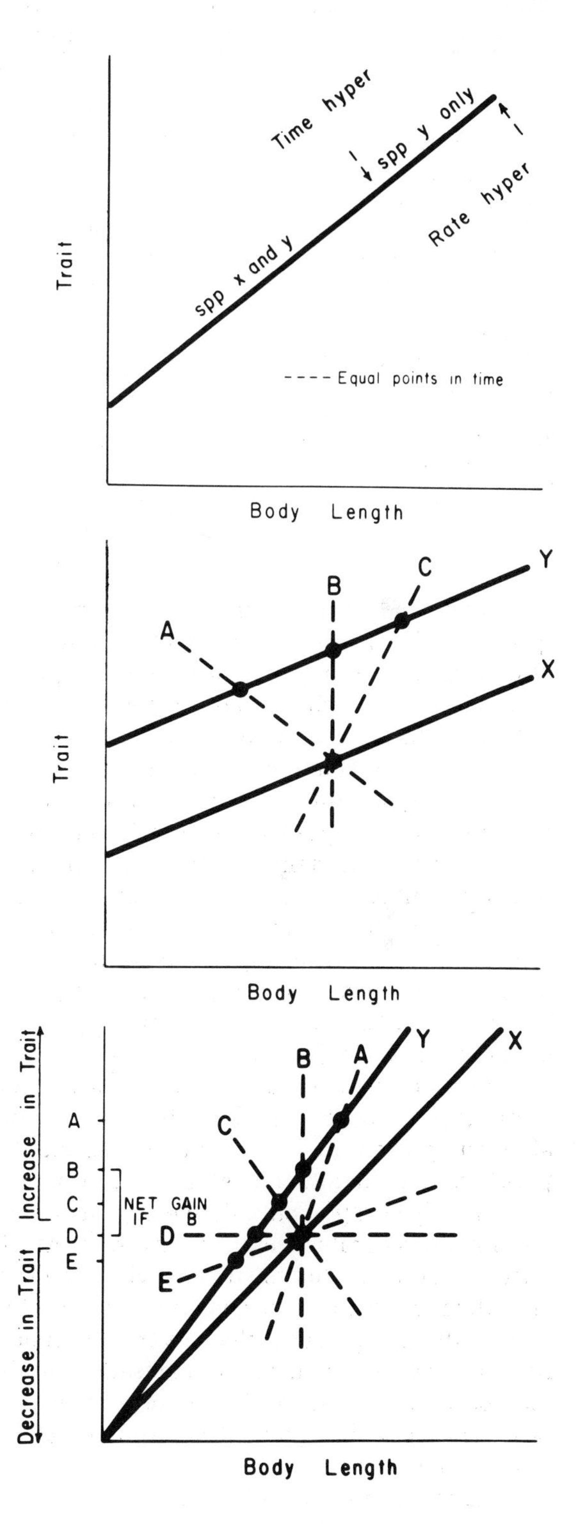

Figure 2. Possible permutations of how different rates of body-size growth can produce the same allometric plot. Dashed line in all three graphs denotes body sizes of same age. See text for explanation.

Table I. Terms of Allometric "Heterochrony" As Defined by Ontogenetic Trajectories of Descendant Species Relative to Ancestral Species

Allometric term	Trajectory of descendant species relative to ancestor[a]		
	Slope (or curve shape)	y-Intercept	Adult body size
Allometric progenesis (= ontogenetic scaling)	Same	Same	Smaller
Allometric hypermorphosis (= ontogenetic scaling)	Same	Same	Larger
Allometric neoteny (= shape dissociation)	Lower	Same	May vary
Allometric acceleration (= shape dissociation)	Higher	Same	May vary
Allometric predisplacement (= shape dissociation)	Same	Greater	May vary
Allometric postdisplacement (= shape dissociation)	Same	Less	May vary

[a] These are illustrated in Fig. 1. Synonyms in parentheses are from Shea (this volume).

itself of considerable use in analyzing ontogeny, evolution, and many of the same questions that heterochronic analysis addresses. This is very important because allometry is one of the most common types of data available, especially in paleontology, where the added dimension of age is often simply unavailable. (Consider also the huge amount of allometry found in bivariate plots to separate groups in many purely taxonomic studies.)

More important is the fact that looking at trait change as a function of size is itself of great interest. Even if size is a differing function of time among many species, it is important to know what trait/body proportions the organism had during its ontogeny. This is useful not only in functional–mechanical ways (Gould, 1968, 1970), but in a more fundamental sense. There is a growing awareness in biology of the critical importance of body size as a primary determinant of not only physiological, but ecological, social, and many other traits as well (Peters, 1983; Calder, 1984). Indeed, it generally accounts for about 75% of the variation in these traits (Calder, 1984).

Further, in some cases body size is a better predictor of an individual's traits than age (Hughes, 1984). This is because "ontogenetic" or "intrinsic" time is often affected by external events so that absolute ("extrinsic") time often corresponds poorly to it. Metabolic rate seems to approximate most closely this internal clock, so that body size (the cumulative product of how much metabolism has occurred) often more accurately reflects intrinsic time (Reiss, in press). Thus, environmentally deprived smaller organisms will often show "compensatory" higher rates of growth later if conditions change (Atchley, 1984): more "intrinsic" time is passing during these high rates than in the "normal" individuals, which experienced higher rates earlier. If no compensation is allowed, the slowed "intrinsic" age may become permanent: hence, organisms [such as gastropods (Kemp and Bertness, 1984)] with sustained environmentally induced differing rates of growth will evince differing allometries.

Table I presents a proposed list of these allometric terms. Thus, we have "allometric progenesis," "allometric predisplacement," and so on. Note that Shea (1983, and this volume) uses a different terminology. Either way, such terms are strictly a matter of expediency and it must always be clear that we are *not* referring to the manifold possible timing or rate processes (as indicated by the qualifying "allometric" adjective) which may have led to any given allometry. It just seems a useful tool to be able to note, for example, that species Y is allometrically accelerated relative to species X.

4. When Age Data Are Present: Heterochrony and Size

4.1. Introduction

Complications of intrinsic age aside, even basic absolute age data in living animals has, surprisingly, not often been used in heterochronic studies. For the fossil record, the lack is more understandable, but certainly no less important. There are a number of ways of obtaining such data from fossils, especially by the use of growth zones. While greatly understudied, these occur in a surprisingly wide variety of organisms (Jones, this volume).

The next question: once obtained, how does one combine the age data with allometric data to get at heterochrony? In the original scheme of Alberch *et al.* (1979), the three factors of heterochrony, age, size, and shape, are visualized by a three-dimensional graph. Such a graph is very informative theoretically, but somewhat difficult in application. I propose that we might be better off by considering shape to be a subset of size. To begin, this is theoretically more honest because in physical terms all that really exists is time and space, so that what changes as a function of time is space occupied. We usually think of size as the quantity of space occupied (or some proxy: mass, length, etc.) and shape as the quality. However, like most dichotomies, this one is specious because what we in practice define as shape is actually the ratio (or relative states) of *sizes* (e.g., "oval" shape as a derivative of the relative magnitudes of length and width). Thus, if we measure rates of growth of trait x and trait *y*, the resulting allometric plot of "shape" change is really just a comparison of the relative "size" change of organs or other morphology. All shape change ultimately reduces to size change(s) in one or more dimensions.

More importantly, this approach is easier to apply. We directly observe the change in rate or timing of a trait (be it biochemical, an organ, or body size) by observing the change in its magnitude over time. Figure 3 shows this treatment in terms of the Alberch *et al.* (1979) scheme, tracing heterochronic events as they are manifested in morphological magnitude (linear, areal, volumetric, or other) through time. The three major changes are growth rate, onset time, and offset time (see caption for explanation and terms). Note that the terms "acceleration" and "neoteny" are truly correct in these graphs since the slope change is really Δ(dimension)/Δ(time), not Δ(dimension)/Δ(size) as in the allometric approach of Fig. 1. Similar logic applies to the timing terms. The sigmoidal pattern is of course not necessary, but merely reflects usual growth patterns.

This scheme has wide application, since most things have a measurable mag-

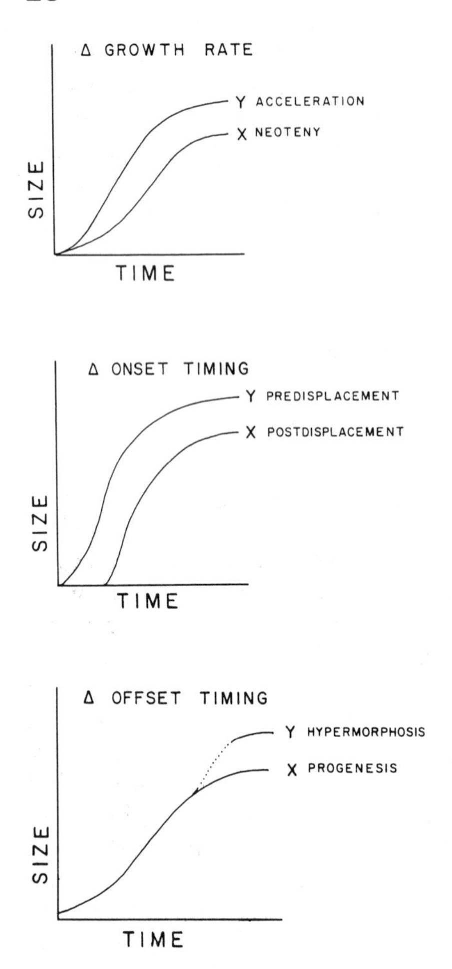

Figure 3. Six major types of heterochrony in terms of size versus age plots. Acceleration: increased growth rate relative to ancestor X; neoteny: decreased growth rate relative to ancestor Y. Predisplacement: early onset of growth relative to ancestor X; postdisplacement: late onset of growth relative to ancestor Y. Hypermorphosis: late offset of growth relative to ancestor X; progenesis: early offset of growth relative to ancestor Y.

nitude and ontogeny (i.e., few things are created *de novo* and last forever) and so have characteristic beginning and ending times and rates, e.g., heterochrony in biochemical processes (Gerhart *et al.*, 1982), hair density changes (Schwartz and Rosenblum, 1981), and even discrete traits such as segmentation in trilobites (Edgecombe and Chatterton, 1987). This last is not surprising, since even discrete traits actually result from threshold effects of continuous processes (Alberch, 1982, 1985b). Even the most common developmental change, the loss of a trait (Raff and Kaufman, 1983), is heterochronic in this light because it represents the endpoint of a gradation where rate (or duration if timing is involved) of growth is reduced to zero. An example might be pigment production in albinos as end member "neoteny" for rate production (or "progenesis/postdisplacement" for zero duration of production), with dark negroids as end member "acceleration, hypermorphosis, or predisplacement," depending on whether rate or timing is involved in pigment differences.

Another important aspect of heterochrony, paedomorphosis and peramor-

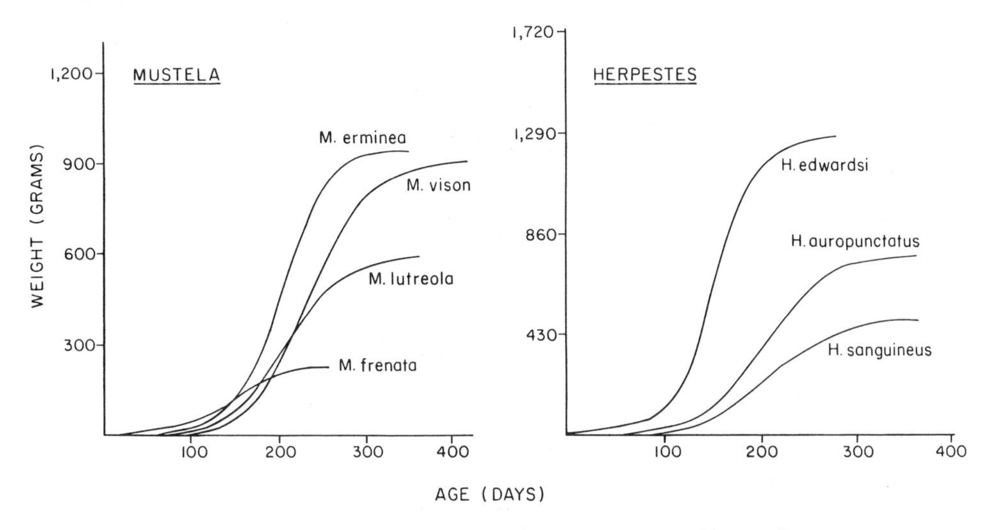

Figure 4. Carnivore body size growth patterns reflecting a variety of heterochronic processes.

phosis, is seen more clearly in this light. Gould (1977) defines these terms as the production of descendants with lesser or greater (respectively) trait development than their ancestors at a given size. Thus, it is often said (Alberch *et al.*, 1979; McNamara, 1986) that neoteny, postdisplacement, and progenesis (Fig. 3) produce paedomorphs, while the converse processes produce peramorphs. However, we see here that it is reduced trait development at a given *age* (through neoteny and the other factors listed) that produce paedomorphosis. Note that this also means that paedomorphic adults do not necessarily resemble ancestral juveniles in this reduced trait development (as is often stated); paedomorphs may only resemble less developed adults.

Finally, I note that body size can itself be considered a "trait" in Fig. 3, subject to acceleration, hypermorphosis, and so on. In this view, Shea's (1983) above-mentioned "rate hypermorphs" refer to animals in which body size (and sometimes, but not necessarily, as discussed below, all other traits) is accelerated (with the converse for rate hypomorphs or progenetics); time hypermorphs refer to those that simply continued to grow for a longer time. It has been stated that larger animals usually grow for a longer time (Bonner and Horn, 1982; Gould, 1974). On a "mouse to elephant" comparison, this is no doubt true (McMahon and Bonner, 1983). However, as Shea (1983, and this volume) has shown, this is not a valid generalization on the level of most heterochronies: closely related species often just grow faster, not for a longer time, in getting bigger. A recent examination of maturation times and body size in carnivores (Gittleman, 1986) shows that larger forms very often just grow faster and may even grow faster *and* for a longer time (sample in Fig. 4). Just why a particular route is chosen may usually be purely a matter of selection and not genetic resources, since body size change is apparently an easy thing to do (Anderson, 1987) with a simple genetic basis (Prothero and Sereno, 1982). Possibly, where selection is on body size alone, it is best to get large quickly, but where other aspects of life history are involved (e.g., prolonged youth for learning in humans, the second largest primate), it may be

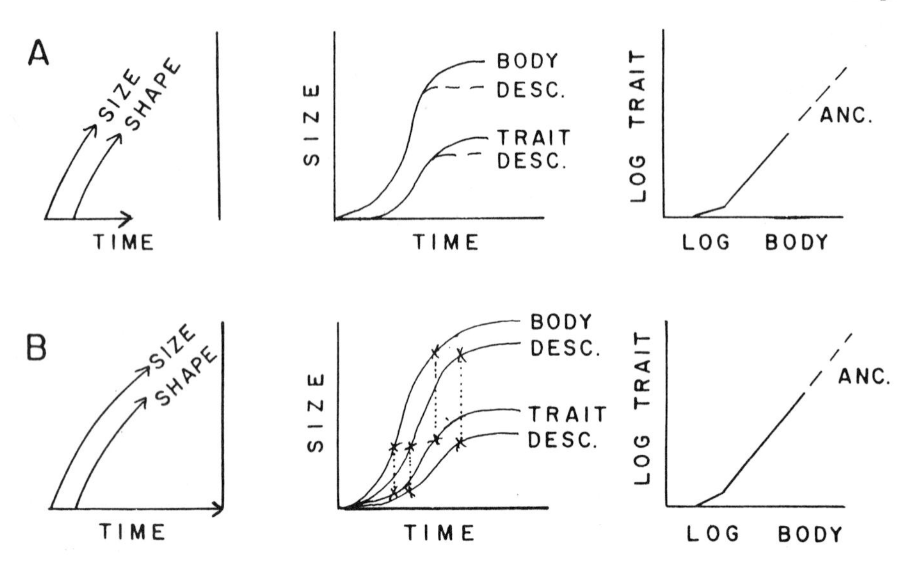

Figure 5. (A) Clock model, size–time, and allometric manifestations of body and trait time progenesis. (B) Same depictions of rate progenesis. Both processes yield the same allometries. Desc., descendant; Anc., ancestor.

more adaptive to increase size by growing for a longer period. Determining whether large animals get there through rate or time changes and what ecological correlates are involved are major prospects for heterochronic research.

4.2. Heterochronic Hierarchy: Trait Coupling and Decoupling

So far we have looked at rate and timing changes in only one trait at a time. But obviously we can make such graphs for many traits in the same individual; the real question is, how often do the traits, including body size, show the same heterochronies? How much independence is there among traits in growth and timing changes? A precise answer awaits empirical data, but "dissociation" is very common, as discussed in detail below.

First, I illustrate a case where the traits under observation (in this case, body and some organ size) show the same heterochrony. Figure 5A is an example of simple time progenesis where the descendant follows the same trajectory as the ancestor but stops growing sooner. As it does this for both traits, in the same pattern, a simple allometric progenesis results. Gould's (1977) "clock model" (Fig. 5A) will work for these cases where both traits show the same heterochrony (for the clock, "size" is body size only, but, as noted above, both body size and shape are reducible: body size is a trait and shape is the ratio of trait sizes). However, the clock fails where the two traits differ. How, for example, can we show cases where one trait grows for a longer time than the other when there is only one time setting? In Fig. 5B we see a case of "rate progenesis" involving both traits and the resultant danger of allometric inference. Both Fig. 5A and Fig. 5B yield the same allometric plot from completely different heterochronies: in the first the

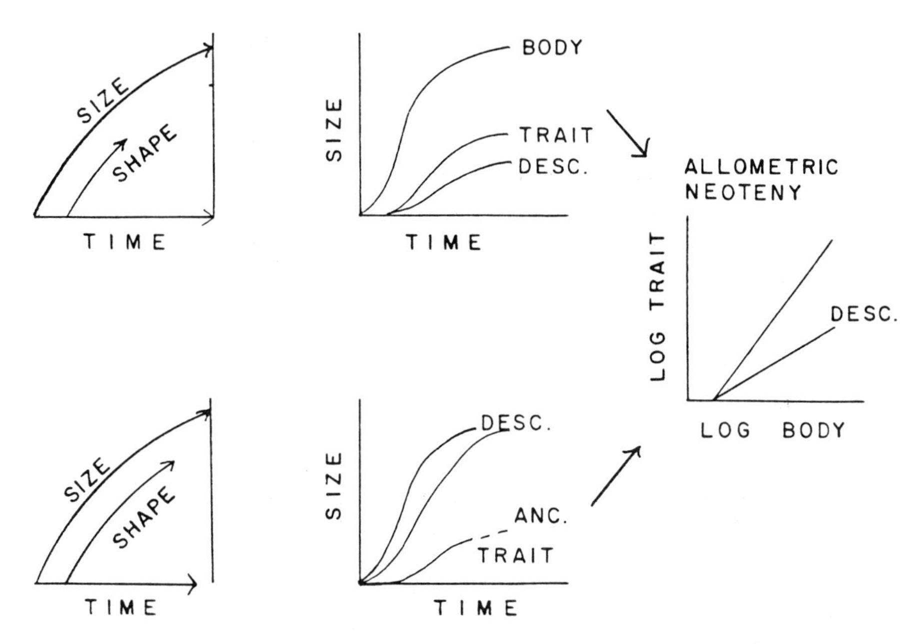

Figure 6. Top: dissociated neotenic trait development in clock and size–time formats. Bottom: accelerated body size in descendant in same formats. Both yield the same allometric plot. Desc., descendant.

descendant grows at the same rate for a shorter time; in the second it grows more slowly for the same time. If either progenesis were truly global in its effects, then all traits would show the patterns of these two.

Figure 6 shows an example where both traits (again organ and body) do not show the same heterochrony. In the case of trait neoteny, the descendant is truly retarded in the growth of the trait and body size is unaffected in both. In the case of body acceleration, the trait is unaffected. However, in both cases, the same allometric plot results, showing yet again the need for time information.

It is not hard to see how such "dissociation" of heterochronies (and hence allometries) can be studied in using this approach. "Mosaic evolution," whereby traits evolve at different rates, has long been a part of the evolutionary vocabulary. In the heterochronic framework, this would be seen as a product of time and rate changes affecting local growth fields (i.e., suite of covarying traits, not necessarily morphologically adjacent) at a variety of magnitudes. A number of evolutionary studies have already been couched in such terms (e.g., Alberch and Alberch, 1981; Guerrant, 1982; Edgecombe and Chatterton, 1987; McKinney *et al.*, in review; Tissot, this volume). As discussed in Chapter 16, this hierarchy of breadth (number of traits affected)—ranging from biochemistry to organs, limbs, and whole body—and depth (magnitude of effect) has major evolutionary implications.

5. A Synthesis: Heterochrony, Allometry, and Ontogeny

What is the biological reality of all these graphs and notions? I have made an effort at a representation in Fig. 7. This relatively simple model is based on a

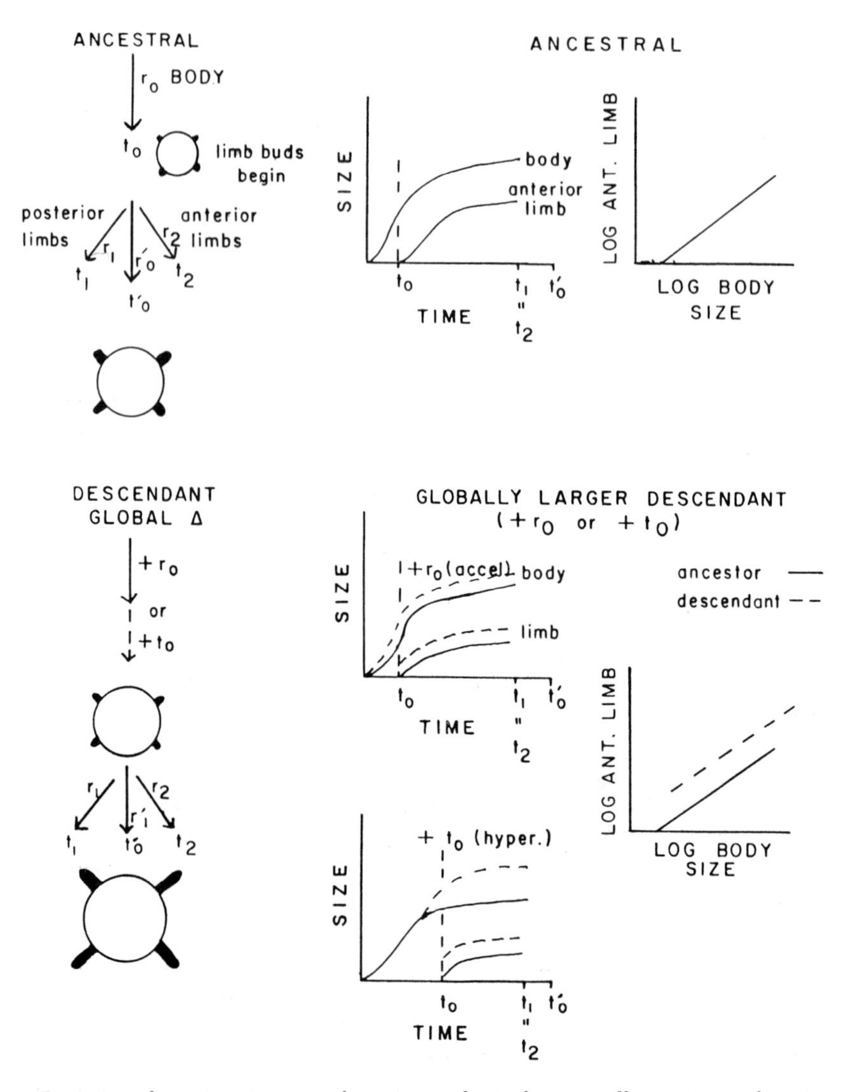

Figure 7. Depiction of ontogenetic events, from time and rate change to allometric manifestation. Upper left: ancestral pattern; *r* and *t* refer to growth rate and offset times of body and organs shown. In global change (lower left) *r* or *t* of undifferentiated embryo is increased, resulting in morphologies (everything is larger) and size–time and allometric plots shown. On right side, *r* and *t* are altered after traits (limbs) have differentiated (i.e., dissociated heterochrony), allowing those traits to be altered individually. Upper right: limb change; lower right: "body" increase without concomitant increase in limb size. Note how allometric curves reflect changes, e.g., body size acceleration (lower right) looks like trait neoteny because at a given body size, limb size is smaller in descendant.

sequence of bifurcating events according to which traits develop from the same tissue until a transition occurs (Oster and Alberch, 1982). It draws heavily from Slatkin's (1987) enlightening paper on quantitative genetics and heterochrony (also see Riska, 1986).

Details are discussed in the caption, but in general two basic kinds of changes are shown. In the global change, a rate or time change occurs early, perhaps when

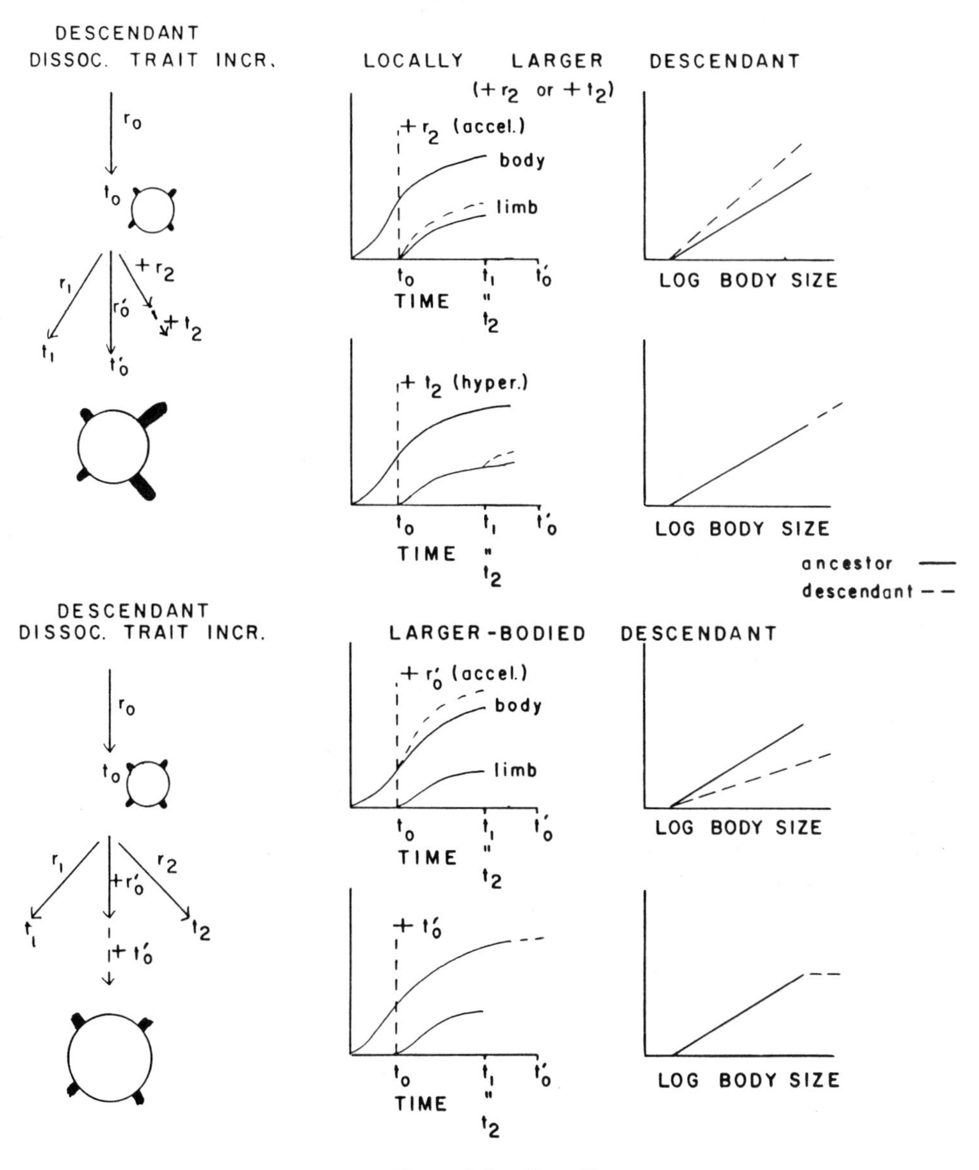

Figure 7. *(continued)*

there is only one undifferentiated cell mass. This means that size decrease or increase cascades through the system because of smaller or larger stem cell size of organs, which affects their ultimate size (e.g., Alberch, 1985b).

In contrast, changes that occur after differentiation can affect only those tissues involved in one area. The later the change is manifested, the more differentiation has occurred and therefore the more local is the impact (as Von Baer noted long ago: the earlier in ontogeny, the more general the effect). This is shown in Fig. 7, which shows a "dissociation" of trait and body size in the sense that after some of the organs have differentiated, a change in timing or rate occurs.

There is ample empirical evidence for this. For example, Riska and Atchley (1985) have demonstrated that brain and body size often share the same growth-regulating systems early in ontogeny. However, changes occurring later in ontogeny affect only body size, since brain tissue has not only differentiated, but finished much of its growth.

We see here, then, that much of the "breadth" of change is determined by when the alteration occurs. In contrast, what determines how much the rate or timing is altered (i.e., "depth" of change) when change occurs?

The proximate cause of course is amount of change in the rate and duration of mitosis (Bonner, 1982; Atchley, 1987). The tie with organ (tissue) allometry is therefore that the constant k apparently represents the relative rate of cell divisions and b is the relative number of stem cells (Katz, 1980). However, this begs the question of what causes these changes in rate and timing of mitosis. A molecular genetic mechanism controlling rate is fairly easy to infer; one for timing is more difficult (Gerhart *et al.*, 1982). In the best-studied examples of heterochrony, e.g., salamanders, growth is controlled by chemical regulators. Thus, genes affecting hormonal production (or reception) can affect rate and timing of growth. In *Ambystoma* a drastic change can be effected by a simple allelic substitution in one gene (Tompkins, 1978). In other cases, e.g., height in humans, the duration and rate of growth are clearly polygenically controlled for the most part, but can be greatly affected by single-gene mutations affecting growth hormones (Shea, this volume). This is also the case in other animals (Prothero and Sereno, 1982). Further discussion is found in Chapter 16, but the main point is that the degree of rate and time change can vary from mild to extreme, depending perhaps in part on whether the change is under polygenic or single-gene control. Either way, the change in rate or time [and therefore allometry (Cock, 1966)] is heritable and is therefore open to fixation in a population. [However, the mechanics of how these genes carry out the rate and timing changes are among the least known aspects of heterochrony (Bonner, 1982; Raff and Kaufman, 1983).]

ACKNOWLEDGMENTS. I thank Bill Calder, Craig Oyen, and Pere Alberch for comments.

References

Alberch, P., 1982, Developmental constraints in evolutionary processes, in: *Evolution and Development* (J. T. Bonner, ed.), pp. 313–332, Springer-Verlag, Berlin.

Alberch, P., 1985a, Problems with the interpretation of developmental sequences, *Syst. Zool.* **34**:46–58.

Alberch, P., 1985b, Developmental constraints: Why St. Bernards often have an extra digit and Poodles never do, *Am. Nat.* **126**:430–433.

Alberch, P., and Alberch, J., 1981, Heterochronic mechanisms of morphological diversification and evolutionary change in the neotropical salamander, *Bolitoglossa occidentalis*, *J. Morphol.* **167**:249–264.

Alberch, P., Gould, S. J., Oster, G. F., and Wake, D. B., 1979, Size and shape in ontogeny and phylogeny, *Paleobiology* **5**:296–317.

Anderson, D. T., 1987, Developmental pathways and evolutionary rates, in: *Rates of Evolution* (K. S. Campbell and M. F. Day, eds.), pp. 143–155, Allen and Unwin, Boston.

Atchley, W. R., 1984, Ontogeny, timing of development and genetic variance–covariance structure, *Am. Nat.* **123**:519–540.
Atchley, W. R., 1987, Developmental quantitative genetics and the evolution of ontogenies, *Evolution* **41**:316–330.
Atchley, W. R., Rutledge, J. J., and Cowley, D. E., 1981, Genetic components of size and shape, *Evolution* **35**:1037–1055.
Bonner, J. T. (ed.), 1982, *Evolution and Development*, Springer-Verlag, Berlin.
Bonner, J. T., and Horn, H. S., 1982, Selection for size, shape, and developmental timing, in: *Evolution and Development*, (J. T. Bonner, ed.), pp. 259–277, Springer-Verlag, Berlin.
Calder, W., 1984, *Size, Function, and Life History*, Harvard University Press, Cambridge.
Cheverud, J., 1982, Relationships among ontogenetic, static, and evolutionary allometry, *Am. J. Phys. Anthropol.* **59**:139–149.
Cock, A. G., 1966, Genetical aspects of metrical growth and form in animals, *Q. Rev. Biol.* **41**:131–190.
Davis, J. C., 1986, *Statistics and Data Analysis in Geology*, Wiley, New York.
Edgecombe, G. D., and Chatterton, B. D. E., 1987, Heterochrony in the Silurian radiation of encrinurine trilobites, *Lethaia* **20**:337–351.
Emerson, S. B., 1986, Heterochrony and frogs: The relationship of a life history trait to morphological form, *Am. Nat.* **127**:167–183.
Gerhart, J. C., *et al.*, 1982, The cellular basis of morphogenetic change (Group Report), in: *Evolution and Development* (J. T. Bonner, ed.), pp. 87–113, Springer-Verlag, Berlin.
Gittleman, J. L., 1986, Carnivore life history patterns: Allometric, ecological, and phylogenetic associations, *Am. Nat.* **217**:744–771.
Gould, S. J., 1966, Allometry and size in ontogeny and phylogeny, *Biol. Rev.* **41**:587–680.
Gould, S. J., 1968, Ontogeny and the explanation of form: An allometric analysis, *J. Paleontol.* **42**:81–98.
Gould, S. J., 1970, Evolutionary paleontology and the science of form, *Earth Sci. Rev.* **6**:77–119.
Gould, S. J., 1974, The evolutionary significance of 'bizarre' structures: Antler size and skull size in the 'Irish Elk', *Megaloceras gigantans*, *Evolution* **28**:191–220.
Gould, S. J., 1977, *Ontogeny and Phylogeny*, Harvard University Press, Cambridge.
Guerrant, E. O., 1982, Neotenic evolution of *Delphinium nudicaule* (Ranuculaceae): A hummingbird-pollinated larkspur, *Evolution* **36**:699–712.
Hughes, T. P., 1984, Population dynamics based on individual size rather than age: A general model with a coral reef example, *Am. Nat.* **123**:778–795.
Huxley, J. S., 1932, *Problems of Relative Growth*, Methuen, London.
Imbrie, J., 1956, Biometrical methods in the study of invertebrate fossils, *Bull. Am. Mus. Nat. Hist.* **108**:219–252.
Katz, M. J., 1980, Allometry formula: A cellular model, *Growth* **44**:89–96.
Kemp, P., and Bertness, M. D., 1984, Snail shape and growth rates: Evidence for plastic shell allometry in *Littorina littorea*, *Proc. Natl. Acad. Sci. USA* **81**:811–813.
Laird, A. K., 1965, Dynamics of relative growth, *Growth* **29**:249–263.
Lande, R., 1979, Quantitative genetic analysis of multivariate evolution, applied to brain:body size allometry, *Evolution* **33**:402–416.
McKinney, M. L., 1984, Allometry and heterochrony in an Eocene echinoid lineage: Morphological change as a byproduct of size selection, *Paleobiology* **10**:407–419.
McKinney, M. L., 1986, Ecological causation of heterochrony: A test and implications for evolutionary theory, *Paleobiology* **12**:282–289.
McKinney, M. L., and Schoch, R. M., 1985, Titanothere allometry, heterochrony, and biomechanics: Revising an evolutionary classic, *Evolution* **39**:1352–1363.
McKinney, M. L., McNamara, K. J., Zachos, L. G., and Oyen, C. W., Heterochrony of growth fields: Evolution by less than monstrous hopefuls, *Paleobiology* (in review).
McMahon, T. A., and Bonner, J. T., 1983, *On Size and Life*, Freeman, New York.
McNamara, K. J., 1986, A guide to the nomenclature of heterochrony, *J. Paleontol.* **60**:4–13.
Oster, G., and Alberch, P., 1982, Evolution and bifurcation of developmental programs, *Evolution* **36**:444–459.
Peters, R. H., 1983, *The Ecological Implications of Body Size*, Cambridge University Press, Cambridge.
Prothero, D. R., and Sereno, P. C., 1982, Allometry and paleoecology of medial Miocene dwarf rhinoceroses from the Texas Gulf Coastal Plain, *Paleobiology* **8**:16–30.

Raff, R. A., and Kaufman, T. C., 1983, *Embryos, Genes, and Evolution*, Macmillan, New York.
Reiss, J. O., 1988, The meaning of developmental time: Attempt at a metric for comparative embryology, *Am. Nat.* (in press).
Riska, B., 1986, Some models for development, growth, and morphometric correlation, *Evolution* **40:**1303–1311.
Riska, B., and Atchley, W. R., 1985, Genetics of growth predict patterns of brain size evolution, *Science* **229:**668–671.
Schwartz, G. G., and Rosenblum, L. A., 1981, Allometry of primate hair density and the evolution of human hairlessness, *Am. J. Phys. Anthropol.* **55:**7–12.
Shea, B. T., 1983, Allometry and heterochrony in the African apes, *Am. J. Phys. Anthropol.* **62:**275–289.
Shea, B. T., 1985, Bivariate and multivariate growth allometry: Statistical and biological considerations, *J. Zool. Lond. A* **206:**367–390.
Slatkin, M., 1987, Quantitative genetics of heterochrony, *Evolution* **41:**799–811.
Thompson, D'A. W., 1917, *On Growth and Form*, Cambridge University Press, Cambridge.
Tompkins, R., 1978, Genetic control of axolotl metamorphosis, *Am. Zool.* **18:**313–319.
White, J. F., and Gould, S. J., 1965, Interpretation of the coefficient in the allometric equation, *Am. Nat.* **99:**5–18.

Chapter 3

Multivariate Analysis

BRIAN N. TISSOT

1. Introduction

This chapter focuses on the multivariate analysis of morphological variation resulting from heterochrony, or changes in the timing of developmental events during ontogeny (de Beer, 1958). My goals are threefold: (1) outline the steps in a multivariate analysis, (2) illustrate the methodology with data on a recent marine gastropod, the black abalone, *Haliotis cracherodii*, and (3) discuss conceptual difficulties involving heterochrony.

Heterochrony invokes multiple changes in organismal morphology. Most traits change ontogenetically and are functionally, developmentally, or genetically correlated with other traits (Falconer, 1981; Atchley, 1984; Maynard Smith *et al.*, 1985). Thus, heterochronic change is multivariate and the measurement and simultaneous description of these changes require a multivariate approach. Bivariate techniques are an additional tool for examining heterochrony; they are useful for dissecting out changes in individual characters and describing dissociations among traits (McKinney, this volume).

BRIAN N. TISSOT • Department of Zoology, Oregon State University, Corvallis, Oregon 97331.

1.1. Historical Perspective

A history of the ideas leading to our current understanding of heterochronic processes was reviewed by Gould (1977). My goal here is to review the development of quantitative multivariate descriptions of ontogeny and their application to problems involving heterochrony.

Descriptions of ontogenetic change began with Thompson (1917) and Huxley (1932) (see McKinney, this volume). Examination of covariation among characters did not occur until the advent of computers in the late 1950s. Early multivariate studies examined polymorphism (Blackith, 1957), growth (Rao, 1958), geographic variation (Jolicoeur, 1959), sexual dimorphism (Jolicoeur and Mosimann, 1960), and interspecific variation (Blackith, 1960) using discriminate analysis and principal component analysis [reviews by Cock (1966) and Gould (1966)]. Links between bivariate and multivariate allometry were provided by Jolicoeur (1963a,b) [see Shea (1985) for examples].

Gould's (1969) factor analysis of paedomorphosis in Bermudian land snails was perhaps the first explicit application of multivariate techniques to problems of heterochrony. His *Ontogeny and Phylogeny* (Gould, 1977) laid the foundations for a synthesis of the fields of heterochrony and the quantitative allometric apraoch. The clock model of heterochronic change, although qualitative, was the basis for the semiquantitative model of Alberch *et al.* (1979) and most current studies of heterochrony are extensions and elaborations of their model [bivariate: Alberch and Alberch (1981); Shea (1983); McKinney (1984, and this volume); multivariate: Shea (1985); Schweitzer *et al.* (1986); Tissot (1988)].

1.2. Metrics of Heterochrony

Alberch *et al.* (1979) proposed a model of heterochrony involving size, shape, and age as independent variables. Relying on the quantification of a character's *ontogenetic trajectory*, a path depicting the development of a character, the model is intended to distinquish between different forms of heterochrony. Within a space defined by age, size, and shape, different trajectories describe variation in three parameters: α, the size (or age) of onset of growth of a character; β, the size (or age) of termination (offset) of growth of a character; and k, the rate of change of shape relative to size (or age). A major problem with the use of their model is a general lack of data on age. As heterochrony, by definition, describes shifts in developmental timing, examination of size and shape describes *allometric heterochrony* rather than pure heterochrony (see McKinney, this volume).

In order to measure quantitative heterochronic changes, analytical methods must supply the following: (1) multivariate metrics of size and shape; (2) calculation of ontogenetic trajectories; (3) discrimination among ancestor–descendant trajectories with respect to slopes, intercepts, and developmental events; and (4) a technique to incorporate age into the morphological framework. Below I describe methods that fulfill objectives 1–3; objective 4 will be treated in the discussion.

2. Multivariate Methodology

In this section I outline the steps in the multivariate analysis of ontogenetic variation. Problems associated with samples, standardizing data, and choosing an appropriate dispersion matrix for analysis are discussed. I describe the use of principal component analysis (PCA) for deriving multivariate measures of size and shape and their formulation into ontogenetic trajectories.

2.1. Samples

The goal is to sample individuals from a population or species from which information relating to ontogenetic variation can be obtained. Typical samples in morphometric studies constitute mixed cross sections of populations: a series of individuals of varying sizes (Cock, 1966). As the age of individuals is unknown, size and shape are usually confounded with age due to temporal variation in growth. A longitudinal sample, data on the growth of individuals throughout their ontogeny, is required to measure the influence of age on size and shape (Cock, 1966). Longitudinal samples, however, are difficult or impossible (in the case of most fossils) to obtain. Mixed cross-sectional samples are frequently used in morphometric studies, and represent average growth patterns within populations (Cock, 1966). Based on information presented by Dudzinski *et al.* (1975), a minimum sample size for PCA is 30–40 individuals.

2.2. Data

Data are grouped by their functional level of representation, that of the operational taxonomic unit (OTU), which is the lowest ranking taxon employed in a study (Sneath and Sokal, 1973). OTUs may represent individuals, populations, or species and are grouped by general factors, such as age, sex, population, or species.

Before analysis by PCA, raw data are frequently transformed to alternate scales of measurement in order to equalize variances, remove scaling effects, and promote linearity (Sokal and Rohlf, 1981). Raw data are rarely used in PCA, due to scale effects. As PCA describes patterns of covariation among variables based on their magnitude, the raw variances of variables directly determine the resulting components (Pimentel, 1979). Log data may circumvent this problem, as logarithmic transformation promotes independence of the variance and the mean (Sokal and Rohlf, 1981; Bryant, 1986).

Many data sets constitute a mixed array of quantitative and qualitative data types of different units (e.g., millimeters, degrees, counts, character state codes) (Sneath and Sokal, 1973). As with raw data, log-transformed mixed-mode data introduce scale effects that influence subsequent principal components. Separate analysis of continuous and discontinuous variables can circumvent this problem

(Humphries *et al.*, 1981). If the number of traits is small, however, this approach may be unfeasible. Moreover, separate analyses of different types of data expand rather than simplify our descriptions of ontogeny. Transformation of the data to z-scores eliminates problems of mixed-mode variables. Variances are standardized by dividing data values by standard deviations (Sokal and Rohlf, 1981). PCA of standardized data results in a more even weighing of variables and has proven useful in taxonomic studies (Thorpe, 1980; Pimentel, 1981).

PCA examines patterns of variation among measures of dispersion of a sample. Two general types of dispersion matrices are defined, based on whether OTUs or OTU traits provide the variation (Pimentel, 1981). In *R*-mode analyses, the dispersion matrix describes variation among OTU traits. In *Q*-mode analyses, the dispersion matrix measures variation among OTUs. Both modes are commonplace in morphometric studies and may produce similar results; *R*-mode techniques are more common in the literature (Thorpe, 1980; Pimentel, 1981) and will be the focus of my analysis.

In *R*-mode PCA, the dispersion matrix is based on variances and covariances, or on correlations, among traits. When variables are linear distance measures, the covariance matrix is preferred, as it is supported by a large amount of statistical theory (Anderson, 1963; Dillon and Goldstein, 1984) and by links between bivariate and multivariate allometry (Jolicoeur, 1963a). The correlation matrix is used when variables are mixed mode, as correlations are independent of scale. When data are standardized, the covariance matrix is equivalent to the correlation matrix (Pimentel, 1981).

2.3. Analytic Methods: Size and Shape

Our goal is to obtain metrics of size and shape that we can combine into ontogenetic trajectories. Factor analysis, or PCA in the case where factors are orthogonal, is a multivariate technique that is concerned with the identification of structure within a set of observed variables (Dillon and Goldstein, 1984). In data sets with OTUs that vary in size, PCA almost invariably produces components that describe variation in size and shape.

PCA represents a transformation and rotation of original variable axes to new variable axes, or components, that describe successively smaller partitions of the variation in the dispersion matrix, subject to the constraint that axes are orthogonal or uncorrelated. The amount of variation described by each axis is measured by the component's length, or eigenvalue. Eigenvector coefficients, or directional cosines, are the cosines of angles between components and original variable axes. Eigenvector loadings are equivalent to the allometric loading of variables on components. Component scores represent values for each OTU based on relationships to eigenvectors; scores are derived from a linear combination of the raw data and component loadings. Component correlations represent product-moment correlations between principal component scores and the original data. Additional PCA metrics that aid in interpretation of components are the amount of variation attributed to each trait or OTU on each component (Pimentel, 1979).

Although PCA has been a common approach in morphometric studies (Jolicoeur, 1959, 1963b; Jolicoeur and Mosimann, 1960; Gould, 1969; Tissot, 1984;

Shea, 1985), it has three problems. First is the observation that the first principal component (PC1) does not always measure pure (or isometric) size. Because traits have different allometric loadings with general size, variation in shape is inextricably associated with changes in size (see Discussion). Much of the confusion associated with the use of PC1 as a metric of size is semantic: PC1 derived from ontogenetic data does not simply describe variation in size, but in addition, variation in shape related to allometric growth (Shea, 1985). PC1 should therefore be labeled a size-allometry axis (hereafter referred to as "*allometric size*"). Second, the mathematical constraint of orthogonality between components removes shape variation contained in PC1 from subsequent components, which are thus not measures of pure (or allometric) shape. Components subsequent to PC1 describe variation in shape that is mathematically independent of allometric growth. These shape axes describe differences in growth attributable to changes in the slope and intercepts of ontogenetic trajectories (Shea, 1985; Tissot, 1988). A third problem pertains to the extrapolation of PCA, originally intended for single-group analysis, to the multigroup case. Differences in size among groups can confound size and shape variation among principal components, in addition to their functional and mathematical mingling above.

The shear-PCA was proposed by Humphries *et al.* (1981) as a solution to the intergroup size problem. Shear-PCA measures the confounding of intergroup size and shape on PC1 by comparing unmodified principal components to those based on group-free dispersion: a pooled within-group dispersion matrix in which intergroup size differences have been eliminated by mean-centering data within group. Using PC1 from the group-free dispersion matrix (the within-group size-allometry axis), residual shape variation in the original PC1 is measured and placed (or sheared) into subsequent components (Humphries *et al.*, 1981). This procedure amounts to calculating a vector orthogonal to the group-free PC1, but within the plane of the original components. The shearing method ignores divergence among groups in PC1, a limitation that should be explored by separate PCAs of each group (see Section 4). When compared to PCA from unmodified multigroup data, shear-PCA shows enhanced group discrimination (Humphries *et al.*, 1981; Bookstein *et al.*, 1985) and clear separation of allometric size and size-independent shape.

Computer algorithms for PCA are listed by Blackith and Reyment (1971), Cooley and Lohnes (1971), Mather (1976), and Bookstein *et al.* (1985). Several statistical packages offer conventional PCA: BIOSTAT (R. A. Pimentel and J. D. Smith, 1985, Sigma Soft, Placentia, CA), BMDP (Dixon and Brown, 1979), SAS (SAS Institute, Cary, NC), SPSS (SPSS Inc., Chicago, IL), and SYSTAT (SYSTAT Inc., Evanston, IL). A SAS procedure for performing a shear-PCA is listed by Bookstein *et al.* (1985). Programs that perform both conventional and shear-PCA are available from the author for use of IBM personal computers.

After a PCA or shear-PCA run, one is faced with the problem of determining which components to examine and the biological interpretation of axes. Eigenvalues, eigenvectors, component correlations, component scores, and percent variation due to variables and OTUs are helpful interpretative aids. Of course, the ultimate basis for interpretation of components is whether they make biological sense (Oxnard, 1978; Pimentel, 1979).

Eigenvalues are measures of the amount of variability described by each component and aid in determining which components describe significant amounts of variation. Several techniques are available: (1) examining axes until 90% of the total variation is described, (2) examining axes with eigenvalues that are larger than average, (3) scree tests in which eigenvalues are plotted against axis number and visually or statistically examined for breaks in their pattern of decline, and (4) comparisons between real and random eigenvalues (Anderson, 1963; Horn, 1965; Cattell, 1966; Dillon and Goldstein, 1984). In most cases the 90% rule combined with biological interpretation is sufficient for separation of real from trivial components.

Of greater difficulty is the objective interpretation of components. In general, significant component correlations indicate which eigenvector loadings to interpret, while the percent variation attributed to each variable explains the amount of variation being described on a variable-by-variable basis. Isometry tests that compare eigenvectors with a theoretical component having equal loadings are useful for detecting isometric size components (Jolicoeur, 1963a, 1984). Since increases in general size almost invariably involve allometric change among variables, deviation from isometry is expected (Gould, 1966; Sprent, 1972). Multivariate methods that force the first principal axis to isometry create relationships among characters that are not likely to exist (Somers, 1986). The correlation between eigenvector loadings and the allometric loadings of variables on a measure of body size, such as length, mass, or volume, is a way of examining PCA size in relation to bivariate allometric size (Tissot, 1988). Significant correlation between eigenvector and allometric loadings will identify the size-allometry axis, which is invariably the first component in ontogenetic samples.

Component scores are useful in several ways: they aid in the interpretation of components via group ordination, and serve as measures of multivariate variation in the calculation of ontogenetic trajectories. Examination of group ordinations based on component scores is helpful toward identifying outliers and detecting nonlinearity in the data (Pimentel, 1979).

2.4. Analytic Methods: Multivariate Ontogenetic Trajectories

After interpretation of components as contrasts of allometric size and statistically independent measures of shape, component scores serve as metrics of size and shape in the calculation of ontogenetic trajectories. By using component scores as linear combinations of the original variables, PCA reduces a multivariate system to a series of bivariate relationships. Linear regression is used to measure relationships between allometric size and shape. The resulting linear equation predicts shape on the basis of size and serves as a depiction of the average path of ontogeny of OTUs within groups in the size–shape space defined by component axes.

As neither size nor shape can be considered a dependent variable, the appropriate method of analysis is model II regression. As principal component scores are in the same units (those of the combined variables), the slope of the major (or principal) axis for each group is the best estimate of the interrelationships between size and shape (Sokal and Rohlf, 1981). The problem with using model II regression is that methods for discriminating among group slopes and intercepts (analysis of covariance) are based on model I regression, in which one variable is considered dependent on the other. Thus, in order to discriminate among group trajectories, model I techniques, such as least-squares regression, must be used.

In reference to the allometric outcomes of the Alberch *et al.* (1979) model, the slope of the regression is an estimate of shape change relative to size (k), while the intercept serves to delineate onset or offset signals (α and β). Intercepts on ontogenetic trajectories represent average shape scores at size scores of zero. In order to make intercepts biologically meaningful, PCA scores need to be centered based on consistant criteria (e.g., the size score at which length is zero). Developmental events are incorporated into ontogenetic trajectories by calculating the intercept of their regression on size components (i.e., predicting the size at which the event occurs) (Tissot, 1988).

Ontogenetic trajectories are examined for complex allometry by comparing F-ratios of regressions using higher order size and shape terms (McKinney, 1984). Once trajectories are linear, discrimination among groups can be accomplished by analysis of covariance, which tests for differences among slopes and intercepts of regressions (Sokal and Rohlf, 1981).

3. An Example: Intraspecific Divergence and Allometric Heterochrony in the Black Abalone, *Haliotis cracherodii*

To illustrate the multivariate measurement of heterochronic change, I present an analysis of ontogenetic variation in a Recent marine gastropod. The data analyzed are typical of most morphometric studies: simple, mixed-mode traits describing, in this case, shell morphology.

3.1. Introduction

The black abalone, *Haliotis cracherodii,* is a common gastropod of the intertidal and subtidal zones of the eastern Pacific coast from San Francisco to southern Baja California (Cox, 1962). Developmental variation in the black abalone involves increases in the number of respiratory pores, or tremata, on the shell with increasing size (Fig. 1). Associated with increasing tremata number are changes in tremata diameter and spacing (Fig. 1). In addition, black abalone ontogeny is marked by a well-defined developmental event: the attainment of a muscle scar on the shell. Development of the muscle scar coincides with a slowing or cessation in shell growth.

Intraspecific divergence among populations of the black abalone involves ontogenetic processes. Individuals from populations at lower latitudes possess more numerous tremata at a given size than individuals from populations at higher latitudes. As the number of tremata varies with ontogeny, geographic variation in size–tremata relationships is *a priori* evidence that heterochrony is promoting intraspecific divergence.

3.2. Materials

Samples of black abalone were obtained by haphazardly collecting ontogenetic series of animals from several localities. Samples were obtained from four

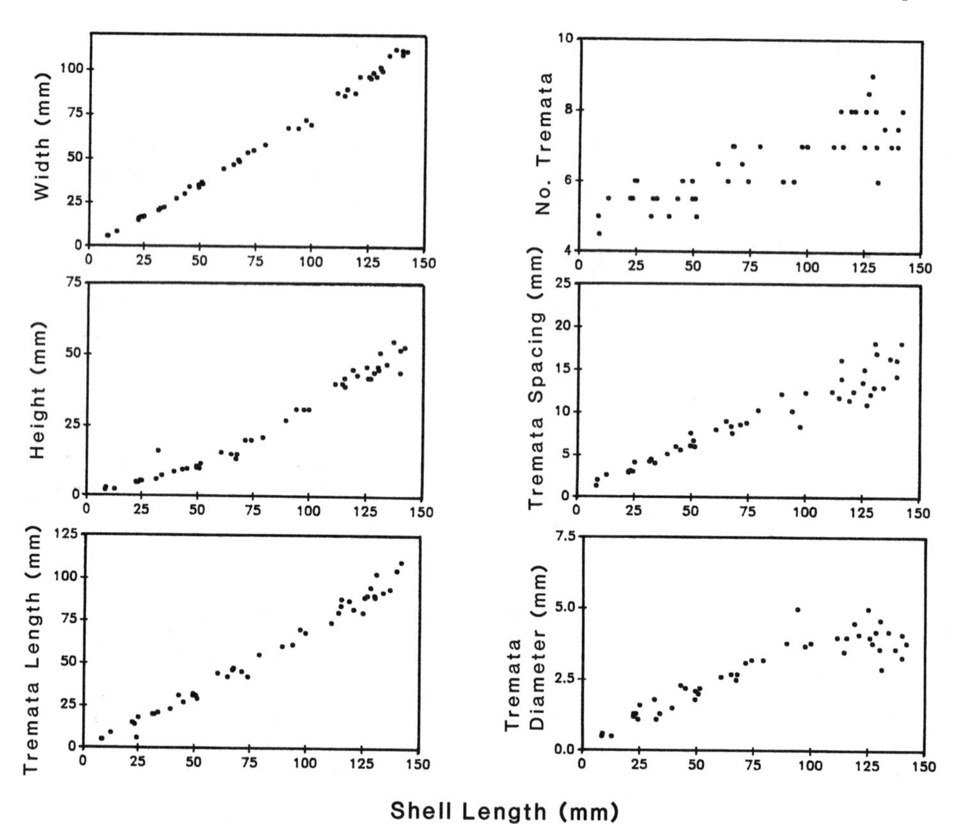

Figure 1. Allometry in black abalone from Santa Cruz Island, California. The relationship between variation in shell length and morphological variables is presented. See Fig. 2 for a diagram of variables. Table II lists the allometric loadings of each variable on shell length.

geographically separate populations: Año Nuevo Island, San Mateo County, CA (n = 54; 37° 6′ N, 122° 20′ W); Valley Anchorage, Santa Cruz Island, CA (n = 47; 34° N, 119° 40′ W); Natividad Island, Baja California (n = 41; 27° 40′ N, 115° 20′ W); and Guadalupe Island, Baja California (n = 64; 29° N, 118° 15′W). As the age of specimens was unknown, collections represented mixed, cross-sectional samples of populations.

Seven traits were measured on the shell of each individual (Fig. 2). Length, width, tremata length, and height describe the overall size of the shell; tremata number, size, and spacing describe developmental variation in shell shape (Fig. 1). The muscle scar was recorded as the proportion of the muscle attachment area covered by scar nacre.

3.3. Multivariate Analysis

As the data were of mixed mode (millimeters and counts), the raw data were transformed to z-scores prior to analysis. The pooled correlation matrix of vari-

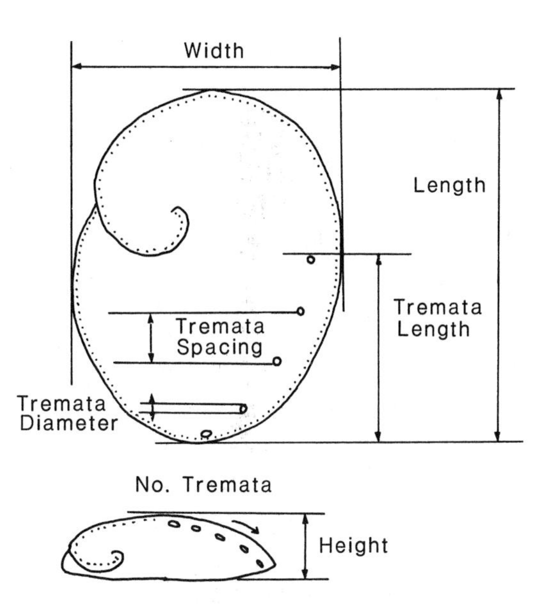

Figure 2. Morphometric variables used to describe the size and shape of black abalone shells.

ables was analyzed using both PCA and shear-PCA (*R*-mode analyses). As the results of both techniques were similar, I will present the shear-PCA and indicate differences between analyses where they occurred.

Two components from the shear-PCA described greater than 90% of the total variation among variables and had eigenvalues close to or greater than average eigenvalues ($\lambda = 1$ for correlation matrix analyses) (Table I). Based on component correlations and the percent variation due to each trait, the first two components described most of the variation among variables and were significantly correlated with the original data (Table I).

The first principal component described 78% of the total variation in the correlation matrix. Eigenvector loadings for PC1 deviated significantly from isometry ($p < 0.001$) and were significantly correlated with the allometric loadings of each variable on shell length ($r = 0.94$, $p < 0.05$). I thus interpreted the first

Table I. Results of Shear-PCA on 206 Individual Black Abalone from Four Localities[a]

	Eigenvectors			Component correlations			Percent variation		
Variable	PC1	PC2	PC3	PC1	PC2	PC3	PC1	PC2	PC3
Length	0.42	0.09	−0.05	**0.99**	−0.08	−0.03	97.3	0.2	0.1
Width	0.42	0.04	−0.08	**0.99**	−0.04	−0.05	97.2	0.0	0.2
Height	0.41	0.20	−0.10	**0.96**	**−0.19**	−0.06	92.2	0.5	0.4
Tremata number	0.23	0.74	0.16	**0.54**	**−0.69**	0.09	28.9	67.3	0.9
Tremata length	0.42	0.14	−0.13	**0.98**	−0.13	−0.07	96.0	0.5	0.5
Tremata spacing	0.36	−0.31	−0.48	**0.85**	**0.29**	**−0.28**	72.7	9.6	7.9
Tremata diameter	0.34	−0.52	0.84	**0.80**	**0.49**	**0.49**	64.6	10.2	24.2
Eigenvalues	5.49	0.88	0.34						
Percent variation	78.4	12.6	4.9						

[a] Component correlations significantly different from zero are indicated in boldface ($p < 0.05$). Measured variables are diagrammed in Fig. 2.

Table II. Allometric Loadings of Variables on Shell Length for Four Samples of Black Abalone[a]

Variable	Año Nuevo Island	Santa Cruz Island	Natividad Island	Guadalupe Island	All data pooled
Width	1.11	1.07	1.01	1.07	1.08
Height	1.25	1.21	1.35	1.38	1.24
Tremata number	0.17	0.18	0.26	0.40	0.26
Tremata length	1.08	1.10	0.99	1.12	1.11
Tremata spacing	0.94	0.83	0.71	0.67	0.67
Tremata diameter	0.59	0.75	0.52	0.45	0.61
Sample size	54	47	41	64	206

[a] Divergence in allometric size among groups is due to interactions among allometric relationships of traits within samples. All loadings are significantly different from zero at $p < 0.01$.

component as the size-allometry axis: variation in shape attributable to allometric growth.

Multigroup PCA revealed small but significant divergence among samples in the size-allometry axis. The angles between PC1 from pooled shear-PCA and PC1s based on single-group analyses were 2.6° (Año Nuevo Island), 7.3° (Santa Cruz Island), 9.3° (Natividad Island), and 5.7° (Guadalupe Island) (all angles $p < 0.01$ from zero). Although group divergence in size allometry was statistically significant, the magnitude of these differences was small and not likely to confound the use of PC1 as a measure of size. Group differences in allometric size resulted from statistical interactions among the allometry of traits within samples (Table II).

Component two described 13% of the total variation (Table I). Eigenvector loadings and component correlations indicated that this axis contrasted variation among individuals in the number, diameter, and spacing of tremata. Variable loadings indicated that as tremata increase in number, tremata diameter and spacing decrease. As traits covaried consistently with initial observations on the ontogeny of black abalone, I interpreted PC2 as describing ontogenetic variation among individuals.

Multigroup PCA revealed significant group divergence in the second principal component (PC2). The angles between PC2 derived from pooled and single-group analyses were 16.7° (Año Nuevo Island), 22.6° (Santa Cruz Island), 19.4° (Natividad Island), and 18.4° (Guadalupe Island) (all angles $p < 0.01$ from zero). This pattern illustrates two points: (1) the pooled analysis describes variation due principally to the extreme samples (high-latitude Año Nuevo Island and low-latitude Guadalupe Island), and (2) PC2 is not simply a measure of size-independent shape, but also describes variation *among* samples. As the shear-PCA removes intergroup size differences, PC2 describes divergence in shape that is independent of allometric growth.

The shear-PCA produced greater group discrimination than conventional PCA. Based on analysis of variance of second-axis principal component scores, shear-PCA produced larger F-ratios based on size-independent shape ($F = 222.0$, $p < 0.01$) than conventional PCA ($F = 219.5$, $p < 0.01$). In essence, the shear-PCA removed intergroup differences in shape on the first axis ($F = 10.7$,

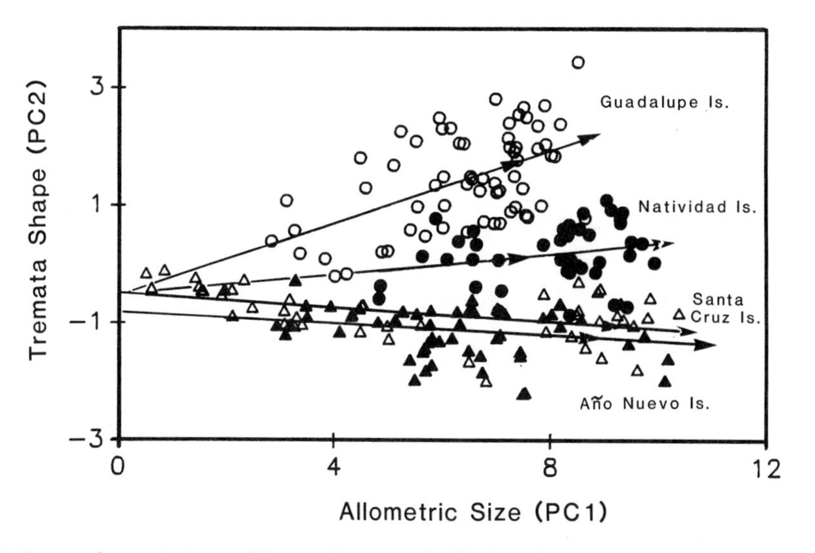

Figure 3. Geographic variation and heterochrony in the black abalone. Ontogenetic trajectories for four populations based on first- (PC1) and second-axis (PC2) principal component scores are presented in relation to the size at the onset of muscle scar development (see text), which is denoted by arrowheads on the length of trajectories. Population samples are denoted by symbols: (○) Guadalupe Island, (●) Natividad Island, (△) Santa Cruz Island, and (▲) Año Nuevo Island. Analysis of covariance revealed significant divergence among populations in trajectory slopes: the rate of change of tremata development relative to general size. Low-latitude populations (e.g., Guadalupe Island) display accelerated morphological development relative to high-latitude populations (e.g., Año Nuevo Island). Dissociation of allometric size and tremata shape covaries with maximum size and a shift in the size at which muscle scar development begins.

$p < 0.01$ on PC1; $F = 0.0$, $P = 1.0$ on shear-PC1) and placed this residual variation into the second axis, thus enhancing group discrimination.

Multivariate ontogenetic trajectories were calculated using model I linear regression analysis on first- and second-axis component scores as measures of allometric size and shape, respectively (Fig. 3). Shape was assumed to be dependent on size. Within groups, trajectories described the average path of tremata ontogeny with increasing allometric size. As individuals were derived from mixed cross-sectional samples, trajectories contrasted variation in shape resulting from variation in allometric growth and age.

The allometric size at the onset of muscle scar development was estimated by a series of second-order curvilinear regressions of PC1 scores on the proportion of scar nacre in the muscle attachment area (all $R^2 \geq 40\%$). Using the regression equations, I estimated the size at which the proportion of scar cover was equal to 0.5 in each sample (Fig. 4).

Analysis of covariance of ontogenetic trajectories indicated significant divergence among trajectory slopes. All slopes were significantly different from one another (all $p < 0.001$) and varied inversely with latitude. Ontogeny among individuals from lower latitudes (e.g., Guadalupe Island) resulted in more small, closely spaced tremata when compared to the ontogeny of individuals at higher latitudes (e.g., Año Nuevo Island), which possessed fewer, larger, more distantly spaced tremata at comparable allometric sizes (Fig. 3). This dissociation of al-

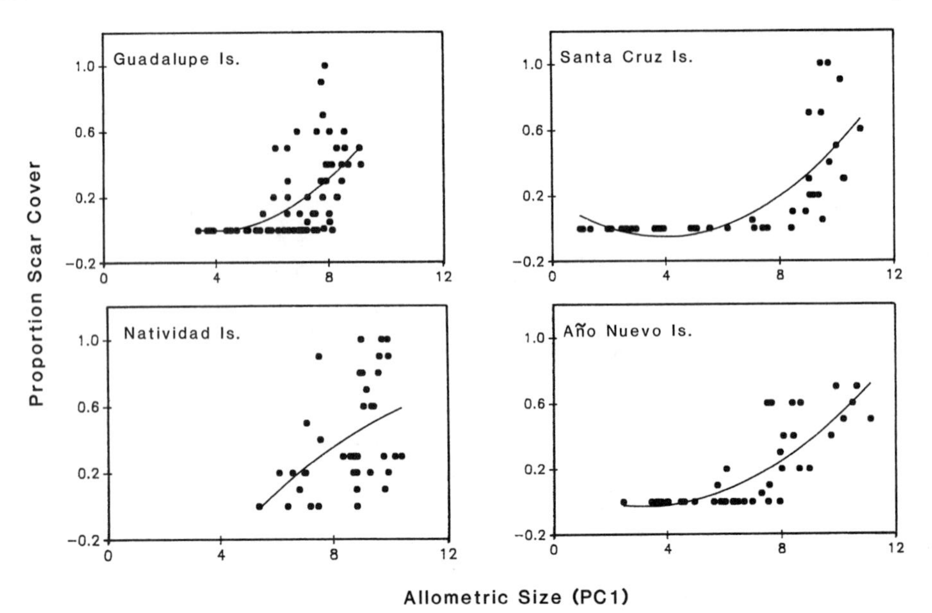

Figure 4. Relationships between allometric size and muscle scar development in four populations of black abalone. The size at the onset of scar development was estimated by second-order curvilinear regression analysis of scar cover on first-axis principal component scores (PC1), which are shown for each sample.

lometric size and shape covaried with a shift in the maximum size of individuals, shells were larger at higher latitudes, and the size (and possibly age) at which muscle scar development begins. The muscle scar develops at smaller sizes in low-latitude populations when compared to high-latitude populations (Figs. 3 and 4).

3.4. Biological Interpretation

Latitudinal dissociation of size and tremata development is associated with a latitudinal habitat shift: individuals in northern populations occur in the mid- and low-intertidal zones, while individuals in southern populations are found in the low-intertidal and shallow subtidal zones. Based on these observations, I propose that heterochronic processes are dissociating tremata development from shell growth rates in response to geographic variation in intertidal habitat.

4. Discussion

The analytical methods presented were successful with respect to their initial goals: the multivariate description of ontogeny and heterochronic change. Using a combination of PCA and linear regression, I obtained metrics of allometric size and developmental shape and described their relationship by calculating ontogenetic trajectories. The methods measured the onset of a developmental event relative to size, and discriminated among group trajectories. Below I discuss the interpretation of PCA size and shape and their usefulness in ontogenetic studies. I conclude by examining future prospects of multivariate techniques as they apply to the measurement of heterochrony.

4.1. Size and Shape

A conceptual difficulty associated with the use of PCA toward the elucidation of ontogenetic processes is the biological interpretation of PCA size and shape. The question is: to what extent is our biological understanding of organismal size and shape consistent with their PCA metrics?

Body size is not a single character, such as length, mass, or volume; size is an unmeasured latent trait. Our goal is to obtain an estimate of size by calculating a general size factor. As organisms increase in size, the absolute magnitude of most morphological traits increases. General size, therefore, is defined as a linear combination of variables that most parsimoniously accounts for observed covariances or correlations among traits of organisms of varying sizes (Bookstein *et al.*, 1985). The *R*-mode PCA extracts linear contrasts of variables that account for the largest proportion of variation among trait dispersion. Thus, with respect to general size, PCA is consistent with the biological definition of the term.

Most of the controversy associated with the use of PCA size as a measure of general size is based on an ignorance of allometry. Few organisms exhibit traits that increase isometrically with size (Sprent, 1972; Schmidt-Nielsen, 1984). As a result, variation in size is inextricably associated with changes in shape (Gould, 1966). Therefore, a measure of general size *by definition* contains variation in shape. Relabeling of PC1 as the size-allometry axis emphasizes this notion (Shea, 1985). In taxonomic studies the inseparability of size and allometric growth on PC1 is regarded as a disadvantage, as evidenced by the multitude of techniques designed to remove size (e.g., Lemen, 1983). From an allometric perspective, however, the merging of size and allometric shape on PC1 is the *goal* of the analysis (Shea, 1985).

Another source of confusion regarding PCA relates to the multigroup problem: intergroup differences in size introduce intergroup variation in shape into the size-allometry axis. When intergroup size differences are significant, results of shear-PCA are distinct from those of conventional PCA and group discrimination is significantly enhanced (Humphries *et al.*, 1981; Bookstein *et al.*, 1985). To the extent that patterns of group size allometries are similar, shear-PCA is an effective technique for removing these intergroup differences. However, the reliance of shear-PCA on similar intergroup size variation is a major problem of the method. When groups diverge significantly in their pattern of size allometry, shear-PCA would calculate nonsensible components. Further, the level of PC1 divergence at which problems would occur is unknown. In my intraspecific example, PCA and shear-PCA produced similar results, largely due to small differences in allometric size among samples and little divergence on the size-allometry axis. Analysis of interspecific divergence is likely to involve more significant differences in size and patterns of within-group size. For these reasons shear-PCA should always be examined in relationship to single-group PCAs.

A more difficult question is to what extent PCA axes beyond the first measure biological attributes of shape. As with size, shape is an unmeasured latent trait: the orientation of a vector in an attribute space after the removal of scale (Mosimann, 1970). Studies of ontogeny require a metric of development, a measure of trait change through time. Because the goal is to separate the timing of trait development from allometric growth (Gould, 1977, pp. 235–238), metrics of size and shape that are biologically uncorrelated are needed. The question is: are there aspects of development uncoupled from allometric growth?

The concepts of dissociation of developmental events (Gould, 1977) and onset–offset signals (Alberch *et al.*, 1979) emphasize that some aspects of development are uncoupled from allometric growth. Moreover, several studies illustrate biological independence of growth and development of traits (Atchley, 1984; Atchley and Rutledge, 1980; Atchley *et al.*, 1981) and heritability of PCA size and shape (Atchley, 1983). The problem with the use of PCA to obtain a metric of shape is due to the required orthogonality of PCA axes: shape variation is extracted independent of allometric growth *regardless* of whether such a relationship exists. That is, if there is little structure in the data beyond that of size allometry, components subsequent to PC1 will describe trivial variation that is not related to biological processes. The only solution to this dilemma is a thorough understanding of the system under investigation and a biological interpretation of axes on a case-by-case basis.

4.2. Future Prospects

Our methods are far from complete with respect to measuring the full gamut of heterochronic relationships. One obvious omission is the incorporation of developmental time into our size–shape relationships. Implicit in the concept of heterochrony is the notion of time. The terms hypermorphosis, progenesis, and displacement *by definition* describe shifts in timing. If we ignore time, we ignore the growth rate of morphological characters, which is another mechanism whereby heterochrony operates (e.g., Shea, 1983). Unless we label our definitions with morphological terms (e.g., allometric neoteny; see McKinney, this volume), we need to incorporate age into our multivariate analyses.

I have presented a technique that incorporates developmental events potentially resulting from timing into a morphological framework, i.e., the location of events on ontogenetic trajectories. On the average, body size is a predictor of 75% of the variation in an organism's morphology, life history, and behavior (Calder, 1984; Peters, 1984). Therefore, in many cases, size *per se* is perhaps of more interest than age, and my approach may be sufficient. Several other approaches are: (1) plotting age against size-allometry and shape axes; (2) combining age, size, and shape into a three-dimensional plot (Alberch *et al.*, 1979); or (3) using age as a variable in a factor-analytic approach in which axes are rotated to describe common associations among traits. In order to assess the importance of timing in heterochrony, empirical studies that describe relationships between developmental time and multivariate form are needed.

Finally, it cannot be overstated that any interpretation of morphological divergence resulting from heterochrony must be viewed in a phylogenetic context. Ancestor–descendant relationships are implicit in the terminology of heterochronic outcomes (Fink, 1982, and this volume). To determine the direction of evolutionary change, the spatial relationships of ontogenetic trajectories need to be supplemented by a phylogenetic analysis of character states (e.g., Wiley, 1981). This procedure amounts to integrating a cladogram with PCA. Bookstein *et al.* (1985, pp. 206–211) present an example of this approach.

In conclusion, I would like to emphasize that a multivariate analysis of form is only a first step in the biological investigation of heterochrony. If heterochrony is limited to morphological terms, then it serves only as a mechanistic redescription of evolutionary change. The ultimate significance of heterochrony must lie in its relation to selective factors promoting dissociations of developmental events and their relationships to ecological processes (Gould, 1977; Fisher, 1985; McKinney, 1986, and this volume).

ACKNOWLEDGMENTS. Abalone were collected with the assistance of Ursula Bechert, Paul Dunn, Susan Gaughan, Mark Hixon, Buzz Owen, Alberto Pompo, and John Steinbeck. Helpful comments on the manuscript were provided by Terry Farrell, Dave Foster, Mark Hixon, Roger Kaesler, Michael McKinney, R. A. Pimentel, and John Steinbeck. For emotional support and inspiration I thank Terri Nunn.

References

Alberch, P., and Alberch, J., 1981, Heterochronic mechanisms of morphological diversification and evolutionary change in the Neotropical salamander, *Bolitoglossa occidentalis* (Amphibia: Plethodontidae), *J. Morphol.* **161**:249–264.

Alberch, P., Gould, S. J., Oster, G. F., and Wake, D. B., 1979, Size and shape in ontogeny and phylogeny, *Paleobiology* **5**(3):296–317.

Anderson, T. W., 1963, Asymptotic theory for principal components analysis, *Ann. Math. Stat.* **34**:122–148.

Atchley, W. R., 1983, Some genetic aspects of morphometric variation, in: *Numerical Taxonomy* (J. Felsenstein, ed.), pp. 346–363, Springer-Verlag, Berlin.

Atchley, W. R., 1984, Ontogeny, timing of development, and genetic variance–covariance structure, *Am. Nat.* **123**:519–540.

Atchley, W. R., and Rutledge, J. J., 1980, Genetic components of size and shape. I. Dynamics of components of phenotypic variability and covariability during ontogeny in the laboratory rat, *Evolution* **34**:1161–1173.

Atchley, W. R., Rutledge, J. J., and Cowley, D. E., 1981, Genetic components of size and shape. II. Multivariate covariance patterns in the rat and mouse skull, *Evolution* **35**:1037–1055.

Blackith, R. E., 1957, Polymorphism in some Australian locusts and grasshoppers, *Biometrics* **13**:183–196.

Blackith, R. E., 1960, A synthesis of multivariate techniques to distinguish patterns of growth in grasshoppers, *Biometrics* **16**:28–40.

Blackith, R. E., and Reyment, R. A. 1971, *Multivariate Morphometrics*, Academic Press, New York.

Bookstein, F. L., Chernoff, B., Elder, R., Humphries, J., Smith, G., and Strauss, R., 1985, *Morphometrics in Evolutionary Biology*, Academy of Natural Sciences, Philadelphia, Special Publication 15.

Bryant, E. H., 1986, On use of logarithms to accommodate scale, *Syst. Zool.* **35**(4):552–559.

Calder, W. A., 1984, *Size, Function, and Life History*, Harvard University Press, Cambridge.

Cattell, R. B., 1966, The scree test for the number of factors, *Multivariate Behav. Res.* **1**:140–161.

Cock, A. G., 1966, Genetical aspects of metrical growth and form in animals, *Q. Rev. Biol.* **41**:131–190.
Cooley, W. W., and Lohnes, P. R., 1971, *Multivariate Data Analysis*, Wiley, New York.
Cox, K. W., 1962, California Abalones, Family Haliotidae, California Fish and Game Bulletin, No. 118.
De Beer, G. R., 1958, *Embryos and Ancestors*, Clarendon Press, Oxford.
Dillon, W. R., and Goldstein, M., 1984, *Multivariate Analysis: Methods and Applications*, Wiley, New York.
Dixon, W. J., and Brown, M. B., eds., 1979, *Biomedical Computer Programs, P-series*, University of California Press, Berkeley.
Dudzinski, M. L., Norris, J. M., Chmura, J. T., and Edwards, C. B. H., 1975, Repeatability of principal components in samples: Normal and non-normal data sets compared, *Multivariate Behav. Res.* **10**:109–117.
Falconer, D. S., 1981, *Introduction to Quantitative Genetics*, 2nd ed., Longman, London.
Fink, W. L., 1982, The conceptual relationship between ontogeny and phylogeny, *Paleobiology* **8**(3):254–264.
Fisher, D. C., 1985, Evolutionary morphology: Beyond the analogous, the anecdotal, and the ad hoc, *Paleobiology* **11**(1):120–138.
Gould, S. J., 1966, Allometry and size in ontogeny and phylogeny, *Biol. Rev.* **41**:587–640.
Gould, S. J., 1969, An evolutionary microcosm: Pleistocene and Recent history of the land snail P. (*Poecilozonites*) in Bermuda, *Bull. Mus. Comp. Zool.* **138**:407–532.
Gould, S. J., 1977, *Ontogeny and Phylogeny*, Harvard University Press, Cambridge.
Horn, J. L., 1965, A rationale and test for the number of factors in factor analysis, *Psychometrika* **30**:179.
Humphries, J. M., Bookstein, F. L., Chernoff, B., Smith, G. R., Elder, R. L., and Poss, S. G., 1981, Multivariate discrimination by shape in relation to size, *Syst. Zool.* **30**(3):291–308.
Huxley, J. S., 1932, *Problems of Relative Growth*, Dial Press, New York.
Jolicoeur, P., 1959, Multivariate geographic variation in the wolf *Canis lupus* (L.), *Evolution* **13**(3):283–299.
Jolicoeur, P., 1963a, The multivariate generalization of the allometry equation, *Biometrics* **19**(3):497–499.
Jolicoeur, P., 1963b, The degree of generality of robustness in *Martes americana*, *Growth* **27**:1–27.
Jolicoeur, P., 1984, Principal components, factor analysis, and multivariate allometry: A small-sample direction test, *Biometrics* **40**:685–690.
Jolicoeur, P., and Mosimann, J. E., 1960, Size and shape variation in the painted turtle: A principal component analysis, *Growth* **24**:339–354.
Lemen, C. A., 1983, The effectiveness of methods of shape analysis, *Fieldiana, Zool.* **15**:1–17.
Mather, P. M., 1976, *Computational Methods of Multivariate Analysis in Physical Geography*, Wiley, London.
Maynard Smith, J., Burian, R., Kauffman, S., Alberch, P., Campbell, J., Goodwin, B., Lande, R., Raup, D., and Wolpert, L., 1985, Developmental constraints and evolution, *Q. Rev. Biol.* **60**(3):265–287.
McKinney, M. L., 1984, Allometry and heterochrony in an Eocene echinoid lineage: Morphological change as a by-product of size selection, *Paleobiology* **10**(4):407–419.
McKinney, M. L., 1986, Ecological causation of heterochrony: A test and implications for evolutionary theory, *Paleobiology* **12**(3):282–289.
Mosimann, J. E., 1970, Size allometry: Size and shape variables with characterizations of the lognormal and generalized gamma distribution, *J. Am. Stat. Assoc.* **65**:930–945.
Oxnard, C. E., 1978, One biologists's view of morphometrics, *Annu. Rev. Ecol. Syst.* **9**:219–241.
Peters, R. H., 1984, *The Ecological Implications of Body Size*, Cambridge University Press, Cambridge.
Pimentel, R. A., 1979, *Morphometrics, the Multivariate Analysis of Biological Data*, Kendall-Hunt, Dubuque, Iowa.
Pimentel, R. A., 1981, A comparative study of data and ordination techniques based on a hybrid swarm of sand verbenas (*Abronia* Juss.), *Syst. Zool.* **30**(3):250–267.
Rao, C. R., 1958, Some statistical methods for comparison of growth curves, *Biometrics* **14**(1):1–17.
Schmidt-Nielsen, K., 1984, *Scaling: Why Is Animal Size So Important?*, Cambridge University Press, Cambridge.
Schweitzer, P. N., Kaesler, R. L., and Lohmann, G. P., 1986, Ontogeny and heterochrony in the ostracode *Cavellina* Coryell from Lower Permian rocks in Kansas, *Paleobiology* **12**(3):290–301.

Shea, B. T., 1983, Allometry and heterochrony in the African apes, *Am. J. Phys. Anthropol.* **62**:275–289.
Shea, B. T., 1985, Bivariate and multivariate growth allometry, *J. Zool. Lond. A* **206**:367–390.
Sneath, P. H. A., and Sokal, R. R., 1973, *Numerical Taxonomy: The Principles and Practice of Numerical Classification*, Freeman, San Francisco.
Sokal, R. R., and Rohlf, F. J., 1981, *Biometry: The Principles and Practice of Statistics in Biological Research*, 2nd ed., Freeman, New York.
Somers, K. M., 1986, Multivariate allometry and removal of size with principal components analysis, *Syst. Zool.* **35**(3):359–368.
Sprent, P., 1972, The mathematics of size and shape, *Biometrics* **28**:23–37.
Thompson, D'A. W., 1917, *On Growth and Form*, Cambridge University Press, Cambridge.
Thorpe, R. S., 1980, A comparative study of ordination techniques in numerical taxonomy in relation to racial variation in the ringed snake *Natrix natrix* (L.), *Biol. J. Linn. Soc.* **13**:7–40.
Tissot, B. N., 1984, Multivariate analysis of geographic variation in *Cypraea caputserpentis* (Gastropoda: Cypraeidae), *Veliger* **27**(2):106–119.
Tissot, B. N., 1988, Geographic variation and heterochrony in two species of cowries (genus *Cypraea*), *Evolution* **42**(1):103–117.
Wiley, E. O., 1981, *Phylogenetics: The Theory and Practice of Phylogenetic Systematics*, Wiley, New York.

Chapter 4

Shape Analysis

Ideas from the Ostracoda

DAVID W. FOSTER and ROGER L. KAESLER

1. Introduction

The availability of computing power to paleontologists has been accompanied by the development of new methods of analysis and increasingly sophisticated software, much of which is readily adaptable to the study of heterochrony. Among these new developments is a suite of techniques that can be grouped under the broad category of *shape analysis*. Here we use the term to refer to any multivariate method that is aimed at evaluating shapes, including special applications of multivariate morphometric methods (Tissot, this volume), as well as new techniques designed with quite special purposes in mind. Shapes to be evaluated may be either such smooth curves as outlines or constellations of landmarks that can be homologized from specimen to specimen or from taxon to taxon.

Our purpose in this chapter is to introduce the methods of shape analysis that we judge to be most applicable to the investigation of heterochrony in the fossil record. We hope to provide sufficient information about the techniques to allow users to select an appropriate method for the task at hand. We have drawn most of our examples from studies of Recent and fossil Ostracoda, but we have not delved deeply into the matter of heterochrony in the evolution of the Ostracoda. In spite of the potential of ostracodes for study of shape analysis as applied

DAVID W. FOSTER and ROGER L. KAESLER • Department of Geology, Museum of Invertebrate Paleontology, and Paleontological Institute, University of Kansas, Lawrence, Kansas 66045.

to heterochrony, little research has been done beyond laying the groundwork (but see Schweitzer *et al.*, 1986).

Evaluation of heterochrony requires study of size, shape, and age of ancestors and descendants (Gould, 1977). A critical element that is often missing from a study of heterochrony in the fossil record is the age at death of the fossils being studied. For most kinds of fossils no means independent of shape or size is available to estimate age. Even relative ages of conspecific fossils are typically difficult to establish, a circumstance that detracts significantly from most investigations of heterochrony in the fossil record.

2. Biology and the Analysis of Shape

2.1. Ideas from the Ostracoda

If one is interested in heterochrony of trilobites or amphibians, studying ostracodes will provide little benefit. For developing ideas about heterochrony and testing quantitative methods that are to be applied to its study, however, the ostracodes offer many interesting possibilities. Ostracodes are crustaceans, a class of complex, highly organized metazoans. They range from Cambrian to Recent and have often been relatively abundant in a wide variety of marine and freshwater environments. Their small size and the heavy calcite carapaces of many species enhance their likelihood of preservation.

As we have mentioned, studies of heterochrony in the fossil record are typically flawed by the inability of paleontologists to estimate ages of fossils independently of size and shape. The stage of growth, however, provides an estimate of relative age, and for fossil ostracodes stage of growth can be determined readily. Unlike some crustaceans that continue to molt throughout life, the ostracodes have a number of instars that is fixed at high taxonomic levels and a terminal adult stage. When an ostracode has reached sexual maturity, it does not molt further. (Postmaturational molting has been reported, but is, at best, quite a rare phenomenon.) Moreover, one can typically determine relatively easily and independently of size and shape from study of meristic morphological characters the instar to which a fossil ostracode belongs. This is especially true of ornamented species. Thus, the ostracodes provide paleontologists interested in heterochrony with a rare opportunity to incorporate all three aspects of heterochrony into their studies—size, shape, and age—and to plot ontogenetic trajectories (Alberch *et al.*, 1979). On the other hand, the last molt involves appreciable morphological change in addition to incremental growth, largely because of the development of large and elaborate reproductive organs, a circumstance that may limit the usefulness of the ontogenetic perspective.

2.2. Data from Organisms

Here we want to discuss briefly aspects of biology that impinge on the analysis of shape and the use of quantitative methods. These include (1) meristic versus

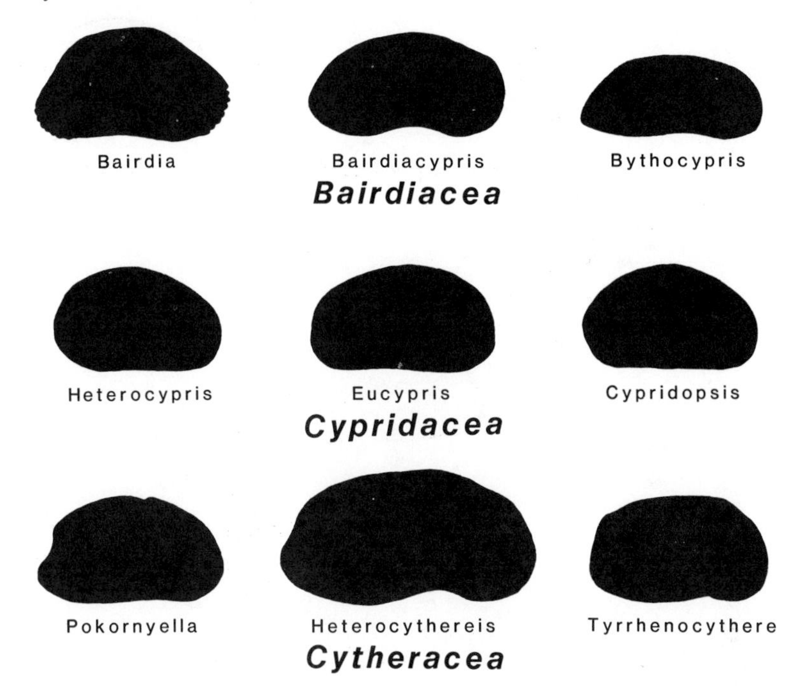

Figure 1. Silhouettes of representative genera of the three major superfamilies of podocopid ostracodes, showing that outlines provide information for discrimination of closely related taxa.

continuously variable heterochronic morphological features; (2) the shape analyst's view of homology; (3) the nature of data, specifically the differences between morphological features of organisms that may show heterochrony and the measurements used to make such heterochronic changes accessible to the computer; and (4) the meaning of tests of statistical significance.

2.2.1. Continuous and Meristic Characters

Gould's (1977) clock model of heterochrony expresses both continuous and meristic, heterochronic change of shape. More recent, hypothetical examples used to illustrate the nomenclature of heterochrony, however, have drawn attention to continuous change of size more than to meristic change of shape (Alberch *et al.*, 1979; McNamara, 1986). Such readily apparent meristic changes of shape as the addition of spines or ribs lend themselves to heterochronic analysis, and their study does not call for elaborate quantification. Similarly, different shapes of such continuously varying morphological features as outlines of specimens from different higher taxa are typically readily apparent without the use of quantitative methods. The differences between the typical representatives of three superfamilies of podocopid ostracodes shown in Fig. 1, for example, are quite readily discernible from inspection of their outlines alone without multivariate analysis of shape. Study of heterochrony, however, focuses on morphological similarities and differences of identifiable growth stages of closely related species. Both the shapes of the outlines of such species and the configurations of constellations of

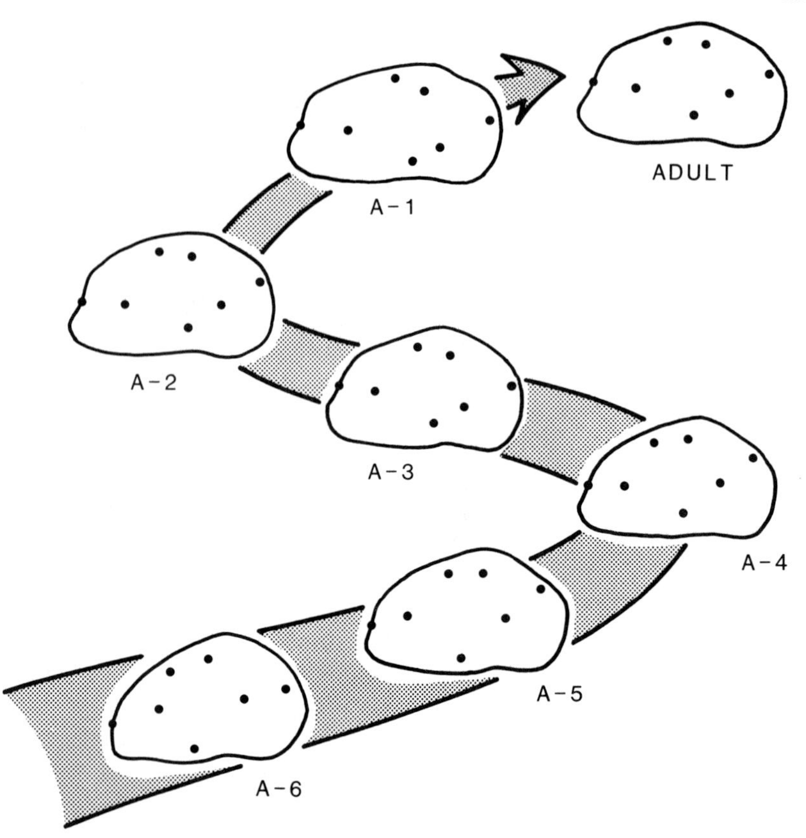

Figure 2. Ontogeny shown by outlines and seven homologous landmarks of six instars and the adult of *Tyrrhenocythere amnicola* with size removed. Successive instars are morphologically quite similar, suggesting the usefulness of pairwise comparisons in the study of ontogeny and heterochrony.

their homologous landmarks are likely to be closely similar (Fig. 2). A paleontologist studying heterochrony of such an evolutionary system, where meristic differences do not tell the whole story, would be well advised to consider using the quantitative methods of shape analysis, which integrate small differences that are expressed by a large number of measurements of morphology.

2.2.2. Homology

Homology is defined as "correspondence between structures of different organisms due to their inheritance of these structures from the same ancestry" (Simpson and Beck, 1965, p. 494). Leaving aside the difficulty in practice of applying this definition, one would like to base investigations of evolutionary pathways through heterochrony on homologous morphological features. In this respect the differences among the various methods of shape analysis are straightforward because either they deal with configurations of homologous landmarks or they deal with outlines (Fig. 3). None of the methods has successfully integrated use of the two kinds of data. Besides their intuitive appeal, the techniques based

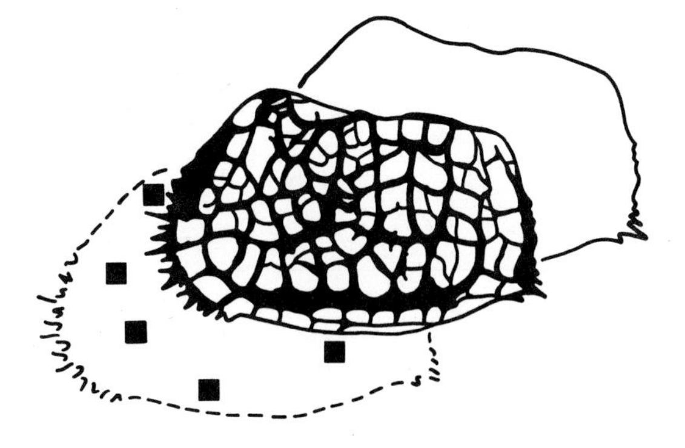

Figure 3. Graphical representation of the surface ornamentation of the ostracode *Bradleya normani*, its outline, and five homologous landmarks. Methods of shape analysis are clearly divided into those that operate on outlines and those that use data on locations of homologous landmarks.

on biological homology that use configurations of constellations of homologs or *homologous landmarks* offer two additional advantages. First, orientation of specimens is not a problem if the morphological features being studied can be homologised. Second, data on the configuration of homologous landmarks may lead immediately to the study of functional morphology, which is often overlooked in using heterochrony to determine evolutionary pathways.

For all their intuitive appeal, however, homologous landmarks by no means tell the whole story of an organism's ontogeny. The outline of the ostracode in Fig. 4, for example, conveys a great deal more of the kind of information that can be used to recognize taxa than does the configuration of the homologous landmarks. This is seen in the way results of the two methods have been illustrated. In analyses of outlines, homologous landmarks are typically not shown (e.g., Lohmann, 1983; Read and Lestrel, 1986; Schweitzer *et al.*, 1986; Kaesler and Maddocks, 1984). On the other hand, in studies based on homologous landmarks, outlines are nearly always shown to give a better impression of the place of the homologs on the organisms (e.g., Chapman *et al.*, 1981; Benson *et al.*, 1982; Siegel and Benson, 1982; Maness and Kaesler, 1987; but see Kaesler and Foster, 1988). When considering outlines, however, one is not dealing with biological homology. Instead, one assumes that regions of the outline are broadly homologous, a characteristic we call *geometrical homology*. For example (Fig. 5), the anterodorsal regions of two ostracode valves may be thought of as geometrically homologous although neither contains homologous landmarks. In the absence of homologs, orientation of specimens is by no means a trivial exercise.

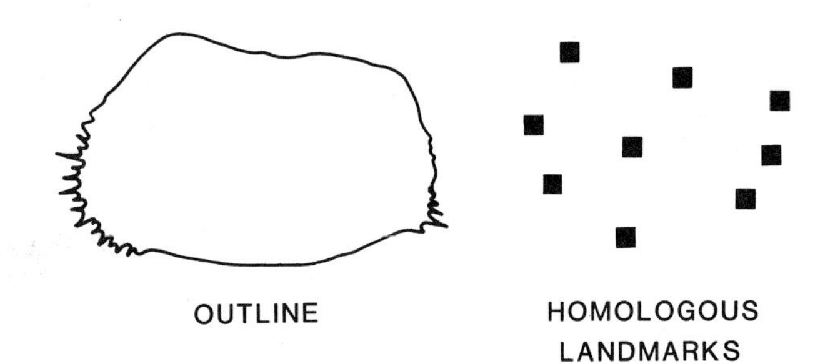

Figure 4. Most investigators prefer to base their studies of heterochrony on biological homology. Nevertheless, outlines convey a great deal of information about shape that configurations of constellations of homologous landmarks do not.

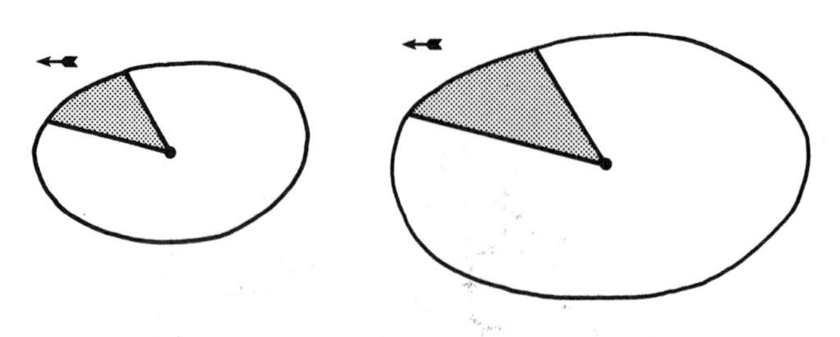

Figure 5. Analysis of outlines is based on geometrical homology rather than biological homology. Here two closely related species of the late Paleozoic ostracode genus *Cavellina* are compared on the basis of the shapes of their outlines. After orientation of the outlines, data from similarly defined segments of the outlines are used as if they were biologically homologous.

2.2.3. Morphological Features and Quantitative Characters

The morphology of some organisms is comparatively simple and can be reconstructed more or less fully by reference to only a few morphological features. This is especially true of organisms that grow deterministically according to a plan that can be described mathematically: spheres, cylinders, and spirals. Raup (1966), for example, used only a few parameters to simulate a wide variety of coiled shells. Real organisms, of course, typically depart from any perceived plan and are especially likely to add during ontogeny such additional morphological features as spines, ribs, and muscle scars. When numerous morphological features are available, it is important to select for study those that vary in biologically interesting ways. One would have little success, for example, using height of sexually dimorphic adult ostracodes if the principal difference between the sexes were in the length of the carapace. Moreover, when morphology is complicated, successful analysis is likely to depend on the study of a large number of morphological characters.

An important distinction here is between morphological features of the organisms and the numerical measurement of characters that are used to describe these features. The location of a homologous landmark—a morphological feature—can be expressed in a number of ways. Among these are x and y coordinates from a common origin, the coordinates from another point of interest, the distance and angle from the origin, and the distance and angle from another point. If the morphological feature of interest is a smooth outline without landmarks, it can be described by measuring x and y coordinates of any number of points along the outline. These may be equally spaced, spaced at equal angles from the centroid, spaced at equal angles from such a landmark as a muscle scar, or variably spaced according to the complexity of the curvature. Naturally one must expect the results of shape analysis of such widely different kinds of data to vary. Several newer methods of shape analysis are of special interest in this regard because they draw attention to those aspects of shape that vary. Some of them focus on those morphological features that change position (Benson, 1982; Siegel and Benson, 1982), whereas others are designed to identify the directions of elongation and shortening (Bookstein, 1978).

2.2.4. Tests of Statistical Hypotheses

In the attempts of scientists to identify underlying processes from observed patterns, one sees increasingly frequent references to the need to test null hypotheses. The term *null hypothesis* has been borrowed from statistics, where it has quite a specific meaning and states that "there is no real difference between the true value of [a parameter] in the population from which we have sampled and the hypothesized value of [that parameter]" (Sokal and Rohlf, 1981, p. 158). The key words in this definition are *real* and *sampled*. Any statistical test "examines a set of sample data and, on the basis of an expected distribution of the data, leads to a decision on whether to accept the hypothesis underlying the expected distribution or whether to reject that hypothesis and accept an alternative one" (Sokal and Rohlf, 1981, p. 157).

A statistical population is "the totality of individual[s] . . . about which inferences are to be made, existing anywhere in the world or at least within a definitely specified sampling area limited in space and time" (Sokal and Rohlf, 1981, p. 9). Thus a statistical test always tests a null hypothesis and is an attempt to understand a statistical population on the basis of a sample drawn from it. Statistical testing of hypotheses is necessary because no sample has parameters that are exactly the same as those of the statistical population from which it was drawn. For this reason the probability always exists of making a *type I error*, in which one rejects a null hypothesis that is true. The *level of significance* of a statistical test is, in fact, the probability of making a type I error. For example, a statistical test with a probability $p < 0.05$ means that the probability of falsely rejecting the null hypothesis about the statistical population on the basis of the sample being studied is less than 1 in 20.

A characteristic of statistical hypothesis testing as practiced by many paleontologists is the tendency to erect utterly trivial null hypotheses. To clarify this point, let us look for a moment at differences between individuals, which are outside the realm of statistical analysis. Two organisms, be they identical twins or members of different phyla, are different from each other in any phenotypic character that can be measured. The differences between genetically identical but ontogenetically different individuals may be readily apparent. On the other hand, differences between two genetically similar individuals of the same age or stage of growth may be extremely difficult to detect. Nevertheless, a null hypothesis about differences of any phenotypic character is trivial. The individuals differ, and failure to detect difference says more about the investigator than about the subject of study. The same can be said of statistical populations. Imagine that an investigator has knowledge of the parameters of two statistical populations. Any differences, however slight, in the parameters of these statistical populations, termed *parametric differences*, are statistically significantly different. The purpose of statistical hypothesis testing is to draw inferences about unknown parameters of statistical populations from study of samples. If one knows the parameters of the statistical populations rather than merely of samples collected from the populations, then testing statistical hypotheses about the populations is obviated. The statistical populations are different from each other. Failure to detect differences on the basis of study of samples of the populations indicates either that the investigator has not studied large enough samples or that measurement was performed at too coarse a scale.

A null hypothesis is trivial that asserts no morphological differences between sexes or, in a study of heterochrony, no differences of shape during ontogeny. Of course any two populations will always be different. A statistical test of such a hypothesis is also trivial. Such populations can be asserted to be different *a priori*. To detect difference sheds no light on the populations. Worse, failure to detect such difference means only that the data were not adequate to the task. What is needed instead of tests of such hypotheses are attempts to discover consistent morphological differences that are likely on other grounds to have biological meaning. This kind of work requires a biological decision: what magnitude of morphological difference is biologically important? Having selected such a difference, the investigator must estimate the variance of the populations. Using the methods of experimental design, he or she may then determine how large a sample is needed in order to detect such a biologically significant difference if it exists (Sokal and Rohlf, 1981, p. 263).

3. Methods of Shape Analysis

Methods of shape analysis may be grouped into categories: multivariate morphometrics, analysis of homologous landmarks, and analysis of outlines. The members of these groups are not distinct, however, because all have a common purpose and many are based on similar procedures.

3.1. Multivariate Morphometrics

Multivariate morphometrics comprises techniques with a broad range of applications to the analysis of shape, including heterochrony and the ontogenetic change of shape. The techniques of multivariate morphometrics are used both as the basis of some methods of shape analysis and, with other procedures, for analyzing outlines or constellations of homologous landmarks. Advantages of multivariate methods stem from their ability to deal with covariation of characters and from difficulties in interpreting bivariate analyses (Willig *et al.*, 1986).

An important goal of the study of heterochrony is to partition the effects of size, shape, and age. To separate size from shape, data are typically standardized for size before further analysis. Humphries *et al.* (1981) separated size and shape with modified principal component analysis (see also Tissot, this volume). Allometric growth is a confounding factor that is especially important in the study of heterochrony and that must be considered when standardizing for size, especially if one does not have a reliable measure of age. Strauss and Bookstein (1982) suggested a method of regressing individual variables onto a measure of general size to account for allometric growth. Study of heterochrony of fossils is severely hampered because the age of a fossil at death is typically unknown.

A major strength of multivariate morphometric methods in shape analysis of heterochrony is the capability they provide of dealing with variations within and among populations. This capability allows for hypothesis testing where appropriate. Some of the methods of shape analysis, however, are limited to pairwise comparisons of shapes and do not take into consideration the variation of shapes within a population. Such methods are nevertheless useful in exploratory data analysis.

Selection of variables is an important part of shape analysis because some characters show differences of shape better than others. Biorthogonal analysis is especially helpful in choosing among characters that have been measured or in suggesting new ones that are suitable metrics for the observed differences of shape (Strauss and Bookstein, 1982).

To one degree or another, multivariate data typically fail to meet the statistical assumptions required, a cause for concern especially when testing hypotheses. Unfortunately, sometimes the assumptions have been treated lightly, so that results of analysis are open to criticism. One approach that is free of assumptions is the use of randomization techniques to compute statistical significance (Sokal and Rohlf, 1981), for which computer programs are available (Sokal and Rohlf, 1981; Edgington, 1980; Romesburg, 1985). The power of such methods in detecting subtle differences in morphological variation was demonstrated by Foster and Kaesler (1983).

3.2. Analysis of Homologous Landmarks

Much of the analysis of homologous landmarks has been done with traditional multivariate morphometrics, although the results of such analyses are sometimes difficult to interpret. More recently, biological interpretation of differences of shape has received a great deal of attention. Thompson's (1917) use of transformation grids was an elegant approach to this subject, but his approach has been difficult to quantify. Sneath (1967) used trend-surface analysis to compare shapes and produce transformation grids by mapping the residuals of Cartesian coordinates. To detect allometric growth, Brower and Veinus (1978) used principal component analysis of Cartesian coordinates of landmarks of aligned specimens. We now discuss two methods that produce especially biologically interpretable results.

3.2.1. RFTRA

Resistant-fit, theta-rho analysis (RFTRA) was introduced by Benson (1982) and Siegel and Benson (1982) for study of localized allometry, such as often occurs between sexes or among adults of the same sex and in which most of the shape remains nearly constant (Benson, R. H., 1987 *personal communication*). Although not originally intended for study of ontogeny, it is well suited for such applications, especially where change is gradual and localized. Using pairwise analysis, RFTRA determines the overall best fit of shapes to each other, producing both a measure of overall similarity and residual vectors that show differences in the configuration of landmarks (Fig. 6). The residuals are especially useful for interpretation of heterochrony and ontogenetic changes of shape. The measure of similarity can be used to construct a distance matrix that can be analyzed using numerical taxonomic techniques. Least squares is a criterion that can be used to determine the overall similarity of two shapes. It suffers, however, when two similar shapes have dramatically differently placed landmarks in a localized region. Under such circumstances, the residual difference between the two shapes is distributed among all landmarks instead of in the area of maximum difference.

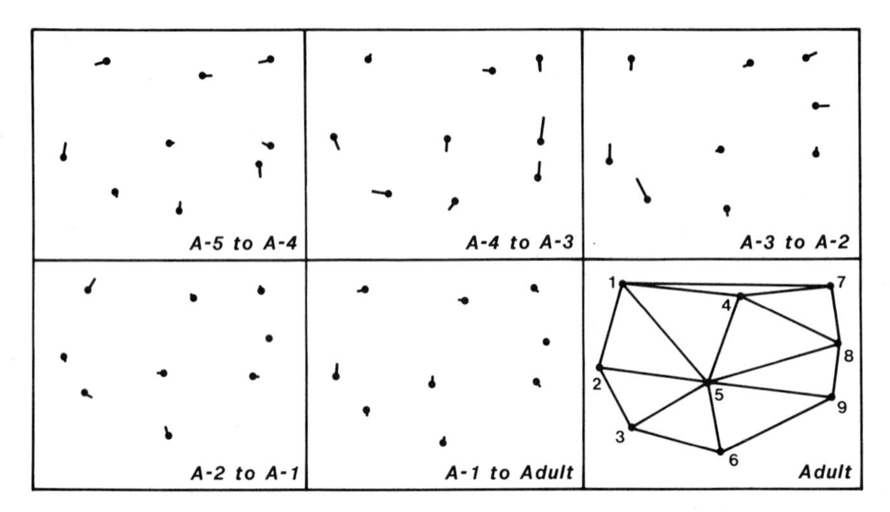

Figure 6. Residual vectors from RFTRA analysis of adults and instars of the ostracode species *Bradleya normani*. Long residual vectors, such as occur in the transformation from A-4 to A-3 and from A-3 to A-2, result from large changes in the relative positions of the homologous landmarks. Transformations from A-2 to A-1 and from A-1 to the adult are marked by short residual vectors as a result of very small changes in the configuration of the landmarks during these transformations.

Using robust regression, RFTRA determines a fit that best aligns landmarks. It is less sensitive to deviations from widely different landmarks and provides a more easily interpreted fit. Siegel (1982) has published a computer program that implements both criteria.

RFTRA is applicable only to pairwise comparison of shapes and does not allow study of variation of populations. To compare shapes of two populations one must first either derive a composite shape for each population or select a representative individual from each. Such an approach discards all information about variation within the populations, with the danger that the composite or exemplar may not be truly representative. As implemented by Siegel (1982), RFTRA is an interactive graphical system of analysis. It can produce a variety of graphics to display differences in shapes and provides an excellent descriptive tool for showing differences graphically. Furthermore, while the actual quantitative comparison of shape is based on landmarks, Siegel's programs can display outlines and other features. Thus, outlines can be incorporated into the analysis of landmarks, as stressed by Read and Lestrel (1986).

3.2.2. Biorthogonal Analysis

Biorthogonal analysis (Bookstein, 1978) is well suited for analyzing constellations of homologous landmarks. The analytical method is a tensor-based approach similar in concept to that used by structural geologists for stress and strain analysis of deformed rocks. The technique compares two shapes at a time and produces a graphical representation showing how one shape may be transformed into another. It focuses on change of shape, being conceptually rather like looking at the first derivative of a mathematical function in addition to the function itself. The technique provides a quantitative derivation of a transformation grid similar

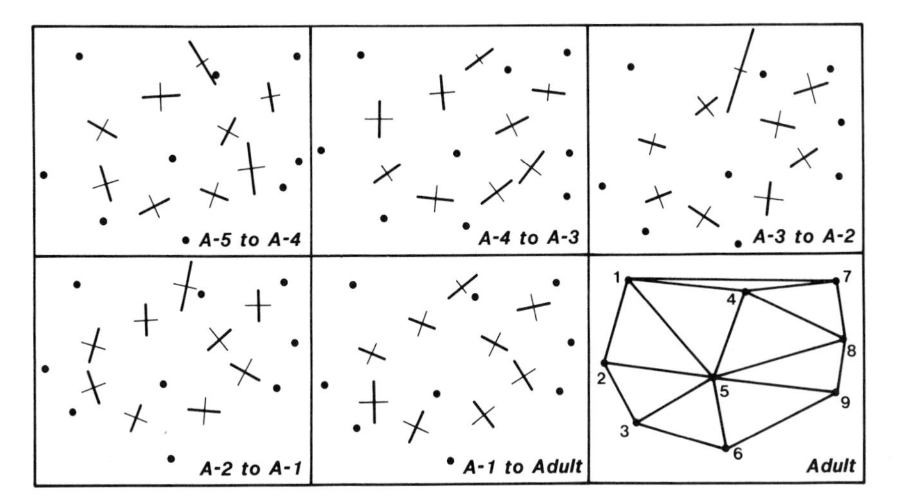

Figure 7. Results of biorthogonal analysis of *Bradleya normani*. Each cross shows the change of shape during ecdysis of a triangle (not drawn) with apices at the three points surrounding the cross. The heavy bar on the cross shows the direction of greatest change. Homologous landmarks are shown in their position before ecdysis. Unlike RFTRA analysis, which shows changes of positions of landmarks, biorthogonal analysis shows the change of shape of the area bounded by landmarks and is thus singularly well suited for expressing allometric growth.

to those of Thompson's (1917) grids. An extension of biorthogonal analysis that is particularly applicable to heterochrony identifies forms with intermediate morphology in a sequence of shapes (Bookstein, 1980).

In our application of biorthogonal analysis, we first construct a network of triangles connecting triplets of homologous landmarks distributed broadly on ostracode valves (Kaesler and Foster, 1987). Shapes of two specimens are analyzed triangle-by-triangle, with magnitude and direction of strain being computed for each triplet (Fig. 7). A similar procedure was presented by Sanderson (1977). The strains may then be presented graphically to show the transformation necessary to derive one shape from the other (Fig. 7).

Biorthogonal analysis compares shapes pairwise, but does not include information about the variation of populations. Thus, the same *caveats* apply to its use as to RFTRA. Bookstein (1982), however, described a procedure for following the growth of single individuals through time and computing average shape of groups. His approach might be useful if skeletal growth is accretionary and the record of early developmental stages of an individual is preserved.

We use biorthogonal analysis as a graphical system of analysis. It can produce a variety of graphics and is useful for describing differences in shapes. One application of biorthogonal analysis is to evaluate metrics of change of shape (Strauss and Bookstein, 1982). It aids in selecting morphological characters that are aligned with directions of change of the organisms to show major differences in shape.

3.3. Analysis of Outlines

Analysis of outlines differs from analysis of homologous landmarks principally in the lack of biological homology. Methods of outline analysis provide a way to describe or compare shapes that might otherwise be difficult. The outline itself is converted to morphological characters suitable for multivariate analysis. In a sense, outline analysis may be regarded as the transformation of data, i.e., the coordinates of points on the outline, into a metric suitable for further multivariate analysis. It is applicable to the study of heterochrony where variation of populations is important.

Analysis of outlines is based on geometrical homology, which requires consistent alignment of outlines. In eigenshape analysis outlines are aligned by comparing shapes with a standard and rotating them until a correlation criterion is maximized (Lohmann, 1983). Full and Ehrlich (1982) suggested methods for locating centroids of shapes for Fourier analysis. Although numerical methods can be used, biologically homologous points should be used for alignment when available. Such features as muscle scars, eye tubercules, and spines are ideal for alignment and also incorporate some homology into the analysis. Note that analyses of identical shapes with different alignments will produce different results.

An advantage of outline analysis is that outlines are easy to measure. Because specimens do not have to be manually oriented, outline analysis can be used to study large populations in a short time. For such ostracodes as *Cavellina* (Schweitzer *et al.*, 1986) with no suitable homologous landmarks, outline analysis may be the only way quantitatively to study heterochrony. Moreover, in our studies of ontogeny of ostracodes, we have had difficulty following homologs through ontogeny because the early instars are quite small and often without the morphological features that develop later. On the other hand, regardless of size or simplicity, each instar always has an outline.

Outline analysis begins with quantification of the outline, usually by using a digitizing device that represents the outline as a series of Cartesian coordinates. An automatic imager can produce large quantities of data rapidly, but for our work with ostracodes we have found a manual digitizer to be adequate. Indeed, much more time is spent in sample preparation than in digitizing, and we have never considered that time for digitizing was limiting to our productivity. After digitizing a specimen, we typically use a procedure, called a splining (Evans *et al.*, 1985), to interpolate the outline between actual digitized points. Splining makes spacing of points less critical and smooths the outline. After splining, the outline is converted into a shape function for analysis.

Shape functions represent the outline as an ordered series of measurements, each of which can be considered to be an individual morphological character. The radius function used in Fourier analysis represents the outline in polar coordinates measured at equally spaced angular increments and requires selection of an appropriate origin of the polar coordinate system. The origin can be a centroid or a biologically homologous landmark, such as a muscle scar. In his original theta-rho analysis, Benson (1967) used polar coordinates to compare graphically outlines of ostracodes. Another shape function, the Phi* function (Zahn and Roskies, 1972), has advantages when the outline is a complex shape that might, for example, have reentrants (Lohmann, 1983). The Phi* function is the tangent angle of the outline measured at equally spaced length intervals along the outline from a starting point. Either shape function represents the outline in a format that is suitable for analysis by the usual multivariate methods. Both methods archive the outline with resolution equivalent to the number of terms in the shape function.

While shape functions can be analyzed directly with most multivariate methods, they are typically subjected to a dimension-reducing technique such as the Fourier transform or eigenshape analysis.

3.3.1. Fourier Analysis

Fourier analysis has been the most widely used method of quantitative outline analysis in geology. It was first applied to analysis of outlines of sand grains by Ehrlich and Weinberg (1970). It has been used for analysis of a variety of fossils, including ostracodes (Kaesler and Waters, 1972; Kaesler and Maddocks, 1984), blastoids (Waters, 1977), bryozoa (Anstey and Delmet, 1973), foraminifera (Healy-Williams and Williams, 1981), and miospores (Christopher and Waters, 1974).

In Fourier analysis, the shape function is decomposed into a series of harmonic amplitudes and phase angles. The radius function has been used most often (but see Rohlf and Archie, 1984). These variables are then further analyzed quantitatively. Phase angles are typically not included in the analysis (but see Ehrlich *et al.*, 1983). Typically a subset of the total number of harmonic amplitudes is adequate to show variation of shape within or among populations. Subsequent analysis may involve principal components or similar procedures. The harmonics have the property of being geometrically interpretable because each harmonic is associated with a specific geometrical shape. This geometrical relationship, however, may not be morphologically interpretable (Rohlf and Archie, 1984), except perhaps with such highly symmetrical organisms as blastoids (Lohmann, 1983).

A major spinoff from Fourier analysis has been increased awareness of the polymorphism of some populations of foraminifera (Healy-Williams *et al.*, 1985). Techniques have been developed to analyze such polymorphic populations (Full *et al.*, 1982).

3.3.2. Eigenshape Analysis

Eigenshape analysis was developed by Lohmann (1983) and later used by Malmgren *et al.* (1983) to study trends in shape of microfossils in stratigraphic sequences. The method involves principal component analysis of Phi* shape functions. The outline as represented by the Phi* function is divided into a fixed number of segments of equal length. The value of the Phi* function at each segment is treated as a morphological character. Eigenshape analysis looks at these characters with *R*-mode principal component analysis to extract the eigenvalues and eigenvectors from the correlation matrix of Phi* functions. The first eigenshape is the eigenvector associated with the largest eigenvalue. It is approximately an average shape of the set of Phi* functions and may be used to reconstruct the outline for illustration. The second and third eigenshapes, analogous to the eigenvectors of the second and third principal components, represent variation of shape within the data. The differences between these eigenshapes may result from such biologically interpretable factors as sexual dimorphism (Maness and Kaesler, 1987) or trends in change of shape during ontogeny (Schweitzer *et al.*, 1986). Eigenshape analysis is often a two-step procedure in which such a natural grouping of specimens as an ostracode instar is first analyzed to obtain the first eigenshape for the group. These eigenshapes, which are in the same format as the Phi* function, are then processed with further eigenshape analysis to explore variation in shape among the groupings. Because it is based on principal component analysis, eigenshape analysis is useful for examining variation in populations.

3.3.3. Discussion

Although the goals of Fourier and eigenshape analysis are the same, they are not directly comparable. The Fourier transform is applied to each specimen and alone does not reduce the dimensionality of the variable space. The reduction is accomplished in another step by examining the variation explained by each harmonic (or phase angle). Only harmonics exhibiting significant variation are retained for subsequent analysis. In eigenshape analysis, principal component analysis is done directly on the shape functions of groups of specimens. This analysis results in an average shape of the group that may be used as data for still further analysis to gain information about variation of shape within the group.

Unfortunately a bitter disagreement has arisen between the proponents of Fourier and eigenshape analysis that is likely to distract investigators from using either method to study heterochrony. Part of this disagreement results not so much from choice of Fourier analysis or eigenshapes analysis as it does from subsequent use of different multivariate methods to complete the analysis. We suggest that the techniques used at various stages of analysis by both factions should be decoupled and considered independently.

Rohlf (1986) suggested that for some studies the use of the Fourier transform may be an unnecessary step in the overall analysis, although he cited the space-reducing virtue of the Fourier transforms for some data sets. Bookstein *et al.* (1982) argued against the application of the Fourier transform where homologous points are present along the outline. They were concerned that biological information contained in homologous points was being ignored and that the Fourier transform was being used where landmark analysis was more appropriate. Read and Lestrel (1986), however, stressed the importance of an integrated approach using both outline analysis and landmark analysis.

Ehrlich and Full (1986) and Full and Ehrlich (1986) suggested problems that may result with eigenshape analysis. They pointed out the fundamental difference between mathematical (geometrical) homology and biological homology that must be considered in interpreting the results. What is perhaps more important, they questioned the stability of eigenshapes as shape estimators when the data contain statistical outliers. They were especially concerned about use of eigenshape analysis with data having morphologically distinct subpopulations. We believe the message of Full and Ehrlich's (1986) paper ought to encourage investigators to use good techniques of data analysis by graphically examining distributions before analysis and not overinterpreting the results after analysis.

We have used both Fourier analysis and eigenshape analysis in various studies and have found that both methods produce biologically reasonable results. While the arguments about methodologies may continue to rage, we believe researchers may safely continue to apply either method, albeit with caution. Similarly, we doubt that a universal rule can be developed for choosing landmark analysis or outline analysis. Instead, we stress that the method must be chosen to fit the problem at hand (Kaesler and Foster, 1988). The problem will include elements of both the biological questions under study and the nature of the organism.

ACKNOWLEDGMENTS. We are grateful to Peter N. Schweitzer and Timothy R. Maness, who participated in discussions during the formulation of our ideas on shape analysis. Rosalie F. Maddocks and Richard H. Benson critically reviewed the manuscript and contributed to its improvement from their special perspectives as specialists on the Ostracoda. A. J. Rowell reviewed part of the manuscript, and colleagues attending the Geology Colloquium in the Department of Geology at The University of Kansas contributed from the viewpoints of diverse backgrounds.

References

Alberch, P., Gould, S. J., Oster, G. F., and Wake, D. B., 1979, Size and shape in ontogeny and phylogeny, *Paleobiology* **5**(3):296–317.

Anstey, R. L., and Delmet, D. A., 1973, Fourier analysis of zooecial chamber shapes in fossil tubular bryozoans, *Geol. Soc. Am. Bull.* **84**:1753–1764.

Benson, R. H., 1967, Muscle-scar patterns of Pleistocene (Kansan) ostracodes, in: *Essays in Paleontology and Stratigraphy* (C. Teichert and E. L. Yochelson, eds.), pp. 211–242, University Press of Kansas, Lawrence, Kansas.

Benson, R. H., 1982, Comparative transformation of shape in a rapidly evolving series of structural morphotypes of the ostracod *Bradleya*, in: *Fossil and Recent Ostracods* (R. H. Bate, E. Robinson, and L. M. Sheppard, eds.), pp. 147–164, Horwood, Chichester.

Benson, R. H., Chapman, R. E., and Siegel, A. F., 1982, On the measurement of morphology and its change, *Paleobiology* **8**(4):328–339.

Bookstein, F. L., 1978, *The Measurement of Biological Shape and Shape Change*, Springer-Verlag, New York.

Bookstein, F. L., 1980, When one form is between two others: An application of biorthogonal analysis, *Am. Zool.* **20**:627–641.

Bookstein, F. L., 1982, Foundations of morphometrics, *Annu. Rev. Ecol. Syst.* **13**:451–470.

Bookstein, F. L., Strauss, R. E., Humphries, J. M., Chernoff, B., Elder, R. L., and Smith, G. R., 1982, A comment upon the uses of Fourier methods in systematics, *Syst. Zool.* **31**:85–92.

Brower, J. C., and Veinus, J., 1978, Multivariate analysis of allometry using point coordinates, *J. Paleontol.* **52**:1037–1053.

Chapman, R. E., Galton, P. M., Sepkoski, J. J., Jr., and Wall, W. P., 1981, A morphometric study of the cranium of the pachycephalosaurid dinosaur *Stegoceras*, *J. Paleontol.* **55**:608–618.

Christopher, R. A., and Waters, J. A., 1974, Fourier series as a quantitative descriptor of miospore shape, *J. Paleontol.* **48**(4):697–709.

Edgington, E. S., 1980, *Randomization Tests*, Dekker, New York.

Ehrlich, R., and Full, W. E., 1986, Comments of "Relationships among eigenshape analysis, Fourier analysis, and analysis of coordinates" by F. James Rohlf, *Math. Geol.* **18**:855–857.

Ehrlich, R., and Weinberg, B., 1970, An exact method for characterization of grain shape, *J. Sed. Petrol.* **40**(1):205–212.

Ehrlich, R., Pharr, R. B., Jr., and Healy-Williams, N., 1983, Comments on the validity of Fourier descriptors in systematics: A reply to Bookstein *et al.*, *Syst. Zool.* **32**(2):202–206.

Evans, D. G., Schweitzer, P. N., and Hanna, M. S., 1985, Parametric cubic splines and geologic shape descriptions, *Math. Geol.* **17**(6):611–624.

Foster, D. W., and Kaesler, R. L., 1983, Intraspecific morphological variability of ostracodes from carbonate and mixed carbonate–terrigenous depositional environments, in: *Applications of Ostracoda* (R. F. Maddocks, ed.), pp. 627–639, University of Houston, Houston.

Full, W. E., and Ehrlich, R., 1982, Some approaches for locations of centroids of quartz grain outlines to increase homology between Fourier amplitude spectra, *Math. Geol.* **15**:259–270.
Full, W. E., and Ehrlich, R., 1986, Fundamental problems associated with "eigenshape analysis" and similar "factor" analysis procedures, *Math. Geol.* **18**(5)451–463.
Full, W. E., Ehrlich, R., and Bezdek, J., 1982, FUZZY QMODEL—A new approach for linear unmixing, *Math. Geol.* **14**:259–270.
Gould, S. J., 1977, *Ontogeny and Phylogeny,* Harvard University Press, Cambridge.
Healy-Williams, N., and Williams, D. F., 1981, Fourier analysis of test shape of planktonic foraminifera, *Nature* **289**:485–487.
Healy-Williams, N., Ehrlich, R., and Williams, D. F., 1985, Morphometric and stable isotopic evidence for subpopulations of *Globorotalia truncatulinoides, J. Foram. Res.* **15**(4):242–253.
Humphries, J. M., Bookstein, F. L., Chernoff, B., Smith, G. R., Elder, R. L., and Poss, S. G., 1981, Multivariate discrimination by shape in relation to size, *Syst. Zool.* **30**:291–308.
Kaesler, R. L., and Foster, D. W., 1988, Ontogeny of *Bradleya normani* (Brady): Shape analysis of landmarks, in: *Proceedings of the Ninth International Symposium on Ostracoda* (T. Hanai, N. Ikeya, and K. Ishizaki, eds.) (in press).
Kaesler, R. L., and Maddocks, R. F., 1984, Preliminary harmonic analysis of outlines of recent macrocypridid Ostracoda, in: *Symposium on Ostracodes, Additional Communications and Discussions,* Serbian Geological Society, pp. 169–174.
Kaesler, R. L., and Waters, J. A., 1972, Fourier analysis of the ostracode margin, *Geol. Soc. Am. Bull.* **83**:1169–1178.
Lohmann, G. P., 1983, Eigenshape analysis of microfossils: A general morphometric procedure for describing changes in shape, *Math. Geol.* **15**:659–672.
Malmgren, B. A., Berggren, W. A., and Lohmann, F. P., 1983, Equatorward migration of *Globorotalia truncatulinoides* ecophenotypes through the late Pleistocene: Gradual evolution or ocean change?, *Paleobiology* **9**(4):377–389.
Maness, T. R., and Kaesler, R. L., 1987, Ontogenetic Changes in the Carapace of *Tyrrhenocythere amnicola* (Sars), a Hemicytherid Ostracode, University of Kansas Paleontological Contributions, No. 118.
McNamara, K. J., 1986, A guide to the nomenclature of heterochrony, *J. Paleontol.* **60**(1):4–13.
Raup, D. M., 1966, Geometric analysis of shell coiling: General problems, *J. Paleontol.* **40**:1178–1190.
Read, D. W., and Lestrel, P. E., 1986, Comment of uses of homologous-point measures in systematics: A reply to Bookstein *et al., Syst. Zool.* **35**(2):241–253.
Rohlf, F. J., 1986, Relationships among eigenshape analysis, Fourier analysis, and analysis of coordinates, *Math. Geol.* **18**:845–854.
Rohlf, F. J., and Archie, J., 1984, A comparison of Fourier methods for the description of wing shape in mosquitoes (Diptera: Culicidae), *Syst. Zool.* **33**:302–317.
Romesburg, H. C., 1985, Exploring, confirming, and randomization tests, *Computers Geosci.* **11**(1):19–37.
Sanderson, D. J., 1977, The algebraic evaluation of two-dimensional finite strain rosettes, *Math. Geol.* **9**(5):483–496.
Schweitzer, P. N., Kaesler, R. L., and Lohmann, G. P., 1986, Ontogeny and heterochrony in the ostracode *Cavellina* Coryell from Lower Permian rocks in Kansas, *Paleobiology* **12**(3):290–301.
Siegel, A. F., 1982, Geometric data analysis: An interactive graphics program for shape comparison, in: *Modern Data Analysis* (R. L. Launer and A. F. Siegel, eds.), pp. 103–122, Academic Press, New York.
Siegel, A. F., and Benson, R. H., 1982, A robust comparison of biological shapes, *Biometrics* **38**:341–250.
Simpson, G. G., and Beck, W. S., 1965, *Life,* 2nd ed., Harcourt, Brace & World, New York.
Sneath, P. H. A., 1967, Trend-surface analysis of transformation grids, *J. Zool. Soc. Lond.* **151**:65–122.
Sokal, R. R., and Rohlf, F. J., 1981, *Biometry,* 2nd ed., Freeman, San Francisco.
Strauss, R. E., and Bookstein, F. L., 1982, The truss: Body form reconstructions in morphometrics, *Syst. Zool.* **31**(2):113–135.
Thompson, D'A. W., 1917, *On Growth and Form,* Cambridge University Press, Cambridge.

Waters, J. A., 1977, Quantification of shape by use of Fourier analysis: The Mississippian blastoid genus *Pentremites*, *Paleobiology* **3**:288–299.

Willig, M. R., Owen, R. D., and Colbert, R. L., 1986, Assessment of morphometric variation in natural populations: The inadequacy of the univariate approach, *Syst. Zool.* **35**(2):195–203.

Zahn, C. T., and Roskies, R. Z., 1972, Fourier descriptors for plane closed curves, *IEEE Trans. Computers* **C-21**(3):269–281.

Chapter 5

Phylogenetic Analysis and the Detection of Ontogenetic Patterns

WILLIAM L. FINK

1. Introduction

Some years ago an article in the journal *Nature* recounted a discussion which had occurred at a meeting of paleontologists at the British Museum (Halstead, 1978; see also Gardiner *et al.*, 1979). Part of the debate concerned the differing classifications systematists of two schools might propose based on the same evidence about relationships. The schools were the phylogenetic (or cladistic) school and the "evolutionary" school. One discussant rose to claim that given a lungfish, a salmon, and a cow to classify, a phylogeneticist would perform the ridiculous action of grouping the lungfish and the cow together, exclusive of the salmon. A phylogeneticist replied "Yes, I cannot see what is wrong in that" (Halstead, 1978). Such exchanges exemplify some of the arguments that, until recently, were the stock in trade of gatherings concerning comparative biology and systematics. The debate has touched many areas of the biological sciences, from biogeography to developmental biology. What follows is a summary of the assumptions and methods of phylogenetic systematics, and how its practice can elucidate the relation-

WILLIAM L. FINK • Museum of Zoology and Department of Biology, University of Michigan, Ann Arbor, Michigan 48019.

ship between ontogeny and phylogeny. I also address the problem of integrating studies of ontogenetic processes such as heterochrony into a modern comparative framework.

Phylogenetics has become the leading school of systematics in the past two decades. This rapid ascendancy can probably be traced to several factors, but the dominant one is that it is a logically consistent, explicit method for finding genealogical relationships among organisms (Hennig, 1966). Although there are debates within phylogenetics regarding some epistemological matters, the majority of working systematists using the methods have moved to practical application in nearly every major plant and animal group. As phylogenetic analyses and classifications become available to other comparative biologists, there is a need to explain some of the assumptions, methods, and applications to these new users. Some of the changes wrought in comparative biology are nontrivial and will greatly alter how some groups of organisms are studied.

2. Principles of Phylogenetic Analysis

The basic assumptions of phylogenetics are few and rather simple. The most economical statement of the underlying theory is "descent with modification" (Darwin, 1859). Phylogenetic methods attempt to uncover the pattern of descent (genealogy) by tracing the modifications that arise during evolution (evolutionary novelties, or synapomorphies). We assume that there is a single phylogenetic tree for life on earth, and that all organisms are related genealogically. This stands as nothing more than a rejection of special creation, since our evidence shows that organisms arise from other organisms. We assume that over time the organisms become modified and that some of these modifications are passed to their descendants. The actual mechanisms of modification and transmission to subsequent generations, generally accounted for by inheritance, natural selection, epigenetics, genetic drift, etc., are not important to phylogenetics. What is important is that change does occur and that its results are passed on through genealogical connections. These genealogical connections are represented by a branching diagram, usually called a phylogenetic tree, or cladogram (but these can be different; see below).

Genealogy and novelty are the cornerstones of phylogenetics. The third important part of the method is the communication of these evolutionary patterns in classifications. Linnean classifications are lists of names arranged in a hierarchical order of groups and their subgroups. In spite of claims to the contrary by some systematists (who consider that information about phenetics or "adaptive zones" can be placed into and somehow retrieved from a classification), the only information that a Linnean classification can transmit is this hierarchy. The hierarchy of phylogeny can easily be communicated by these hierarchical lists of taxa, and that is exactly what phylogenetic classifications are intended to do. A phylogenetic classification and a branching phylogenetic tree are equivalent. One can be directly transformed into the other, with no loss of information. Such transformation is done by including in each named group all the descendants of a common ancestor (and that ancestor if known). Such a classified group is termed monophyletic. It is considered a natural group because it springs directly from evolutionary processes of descent. This is in contrast to other kinds of groups, such as grades, where a systematist removes one or more descendants of the common ancestor and places them in another named group. This subject will be returned to below.

2.1. Synapomorphy

Evolutionary novelties, modifications that occur in a species and are passed to its descendants, are the key to understanding phylogenetic systematics. Hennig realized that similarity within a set of organisms can be partitioned into portions that reflect immediate common ancestry (synapomorphy) or more distant relationship (plesiomorphy). The concept of synapomorphy, or evolutionary novelty, is Hennig's greatest contribution to systematics. Armed with this concept, we can see that some of the characteristics of organisms are hierarchical, and that they reflect the history of relationships of their bearers, a history which is also hierarchical. Imagine a study of relationships within a group of songbirds. An example of a plesiomorphic characteristic in this case would be presence of feathers— they represent an evolutionary transformation which occurred in the common ancestor of all birds and so do not aid us in understanding relationships *within* a subgroup of birds. We already know that the species under consideration have feathers because those structures are a synapomorphy of birds. As guides to more recent evolutionary diversification, we might focus on feather coloration patterns. Any such characteristic shared exclusively by two species would be evidence that they shared an immediate common ancestor (that they are more closely related to each other than to any other species). Such characteristics are synapomorphies.

2.2. The Branching Diagram and Classification

The branching diagram used in phylogenetics can mean several things, depending on the intent of the systematist who constructs it. At its most basic, it serves as a device to summarize character distributions and thus illustrates where a character occurs. In this guise the diagram is usually termed a cladogram, as it implies nothing more than character distributions. For example, Fig. 1 presents a cladogram which summarizes the accompanying matrix of features, where a 1 indicates that the feature occurs.

Assume that taxon X represents the "outgroups," that is, it represents several different clades which are closely related to A, B, and C, and that each state listed as present (1) in one of these taxa is absent (0) in the outgroups. In this example, the distribution of character 1 is summarized by the group ABC, while that of

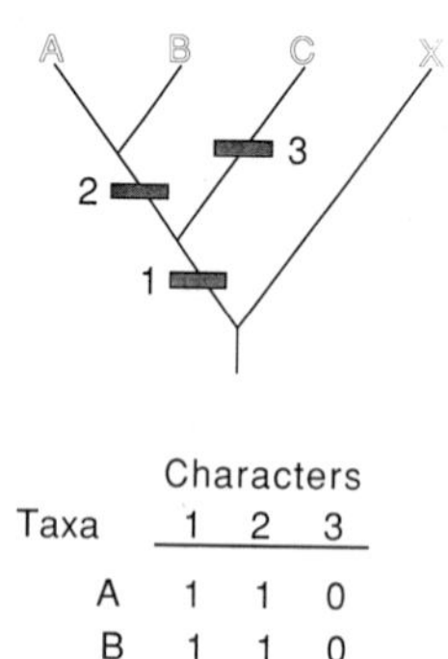

Taxa	Characters 1	2	3
A	1	1	0
B	1	1	0
C	1	0	1
X	0	0	0

Figure 1. A hypothesis of relationships and a data matrix. Here A, B, and C are the taxa of interest and X represents the outgroups. Bars on the branches represent the simplest accounting for the distribution of the characters. See text for discussion.

character 2 is summarized by the group AB. These groups are represented by connections of the branches at nodes. Since A and B share a node not shared by C, the former two taxa represent a group. A classification which contains the same group information as the cladogram is

ABCX
X
ABC
AB
C

or, assuming that the taxa following any particular taxon are the sister group of the latter [a sequencing convention (Wiley, 1979)], the classification is

X
C
B
A

But a branching diagram can imply more information than this. It can also imply phylogenetic relationship, where the branches are taken to indicate evolutionary lineages, and the nodes represent ancestors. Classifications of these phylogenies are different from those of cladograms, because they may have to include positions of ancestors and descendants (Wiley, 1979). For example, consider Fig. 2 with a phylogeny (A) and its cladogram (B). The two classifications based on

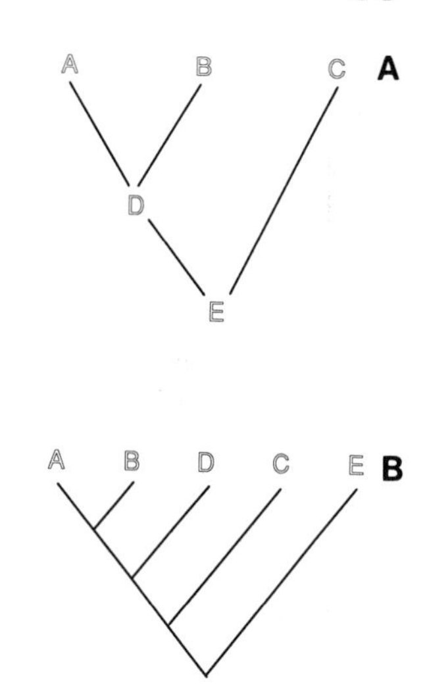

Figure 2. (A) Phylogenetic tree of five taxa, with D and E as ancestors. (B) Cladogram derived from the phylogeny. In this case, all taxa are placed as terminals. Taxa D and E, lacking any unique features, would not be diagnosable.

them are

Phylogeny	Cladogram
ABCD(E)	E
C	C
AB(D)	D
	B
	A

In the case of the phylogenetic classification, the convention (·) is used to indicate that the enclosed taxon is the ancestor of the group to which it is adjacent. Thus, E is the ancestor of the group containing ABCD.

This distinction between phylogenies and cladograms is not necessarily of significance to most comparative biologists, primarily because the evidence for "ancestorship" is usually not available for any given species, and the cladogram is essentially the "default" diagram. One can consider the cladogram as a phylogeny, with the nodes representing hypothetical ancestors.

2.3. Character Analysis

Discovery of the pattern of character transformation is one of the most important aspects of phylogenetics. In order to tell whether a feature is a synapomorphy or a plesiomorphy at a given level of analysis, we need to determine the distribution of that feature or character. If a character is more broadly distributed than in the taxa whose relationships are in question, then it is plesiomorphic for them. This broad distribution is determined by several means, including outgroup comparison, ontogenetic analysis, and stratigraphic sequence. Each of these methods is fraught with problems, in some cases to the extent that they are not used by many systematists.

2.3.1. Outgroup Comparison

This is the most widely used method of determining the generality of a feature (or its "polarity"—the inferred direction of evolutionary change) (Maddison *et al.*, 1984; Stevens, 1980). This is done by comparing the feature in question in the group under analysis with similar features in related taxa. If the feature is restricted to all or some of the taxa in the group under study, it is considered derived and indicative of common ancestry of a lineage comprising just those taxa in which it is found. A feature found in the group of interest as well as in some or all of the outgroup taxa is considered primitive and indicative of relationship at a more general level. The major drawback of outgroup comparison is that it requires at least a broad outline of the relationships of a group. The better the outgroup hypothesis, the more confidence one can have in the subsequent analysis. In some groups, the phylogenies are so poorly supported that the best one can do is employ broad surveys to estimate what the derived states might be (Maddison *et al.*, 1984; Donoghue and Cantino, 1984; Clark and Curran, 1986). Conservative assessments must be done in these cases. As the phylogeny is "fleshed out" by subsequent workers, it is to be expected that mistakes will be found and corrected. This is really the way all science is done, as we build on the structures of past workers and approach, we hope, more and more stable and heuristic hypotheses.

2.3.2. Ontogenetic Criterion

In many cases, the ontogenetic precedence of character states can be used to detect character transformations. This criterion is controversial, with some systematists claiming that it is independent of hypotheses of relationship (Nelson, 1978), while others insist that it must be used in conjunction with outgroup comparison or that it is a form of outgroup comparison (Kluge and Strauss, 1985). As an example of its simplest form, imagine an observable ontogenetic transformation of character state a to state a'. Since a precedes a' ontogenetically, we consider a' to be the derived state of the character. This relationship will hold as long as the ontogeny of the character has been one of terminal addition. The ontogenetic criterion can be especially useful in groups where outgroup comparisons are hampered by sketchy phylogenetic hypotheses.

More recently, some authors have advocated the use ontogenetic transformations as characters themselves (e.g., de Queiroz, 1985), a practice demonstrated here in Section 3.2.3.

2.3.3. Stratigraphy

The stratigraphic criterion of character polarity [also referred to as the "paleontological method" (Nelson, 1978)] involves the assumption that primitive character states will be found in fossils that lie in older strata than do fossils with derived states. This follows from our theoretical expectations that more primitive

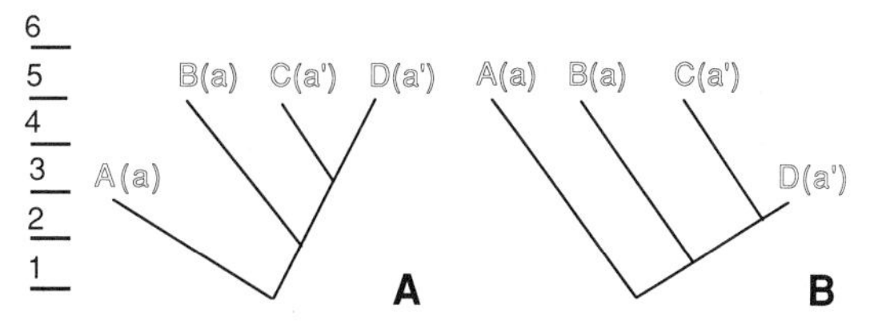

Figure 3. Use of the stratigraphic criterion to determine character polarity. Stratigraphic sequences are represented by the numbers at the left, older strata bearing smaller numbers. As defined, the true polarity of character a is $a \rightarrow a'$. In part A, use of the stratigraphic criterion would yield the proper polarity, since the oldest known taxon (A) has the primitive state. In part B, however, the stratigraphic criterion would yield incorrect polarity, since the earliest known fossil, taxon D, has the derived state.

organisms precede more derived ones in time. However, of the three criteria, this is the least popular, for several reasons. One obvious reason is that many groups have a scanty or nonexistent fossil record and the criterion simply cannot be applied to them. In addition, it is the nature of fossils to present less complete data than Recent organisms, since much fossil material represents only a fraction of the complete animal. Another reason is that to avoid sampling errors, the fossil record must be very dense, assuring that possible character transformations will be seen (the sampling problem). Even in groups with a "reasonable" fossil record there remains the necessary assumption that a fossil will be sampled in a stratigraphic range equal to its relative position in the phylogenetic tree (the positional problem).

The sampling problem is a serious one, since few groups have a virtually continuous fossil record of all the lineages one wishes to study, as well as sufficient geographic coverage so that the sequence of taxa is not confounded by migrations or simple alterations in ranges of the taxa under study. The "completeness' of the record has been dealt with in detail by Dingus and Sadler (1982), who note that the paleontological problem under study will in part determine how complete the record need be. Depositional environment (some environments will offer more complete samples than others) and time (the further back in time, the greater are the chances of encountering gaps that are larger and more frequent) must be acknowledged and accounted for. This means that groups of organisms and the evolutionary problems to be addressed must be chosen and studied in the proper context. It is unlikely, for example, that study of speciation rates in Pennsylvanian amniotes would be a profitable undertaking. The same argument applies to the use of the "stratigraphic criterion" in polarity assessment—one would have to estimate the probability of finding the proper fossils in the proper depositional context to obtain samples that would provide reasonable estimates of character transformations.

The positional problem pertains to the assumption that a fossil organism found at a given stratigraphic range will be primitive in its features relative to a fossil found in subsequent strata. For example, in Fig. 3, imagine that the true phylogeny is (A(B(CD))) and that the character state held by CD (a') is derived. In Case 1 the worker using the stratigraphic criterion correctly assumes that the first fossil has characters more primitive than those of the second; the derived state would indicate that CD is a monophyletic group. In Case 2, however, that assumption is incorrect, and polarity decisions based on it will likewise be incorrect; the derived state would be considered primitive and a group AB would be formed. In part this problem could be ameliorated by a "complete" fossil record, but it also emerges from the common assumption that an older fossil will be more primitive in *all* its features. This assumption ignores the possibility that parts of the body can evolve at different rates (mosaic evolution). The frequency of incorrect character polarity assessments due to the positional problem is difficult to estimate because much of the paleontological literature makes the assumption as a matter of course and thus cannot test it.

In fact, this inability to be independent of and thus to test the stratigraphic record is the best argument against using the stratigraphic criterion. The criterion does not rely on the features of the organisms themselves as much as on their stratigraphic deposition. As a result, there can be no independent tests of phylogeny and stratigraphy.

There has been a history of antagonism between some paleontologists and phylogeneticists, but I see no reason why this must be so. After all, our theoretical expectations are the same: there should be a parallelism between phylogeny and stratigraphic deposition. Disagreements should arise only when the parallelism fails, and one or the other discipline has to account for the failure. Such accounting usually involves reference to character data by phylogeneticists and to the stratigraphic data by paleontologists. Differences in emphasis on the two kinds of data separate phylogeneticists, as typified by Cracraft (1981), from paleontologists, as typified by Paul (1982) or Gingerich (1979). My only comment regarding this split is that interpretations of stratigraphic data are not independent of character data and logically must be subservient to them, because it is character data that allow comparisons between fossils in the first place. In any particular series of strata, many fossils may occur, but we use character data to guide comparisons among them. For example, in a series of strata we might find leaves, lizard bones, and mammal teeth. We do not assume that the leaves represent the ancestor of the lizards or mammals merely because they lie in older strata than the latter. Rather, we use character data to determine which taxa might be related and compared.

2.4. Hypothesis Choice and the Principle of Parsimony

Once we have gathered together a data set and hypothesized the direction of character changes, the next step is to erect and evaluate hypotheses of relationships among the taxa. There are several methods for doing this, but here I will discuss only the dominant one, parsimony analysis. This approach assumes that the better scientific hypothesis is the one that makes the fewest *ad hoc* assumptions, or, in other words, the one that accounts for the data in the most economical fashion. In practice, one usually constructs the tree by determining that a character supports a sister-group relationship of two taxa relative to a third. Another character is then added to the problem, either confirming the first group or suggesting another. For large data sets and large numbers of taxa this process can get rather complicated, and several computer programs are available to automate it (Fink, 1986). However, another, equally valid way to view the problem is to imagine all the alternative trees that represent the interrelationships of the taxa in a given

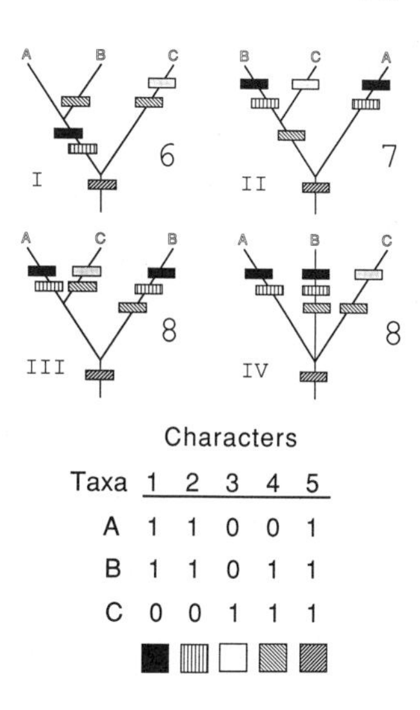

Taxa	1	2	3	4	5
A	1	1	0	0	1
B	1	1	0	1	1
C	0	0	1	1	1

Figure 4. Alternative cladograms of three taxa and a data matrix. Based on outgroup comparisons, the characters are polarized, with 0 representing the primitive state and 1 the derived state. Derived conditions are shown by bars with the appropriate symbol for each character. Note that all four hypotheses of relationship can claim to explain the distributions of the characters. The only difference among the hypotheses, relative to the data, is the efficiency of explanation. Hypothesis I represents the data distribution more economically than II, III, and IV in that only six transformations are required, while II requires seven, and III and IV require eight each. Barring some extra information not included in the analysis, the working hypothesis of choice is hypothesis I.

problem. This gets unwieldy with large numbers of taxa (Felsenstein, 1978), but for small problems it can be a useful way to proceed.

The following example of alternative cladograms will serve to illustrate how one chooses among hypotheses, any of which can explain a given data set (Fig. 4). Take a monophyletic group of species A, B, and C. Assume that we have outgroup information and are able to examine several characters whose polarities are assessable. Additionally assume that each taxon has features unique to itself (autapomorphies), thus indicating that none can be the ancestor of any other. For any such three-taxon group there are four possible cladograms representative of their relationships, and given our evolutionary framework, only one of these can be true. The data matrix represents features we have discovered in the taxa in question. A 1 represents the presence of the derived state and a 0 represents the primitive state. Note that each feature can be placed on each cladogram. In cases where a feature is shared by two taxa that also share a common ancestral node, it can be summarized by placement on the common node. Other features are placed on the terminal node. One can see that each of these alternative cladograms can account for the data in the sense that the presence of each character can be considered an evolutionary novelty wherever it occurs. But there is a definite difference in the efficiency of the accounting. Hypothesis I requires six transformations or evolutionary events (steps) to account for the data, while hypothesis II requires seven, and hypotheses III and IV require eight steps. Given the usual scientific practice of choosing the most parsimonious hypothesis to account for the data at hand, most systematists would consider hypothesis I the best choice. This is not to say that it is necessarily the true one, but only that given the data we now have, it is the most economical way to account for the data. New data could alter our preference.

There are several things to discuss here. First note that each character appears as a synapomorphy—at the group node that it diagnoses. Note also that character 5 is a plesiomorphy relative to any of the possible subgroupings of A, B, and C, while it is a synapomorphy of the entire group ABC. Character 3 is an autapomorphy of C and is never useful in determining the relationships of C. Character 4 is illustrative of the implications of tree topology for the concepts of synapomorphy and convergence. On hypothesis I, 4 is convergently acquired (or homoplasious) in B and in C, while on hypothesis II it is a synapomorphy of group BC. Thus, it becomes clear that hypotheses of synapomorphy and convergence are dependent on the phylogenetic context in which they are interpreted. Neither kind of character can be considered an observation or a datum, but rather both emerge as conclusions from a hypothesis of relationship. As that hypothesis changes, so the status of a feature may change from synapomorphy to convergence. Restated, convergence is detected by incongruent distributions of features on cladograms.

2.5. Multiple Solutions and Character Optimization

The advent of sophisticated computer programs to find phylogenetic trees has brought two important issues to the forefront of phylogenetics discussions. One is the fact that any given data set may support many resolutions equally well. In fact, it is not uncommon for a matrix to support 20 or more equally parsimonious cladograms. There are two reasons for these multiple solutions. One is that there may be no data supporting a particular node. For example, to get a fully resolved cladogram, one must have at least $n - 2$ characters, where n is the number of taxa involved. A matrix of 20 taxa must have at least 18 nonhomoplasious characters to resolve the tree, and fewer characters will result in collapse of one or more nodes into multichotomies, which may be represented by several resolved topologies. Another reason for multiple trees is that the data supporting one or more nodes may be ambiguous, or "noisy." Some characters support one resolution, while others support one or more alternatives. If the total amount of homoplasy required by alternative resolutions is the same, multiple, equally parsimonious trees will exist. It is a humbling experience indeed to analyze a data matrix and find that many equally valid tree solutions are possible. At a minimum, this ease of finding multiple trees must give us caution when we interpret the tree as a phylogeny and try to discuss the evolution of the group.

Even more caution is needed when discussing the evolution of particular features in the context of a cladogram. This is because any ambiguity in a character's distribution can result in its being diagnostic of different groups, depending on how the characters are optimized. Alternate character optimization is a relatively new aspect of phylogenetic analysis, although optimization procedures have been part of Wagner programs for years. Depending on the assumptions one is willing to make about the frequency of parallelisms, convergences, or reversals, the distribution of ambiguous characters can change radically on a single tree topology. This kind of change makes the ability to explain the evolution of such characters very difficult and cautious systematists will be wary of attempting such stories.

2.6. Classification and Naturalness

It was mentioned above that phylogenetic classifications include only monophyletic groups, that is, groups that include a common ancestor and all its descendant species. This is in contrast to some other schools of systematics, which allow grades, groups that do not include all the descendants of the ancestral species. Another term describing such groups is paraphyletic. Monophyletic groups are formed by the process of genealogical descent, rather than arbitrary human decisions. Paraphyletic groups, on the other hand, are human constructs, made by the *removal* of some descendants of the ancestor of a lineage by a taxonomist. Thus, paraphyletic groups cannot be considered the products of genealogical descent; nor has any convincing case been made that any other biological or evolutionary process causes the formation of paraphyletic groups. Since it is the aim of evolutionary biologists to study the processes and products of evolution, it is not paraphyletic but monophyletic groups that will best serve these purposes.

Many of the classically recognized Linnean groups turn out to be paraphyletic groups. The primary reason for this is that species were often grouped together based on primitive similarity rather than on the basis of apomorphic similarities, so that often the primitive members of disparate genealogical lineages were placed together because they looked more or less alike. A prime example of this phenomenon is the class Reptilia. According to many people, reptiles are more or less alike in that they are ectothermic, have scales, are usually quadrupedal and slow-moving, and lack advanced behavioral and communications abilities. Note, however, that none of these features is synapomorphous, but rather all are shared with amphibians and various "fishes." Reptiles share with mammals and birds the presence of many features, including an amnion. But historically they were included in a group with each other because they lack feathers or hair, not because they shared some exclusive attribute. A phylogenetic analysis of tetrapod relationships shows that the Recent reptiles are the sister group of mammals, and that birds are the sister group of crocodilians (Gauthier, 1984; Gauthier *et al.*, 1988). When fossil taxa are included, it is shown that birds are in fact dinosaurs (Gauthier, 1986). And here is a very nice example of the heuristic value of phylogenetic classifications, which reflect the pattern of origination of evolutionary novelties. A closer look at the crocodile + bird relationship reveals some interesting things that most bird and crocodile specialists have overlooked. For example, some traits that these two groups share include: complex social behavior involving vocalizations, nesting and care of the young, imprinting of the young on the female parent, and presence of a gizzard. All of these specializations were or could have been attributed to convergence or overlooked using older classifications. In particular, the older view kept us from insights that shed light on the Mesozoic archosaurian fauna. Instead of a silent era of stupid, hissing dinosaurs, we should expect the Mesozoic to have been noisy with vocalizations of all sorts, marked by large social groups of dinosaurs, which cared for and protected their young. We already know that some dinosaurs had family groups and nesting sites from the fossil record (Horner, 1984) (but this author, for the most part, missed the opportunity for comparisons with birds). Until recently nesting behavior in dinosaurs has been considered a curiosity, but it *should have been expected* in the light of Recent archosaurian behavior.

Another important issue regarding classification pertains to the ontological status of groups. Paraphyletic taxa are names for classes, manufactured by systematists. In contrast, monophyletic taxa are individuals, historical lineages that exist by virtue of the natural processes of evolution—no participation by a systematist is needed to validate their reality. This is more than a trivial point, since the distinction between classes and individuals is a thorny philosophical one (Ghiselin, 1974, 1981; Hull, 1978). For example, it can be argued that classes do not participate in natural processes, only individuals do. Gold does not participate in anything, but individual atoms of gold do. Likewise, Reptilia as a class cannot be a natural group and thus cannot be viewed as being the product of natural processes. A Reptilia that includes birds, however, can be seen as a historical individual that has resulted from the processes of evolution. Taken as a whole, comprising all known and unknown members, it forms a record of those processes.

3. The Discovery of Ontogenetic Patterns

Historically, discovery of evolutionary novelties considered significant or extraordinary by a particular worker were recognized taxonomically by classifying the modified organisms into a new taxon at a categorical rank equal to or higher than that of its closest relatives. This was done because such novelties were considered to be outside the acceptable morphological limits of the preexisting taxon, and to have resulted in the origination of a new genus, family, phylum, or other higher taxon. News of such discoveries was often greeted with interest by other evolutionists or the general public. According to some definitions, the origin of such novelties has been termed macroevolution. I certainly do not deny that marked changes occur, but I deny that there is any substance in the attempt at understanding the "origin" of a new higher taxon *per se* as a natural event, because such origins are, after all, only speciation events and because the categorical level of the taxon is a product of a taxonomist's mind.

The search for mechanisms that can generate major morphological changes has led to the study of ontogeny, in part because some kinds of modifications of ontogenies seem an elegant and economical way to generate major phenotypic change (Gould, 1977). Now the search is on for examples of changes in ontogenies that have indeed been responsible for such novelty. In a fundamental way, of course, such changes must be part of the mechanism for much of the morphological (and other) diversity, since it is ontogeny that actually generates most of the phenotype. But incidences of particular ontogenetic phenomena and their importance to larger scale diversity remain poorly understood, in spite of a growing literature on the subject. One of the reasons for this lack of knowledge is that few researchers have used methods which can unambiguously indicate which mechanisms of ontogenetic modification are involved.

Perhaps the best known mechanism of ontogenetic change proposed as a source for evolutionary novelty is heterochrony. My aim here is not to criticize published studies of heterochrony, but rather to lay out what seems a reasonable way to detect heterochrony and to describe its simpler forms. It seems obvious that the assessment of heterochrony as a historically important process requires that there be historical hypotheses in which to evaluate it. And this historical context requires a phylogenetic basis, a basis that is all too often absent in the literature on the subject.

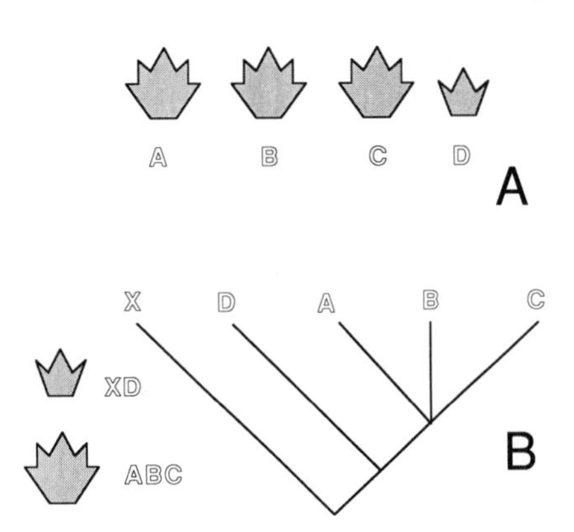

Figure 5. (A) A sample of tooth morphologies representing the taxa A, B, C, and D. With this evidence alone, we are unable to order the evolutionary transformation in tooth cusp number (3→5 or 5→3). (B) Given the knowledge that the morphology of X represents the condition in the outgroups, we can order the transformation to 3→5 and form a monophyletic group ABC.

3.1. Heterochrony

Heterochrony, modifications of developmental rates or timings that alter phenotypes, implies within its definition comparisons against a standard. The standard is generally considered to be the ancestral condition, or the rates and timings that generated the phenotype of the ancestral species morphology. This definition is the one discussed in great detail by Alberch *et al.* (1979) in their classic paper on the subject. Subsequently, I suggested that the cases in which we are able to specify ancestors are rare, and reformulated the model in a phylogenetic context, considering the outgroup condition to be our best estimate of the ancestral condition (Fink, 1982). Thus, a prerequisite for an analysis of heterochrony is a phylogenetic hypothesis of the relevant taxa. With this in hand, the first step in recognizing heterochronic change is to identify character transformations in the context of the phylogeny. The second step is to examine the ontogenies of the features. A third step is to categorize the change. Ultimately, depending on the research program underway, one might wish to determine whether this novelty is associated with other phenomena, such as marked changes in speciation rates or increased resistance to extinction.

3.2. An Example of Heterochronic Analysis

What follows is a simple example of how heterochronic development might be detected. Although the example is hypothetical, it is based in part on the tooth morphology of a group of freshwater fishes from South America.

3.2.1. The Historical and Ontogenetic Evidence

Figure 5A shows a series of teeth, some with three cusps and some with five. The teeth are labeled A, B, C, and D, these names being representative of the species names of the fishes that we think, based on other evidence, form a monophyletic group. As raw data, the tooth samples are unordered. We do not know which is the primitive form. However, if we view the teeth in the context of the larger phylogeny of the group of fishes, we find that tricuspid teeth represent the

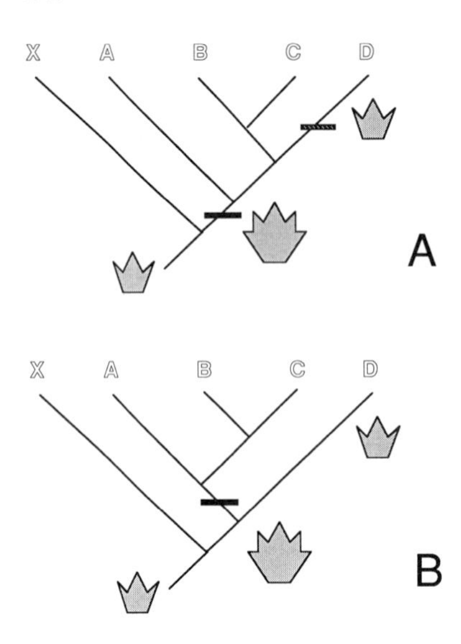

Figure 6. Two resolutions of the interrelationships of taxa A–D, with X representing the outgroup conditions. (A) A resolution based on nontooth information that shows us that D lies within the group containing A, B, and C. In this case the presence of tricuspid teeth in D is seen to be a transformation from the pentacuspid condition. Further examination of the problem may be called for, including a possible analysis of heterochrony. (B) A resolution that excludes D from the group ABC. In this case the presence of tricuspid teeth is simply retention of the primitive condition for the entire group. The problem of greatest interest here is the single transformation of tri- to pentacuspid teeth.

outgroup condition. That is to say, we find that tricuspid teeth are distributed in the phylogenetically primitive members (the outgroups) of the larger group that includes our study group. This relationship is shown in Fig. 5B, where taxon X is considered to be representative of near and more distant outgroups. Taxon D is the sister group of a taxon containing A, B, and C; however, the interrelationships of the latter remain unresolved, as represented by the multifurcation. Nevertheless, we are able to hypothesize confidently that the primitive tooth form is tricuspid, and that there has been an evolutionary transformation to pentacuspid teeth in the study group. We could now do two alternate kinds of analysis. We could use the transformations in teeth to construct a hypothesis of relationships of the taxa, or we could use other features of the species to resolve the cladogram and then use that tree to examine the distribution of the tooth characters. The former analysis would find that ABC form a monophyletic group and that D is their sister group. This follows from a parsimony analysis in which there is a single transformation from the tricuspid to the pentacuspid tooth morphology.

Let us assume now that we have data from other characters to resolve the relationships of the study group. Figure 6 illustrates two possible resolutions, each of which accounts for the tricuspid teeth of taxon D in different ways. Figure 6B is the most parsimonious solution for the tooth data, showing that ABC form a group and that the tooth morphology of D is simply retention of the primitive condition. But another solution is possible, using other evidence, in which tooth character distributions are not as parsimonious. This hypothesis (Fig. 6A) shows two transformations in tooth evolution: from tricuspid to pentacuspid for a group ABCD and then reversal to tricuspid in D. In the context of this tree, the condition in D is anomalous and may warrant further investigation.

So far we have two pieces of information about the evolutionary history of tooth morphology in the group. First, we have the hypothesis that having teeth in the pentacuspid form is a novelty and is preceded phylogenetically by the tricuspid form. Under one hypothesis of relationship, we also know that there

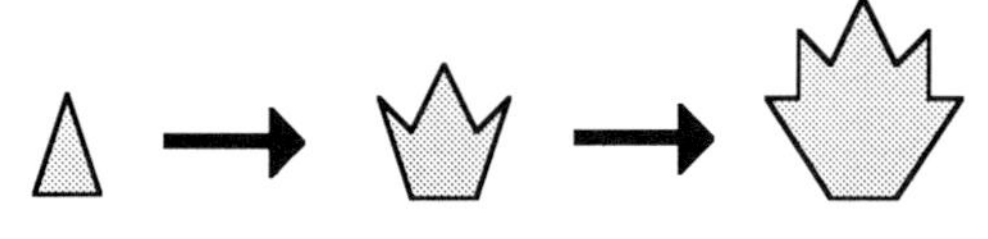

Figure 7. The ontogenetic transformation sequence of teeth from conical, to three cusps, to five cusps. In this case the ontogenetic and phylogenetic transformations are congruent. With this increased information, we can see that the tricuspid condition in taxon D of Fig. 6A is paedomorphosis and the pentacuspid condition of taxon ABC in Fig. 6B is peramorphosis.

has been an evolutionary reversal from pentacuspid to tricuspid teeth in taxon D. Depending on our interests, we might wish to concentrate on the causal mechanisms behind the origin of the novel, pentacuspid, morphology, or we could view the anomalous tricuspid teeth as more interesting and concentrate on them. In either case, or in combination, the next step could be an examination of the ontogenies of the two forms.

As shown in Fig. 7, we find that the tricuspid and pentacuspid teeth are preceded ontogenetically by conical teeth. Further, we find that pentacuspid teeth replace tricuspid teeth during ontogeny. So our ontogenetic sequence, or trajectory, parallels the phylogenetic sequence. This implies that pentacuspid teeth are a terminal addition to the ontogenetic trajectory, and that the reversal to tricuspid teeth in taxon D is a juvenilized, or paedomorphic, state.

3.2.2. A Formalism to Describe Heterochrony

The conclusions deduced above can be further illustrated by showing the developmental stages on a simple graph, with cusp number plotted against developmental time (Fig. 8), as formalized by Alberch *et al.* (1979) and modified by Fink (1982). Those authors noted that simple cases of heterochrony could be described by altering three parameters: onset of growth (α), offset of growth (β), and growth rate (k). In Fig. 8, the outgroup trajectory proceeds from its inception at α, at a rate k, and terminates at β. In this example we assume that k is equal

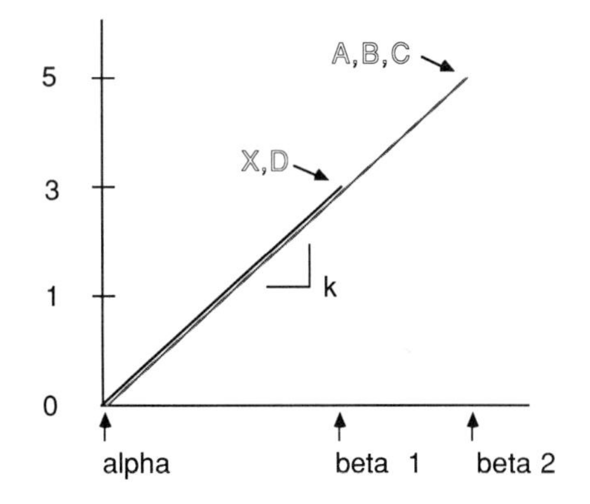

Figure 8. The formalism of Alberch *et al.* (1979). The x axis represents size or time, the y axis shape. Alpha represents onset of growth, beta represents growth offset, and *k* represents growth rate. Using this model, we can describe the heterochronic pattern in the tooth cusp example discussed in the text. The rate of growth is the same in all taxa, as the onset of growth. The variable factor is growth offset time. In the outgroup (X) and taxon D, offset of growth is at beta 1, while in taxa A, B, and C, offset occurs at a later time, beta 2. In the context of the hypotheses of relationships shown in Fig. 6A, pentacuspid teeth in A, B, and C are interpreted as due to hypermorphosis, followed by a return to the primitive offset time (and morphology) in D, aprogenetic event. In Fig. 6B, the presence of similar ontogenetic trajectories in D and X is primitive, and hypermorphosis explains the origin of pentacuspid teeth in taxon ABC.

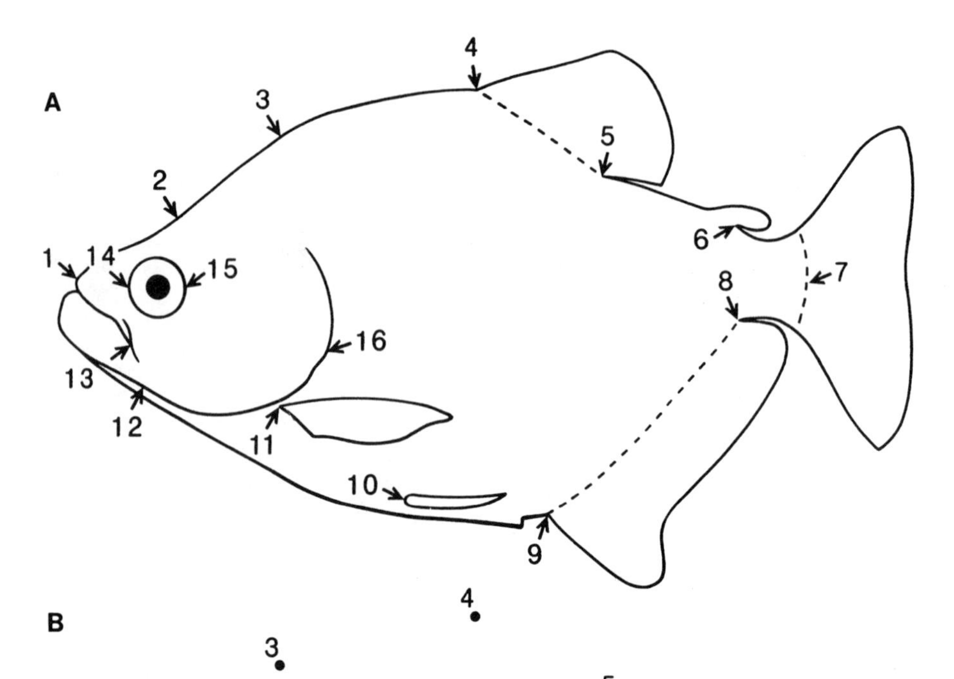
A
1
2
3
4
5
6
7
8
9
10
11
12
13
14
15
16

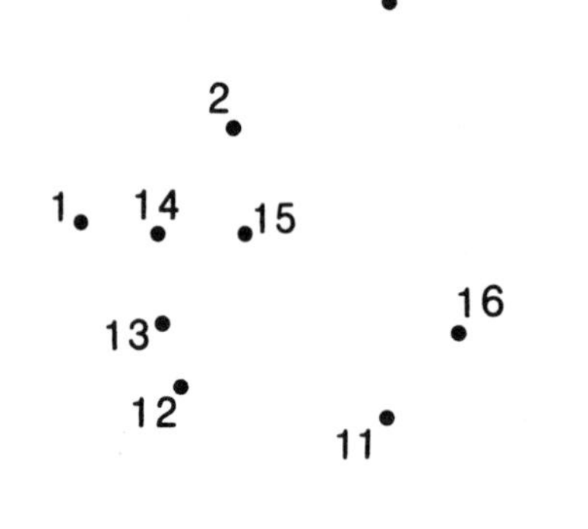
B
1
2
3
11
12
13
14
15
16

10
9

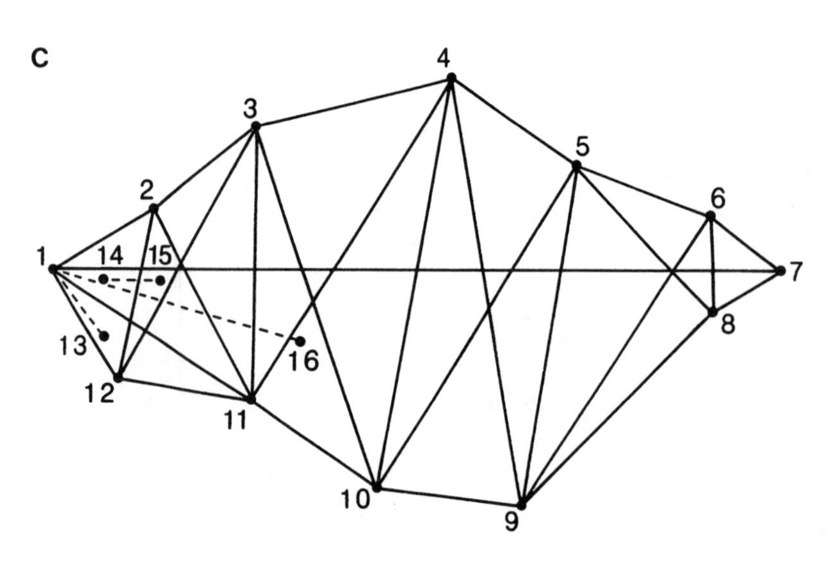
C
1
2
3
4
5
6
7
8
9
10
11
12
13
14
15
16

in all trajectories and that it is offset time β that is altered. In both of our phylogenetic hypotheses (Fig. 6), the condition in D is similar to that of X. In one hypothesis (Fig. 6A), origin of pentacuspid teeth in group ABCD is due to hypermorphosis (delay in offset time), followed by a return to the primitive offset time in taxon D, resulting in progenetic tricuspid teeth. In the other hypothesis (Fig. 6B), the presence of similar ontogenetic trajectories in X and D are simply primitive, and hypermorphosis explains the origin of pentacuspid teeth in the monophyletic group ABC.

3.2.3. An Elaboration of the Formalism

A more complex use of this formalization is the description of the ontogeny of more than one feature. Such description is useful for comparisons of nondiscrete features such as body shape. For example, Fig. 9 shows an outline of a South American piranha fish, with a set of points indicated. These points are digitized and the distances between certain sets of them are calculated, forming a truss system (Strauss and Bookstein, 1982) to estimate body shape. Body width measures are taken as well. When specimens ranging in size from postlarvae to large adults are measured, these distances can be interpreted as informative about the ontogeny of shape in at least two ways. The first method to be discussed is historical, examining shape change in the context of a phylogeny; the other method uses allometric coefficients to describe shape ontogeny in a single species.

Principal components analysis provides a convenient way to summarize shape and size information in both approaches. In the study of piranha ontogeny, I found that PC1 represents body size, while PC2 and PC3 represent various aspects of body shape. By plotting PC2 against PC1 we produce an ontogenetic trajectory of those body elements (in this case all related to body depth) summarized by PC2. To examine how body shape has changed phylogenetically, we analyze the data from several species together, with inclusion of a representative outgroup species. In Fig. 10, there are four ontogenetic trajectories, represented by the scatter of body depth scores plotted against PC1. The trajectory of taxon X is representative of the outgroup condition (based on analysis of several lineages placed successively on the cladogram). Note that taxon X begins with a deeper body than A, B, or C and continues to increase in body depth at a certain rate. Thus, the outgroup condition is to begin development with a relatively deep body and to increase body depth during growth. Taxon A begins development at a shallower body depth, but retains the primitive growth rate, so it gains body depth with age, although it never reaches the depth of taxon X. Taxa B and C begin development near the body depth of A but remain shallow-bodied throughout life. This set of trajectories can be interpreted as resolving a cladogram of the

Figure 9. An example of the data useful in an analysis of shape in an actinopterygian fish (genus *Serrasalmus*). (A) Landmarks chosen are thought to be homologous. Points include some skeletal features as well as origins and insertions of fins. (B) Cartesian coordinate of the digitized body image. Coordinates can be analyzed according to the biorthogonal method, or distances can be calculated to form a truss network (Bookstein *et al.*, 1985). (C) Truss network of distances for reconstructing the body shape of the fish. Dashed lines indicate non-truss measurements.

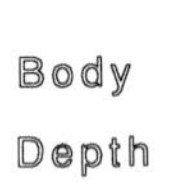

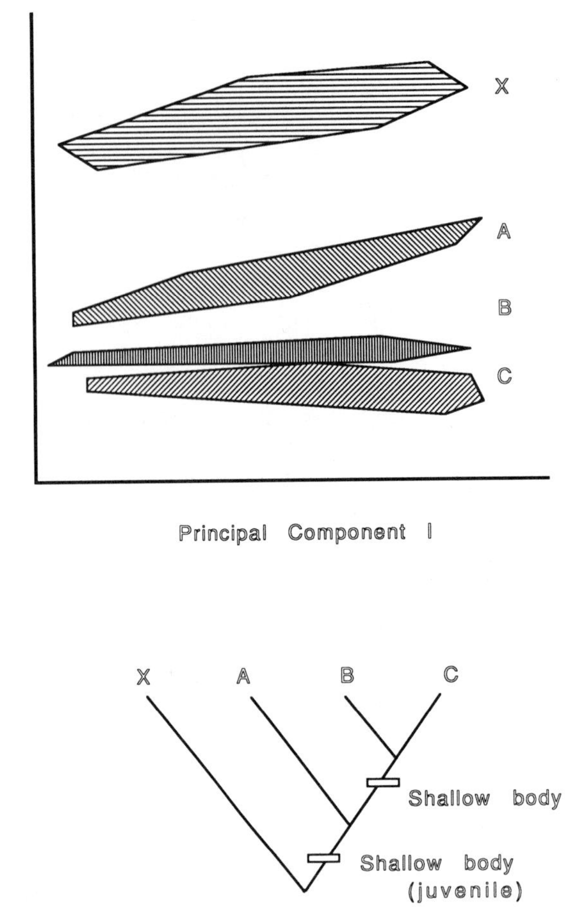

Figure 10. Top. Ontogenetic trajectories of body depth (summarized by PC2) relative to overall size (summarized by PC1) in four taxa. Taxon X is the outgroup. The ingroup includes A, B, and C, all of which have a shallower body depth than X early in ontogeny. Taxon A retains the primitive growth rate, while B and C have a modified rate, which keeps the body shallower then X or A throughout life. Bottom. Cladogram of the four taxa, indicating the generality of the transformations of body depth ontogeny.

interrelationships of A–C (Fig. 10). A change in body depth early in ontogeny is a specialization of A–C, and a retention of the juvenile body depth (paedomorphosis) is characteristic of B and C.

If the cladogram is resolved on other features, we can interpret the ontogenetic data in the context of the cladogram (resolving the cladogram with the same features whose evolution is being examined would be circular). Viewed in this way, the origin of clade ABC was marked by a shallow-bodied juvenile form, produced by an allometric shift before or during the transformation from larval to juvenile stage. Larvae of the fishes in this example are not available, preventing identification of the timing of the shift involved [for other examples see Strauss and Fuiman (1985)]. The origin of BC was marked by a change in the growth rate *k* of body depth, as may be seen by comparison of the growth slopes of B and C with A and X. In BC the juvenile form is retained throughout life, and according to the formalism described above, this is an example of neoteny.

Kluge and Strauss (1985) and Creighton and Strauss (1985) have demonstrated ways one can plot some shape discriminator(s) against overall body size or age, and in addition superimpose the development of particular morphological features. The result is an ontogenetic trajectory based on shape vs. size or age, with discrete morphological events mapped onto the shape changes.

A second, but nonhistorical, kind of ontogenetic information that can be derived from principal components analyses is relative growth rates of different body parts or regions. The scores of individual truss elements are converted to allometric coefficients by calculating an average of all the scores and dividing individual truss scores by the average. Those scores that are greater than 1 indicate positive allometry in that body region (i.e., the region measured by the element grows faster than the average as body size increases), while those below 1 indicate negative allometry. Scores near the average represent isometric growth. One can quickly determine which parts of the body are growing faster or slower than average as body size increases, and thus gain an understanding of how differential growth rates generate the adult body shape within a species.

A more sophisticated method of comparing shape differences, which can be interpreted as ontogenetic when done between specimens of different age classes of a single species, or phylogenetic when done in the context of a cladogram, is biorthogonal analysis. This technique summarizes shape changes in terms of greatest and least deformation when one form is transformed into another. These deformations can be quantified and graphed to resemble Thompson (1961) grids. A description of the procedure and examples of analyses can be found in Bookstein *et al.* (1985).

4. Summary and Conclusions

The historical importance of heterochrony can be determined only by analyzing its frequency and consequences in a historical context. The appropriate historical context is a phylogeny of the group of organisms. Without that phylogeny, one has no evidence concerning the direction of morphological transformation, and thus no problem for the evolutionary biologist to examine. The paucity of well-corroborated phylogenies of most organisms is the major practical obstacle which must be overcome before the actual importance of ontogenetic phenomena in evolution can be assessed. The major conceptual obstacle remains the lack of appreciation of the central role that phylogenies must play in any form of comparative biological analyses. As the conceptual framework changes with increased adoption of phylogenetic methods, the practical problems will decrease. And as detailed phylogenies become more common, we will be able to assess with greater and greater confidence the mechanisms that generate form during the process of evolution. Paleontology potentially has a crucial role to play in the understanding of heterochrony's importance in evolution, since the great bulk of past evolutionary events is contained in its purview. Paleontological material provides a broader picture of the history of life than Recent organisms alone, by giving us a time dimension and many more taxa to incorporate into phylogenies than occur only in the Recent. Carefully built phylogenies of these extinct and often ancient organisms should provide a wealth of information on heterochronic events of the past.

ACKNOWLEDGMENTS. My thanks go to Sara V. Fink for her thoughtful criticisms of the manuscript, to Jacques Gauthier for discussions of stratigraphy and suggestions of references, and to the National Science Foundation for support (DEB-8206939 and BSR-8600115).

References

Alberch, P., Gould, S. J., Oster, G. F., and Wake, D. B., 1979, Size and shape in ontogeny and phylogeny, *Paleobiology* **5**:296–317.

Bookstein, F. L., Chernoff, B., Elder, R. L., Humphries, J. M., Smith, G. R., and Strauss, R. E., 1985, *Morphometrics in Evolutionary Biology*, Academy of Natural Sciences, Philadelphia, Special Publication 15.

Clark, C., and Curran, D. J., 1986, Outgroup analysis, homoplasy and global parsimony: A response to Maddison, Donoghue, and Maddison, *Syst. Zool.* **35**:422–426.

Cracraft, J., 1981, Pattern and Process in paleobiology: The role of cladistic analysis in systematic paleontology, *Paleobiology* **7**:456–468.

Creighton, G. K., and Strauss, R. E., 1985, Comparative patterns of growth and development in criticine rodents and the evolution of ontogeny, *Evolution* **40**:94–106.

Darwin, C., 1859, *On the Origin of Species*, Murray, London.

De Queiroz, K., 1985, The ontogenetic method for determining character polarity and its relevance to phylogenetic systematics, *Syst. Zool.* **34**:280–299.

Dingus, L., and Sadler, P. M., 1982, The effects of stratigraphic completeness on estimates of evolutionary rates, *Syst. Zool.* **31**:400–412.

Donoghue, M. J., and Cantino, P. D., 1984, The logic and limitations of the outgroup substitution approach to cladistic analysis, *Syst. Bot.* **9**:192–202.

Felsenstein, J., 1978, The number of evolutionary trees, *Syst. Zool.* **27**:27–33.

Fink, W. L., 1982, The conceptual relationship between ontogeny and phylogeny, *Paleobiology* **8**:254–264.

Fink, W. L., 1986, Phylogenetics and microcomputers, *Science* **234**:1135–1139.

Gardiner, B. G., Janvier, P., Patterson, C., Forey, P. L., Greenwood, P. H., Miles, R. S., and Jefferies, R. P. S., 1979, The salmon, the lungfish and the cow: A reply, *Nature* **277**:175–176.

Gauthier, J., 1984, A cladistic analysis of the higher systematic categories of the Diapsida, Ph.D. dissertation, University of California, Berkeley [available from University Microfilms, No. 85-12825, Ann Arbor, Michigan].

Gauthier, J., 1986, Saurischian monophyly and the origin of birds, in: The Origin of Birds and the Evolution of flight, (K. Padian, ed.), *Mem. Calif. Acad. Sci.* **8**:1–55.

Gauthier, J., Kluge, A. G., and Rowe, T., 1988, Amniote phylogeny and the importance of fossils, *Cladistics* **4**:105–208.

Ghiselin, M. T., 1974, A radical solution to the species problem, *Syst. Zool.* **23**:536–54.

Ghiselin, M. T., 1981, Categories, life, and thinking, *Behav. Brain Sci.* **4**:269–313.

Gingerich, P. D., 1979, Stratophenetic approach to phylogeny reconstruction in vertebrate phylogeny, in: *Phylogenetic Analysis and Paleontology* (J. Cracraft and N. Eldredge, eds.), pp. 41–77, Columbia University Press, New York.

Gould, S. J., 1977, *Ontogeny and Phylogeny*, Harvard University Press, Cambridge.

Halstead, L. B., 1978, The cladistic revolution—Can it make the grade? *Nature* **289**:759–760.

Hennig, W., 1966, *Phylogenetic Systematics*, University of Illinois Press, Urbana.

Horner, J. R., 1984, The nesting behavior of dinosaurs, *Sci. Am.* **250**:130–137.

Hull, D. L., 1978, A matter of individuality, *Philos. Sci.* **45**:335–60.

Kluge, A. G., and Strauss, R. E., 1985, Ontogeny and systematics, *Annu. Rev. Ecol. Syst.* **16**:247–268.

Maddison, W. P., Donoghue, M. J., and Maddison, D. R., 1984, Outgroup analysis and parsimony, *Syst. Zool.* **33**:83–103.

Nelson, G. J., 1978, Ontogeny, phylogeny, paleontology, and the biogenetic law, *Syst. Zool.* **27**:324–345.

Paul, C. R. C., 1982, The adequacy of the fossil record, in: *The Problems of Phylogenetic Reconstruction* (K. A. Joysey and A. E. Friday, eds.), pp. 75–117, Academic Press, New York.

Stevens, P. F., 1980, Evolutionary polarity of character states, *Annu. Rev. Ecol. Syst.* **11**:333–358.

Strauss, R. E., and Bookstein, F. L., 1982, The truss: Body form reconstruction in morphometrics, *Syst. Zool.* **31:**113–135.

Strauss, R. E., and Fuiman, L. A., 1985, Quantitative comparisons of body form and allometry in larval and adult Pacific sculpins (Teleostei: Cottidae), *Can. J. Zool.* **63:**1582–1589.

Thompson, D'A. W., 1961, *On Growth and Form,* abridged edition (J. T. Bonner, ed.), Cambridge University Press, Cambridge.

Wiley, E. O., 1979, An annotated Linnean hierarchy, with comments on natural taxa and competing systems, *Syst. Zool.* **28:**308–337.

Chapter 6

Sclerochronology and the Size versus Age Problem

DOUGLAS S. JONES

1. Introduction

The rekindling of interest surrounding the role of heterochrony in the evolution of life has produced many examples of paedomorphosis and peramorphosis among fossil taxa, attesting to the ubiquity of this phenomenon (McNamara, 1986, and this volume). Nevertheless, two significant problems are intimately associated with such heterochronic studies: (1) uncertainties of taxonomy and ancestor–descendant relationships (see Fink, this volume); and (2) the problem of assessing absolute age (and hence growth rate) throughout ontogeny. Whereas both problems may be significant when attempting to distinguish among the various heterochronic processes that might have operated in a particular case, it is my contention that sclerochronology can often help resolve the latter.

Life history information such as growth rate and age is almost nonexistent for most fossil organisms. Even for extant material in museum collections or in literature reports, age information is rarely available, making it almost impossible for the paleobiologist or biologist to answer such a question as: Did the descendant achieve its greater size by living longer or by growing faster? Hence, it has proven quite difficult to distinguish between time and rate processes, so that size is often

DOUGLAS S. JONES • Florida State Museum, University of Florida, Gainesville, Florida 32611.

used interchangeably for age in studies of heterochrony. Investigators must then assume that length (or any other metric of body size, such as width, height, thickness, or weight) increases with time at similar rates in the compared taxa. This assumption, however, may not always hold (e.g., Shea, 1983; Emerson, 1986) and could lead to erroneous heterochronic inferences (McKinney, this volume).

Recorded in the hard parts of many fossil and living organisms is life history information that can often help overcome such problems. Age information is preserved as ordered microstructural and/or biogeochemical variations throughout the skeleton, which can form in response to any of a hierarchy of environmental periodicities (e.g., annual, monthly, daily). Proper interpretation of skeletal periodicities permits assessment of growth rate (size vs. time) and age. Such information may then be used to help distinguish among alternative heterochronic processes. The purpose of this chapter is to explore further the size vs. age problem and then consider the diversity of skeletal growth data potentially applicable to the problem.

2. Size As a Proxy for Age

Heterochrony, by definition, refers to processes effecting modifications in ontogeny whereby features of descendant forms are displaced in time relative to the ancestral condition (de Beer, 1930, 1958; Gould, 1977; Shea, 1983). Thus, it should not be surprising that Gould (1977) chose a "clock" model of heterochrony to illustrate the relationships of size, shape, and age while qualitatively defining the principal heterochronic processes. Alberch *et al.* (1979) carried this analysis a step further by developing a quantitative methodology that can be used to relate patterns of morphological change to the processes of heterochrony. Their approach analyzed differences in morphology in terms of changes in a basic set of parameters that control the trajectory of ontogeny: onset of growth, offset signal for growth (either a specific age or a limiting size or shape), growth rate during the growth period, and initial size. Perturbations in these parameters during ontogeny produce the various categories of heterochrony identified by Gould (1977) and formalized by Alberch *et al.* (1979).

However, because our knowledge of growth rates and other life history parameters is typically so incomplete (particularly for the paleobiologist), allometric changes during ontogeny are most often analyzed to explore heterochrony. By analyzing the allometric relationship between shape (or size of some measurable trait or attribute) and body size (e.g., total length, height, weight, or volume) an investigator seeks to circumvent the need for absolute age data. This approach has been widely used on both extinct and extant taxa (e.g., McKinney, 1984; Jungers, 1985). The basic assumption for this scheme to work is that size is an accurate proxy for age, an assumption implicit in the model of Alberch *et al.* (1979).

Shea (1983) and Emerson (1986) have shown that for apes and frogs, the interchangeability assumption between size and age does not always hold. Shea (1983), in particular, showed that the same morphological pattern (i.e., extension or truncation of ancestral allometries to new adult sizes and shapes) may arise

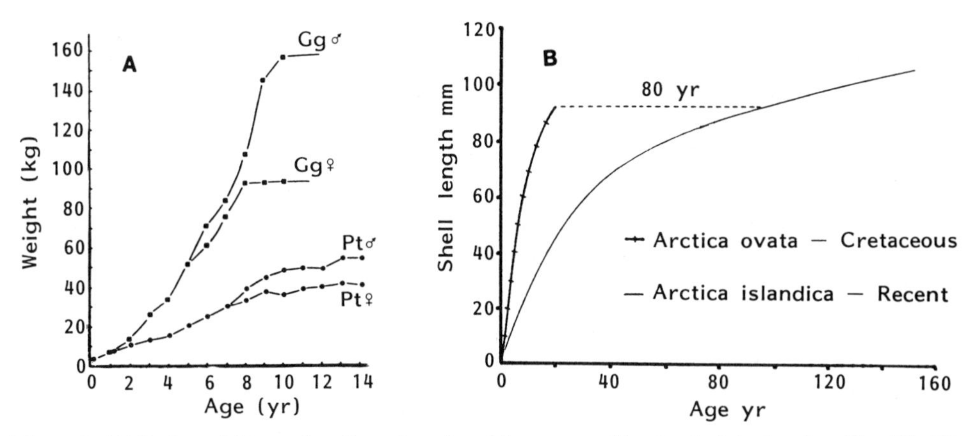

Figure 1. (A) Body weight as a function of age for chimpanzees (*Pan troglodytes*) and gorillas (*Gorilla gorilla*) [from Shea (1983)]. (B) Shell length vs. age comparison for fossil and modern bivalves of the genus *Arctica*.

by two fundamentally different processes: (1) maintaining the ancestral allometric relation and the ancestral *rate* of size increase in time, but lengthening or shortening the duration of ontogeny; or (2) maintaining the ancestral allometric relation and the ancestral duration of ontogeny, but increasing or reducing the rate of size change in the given time. In the absence of age information, it is impossible to distinguish between heterochronic rate processes and time processes. Furthermore, Shea (1983, and this volume) has quite clearly demonstrated that both types of processes may become the separate targets of natural selection in related organisms, even though they produce the same morphological result. Emerson (1986, p. 177) stated, "The major reason for the persistent use of size as an indirect indicator of time is simple: little is known about the actual timing of development for most vertebrate species."

While in some cases size may serve as a valid proxy for time, there are certainly many others in which this assumption is unwarranted (McKinney, this volume). Figure 1 illustrates two such examples, one drawn from among the vertebrates and the other from the invertebrates. Note that the size vs. age relationships for the two species in each diagram are clearly quite distinct. The data in Fig. 1A indicate that gorillas grow faster than chimpanzees, but they do not grow for as long a time. Furthermore, within each species, males continue to grow for a longer period than do females (Shea, 1983). The growth curves of Fig. 1B suggest that the life histories of the bivalves *Arctica ovata* and *Arctica islandica* are also very different, with the former growing more rapidly but not attaining near the longevity of the latter. A comparison of 90-mm specimens of each species suggests an age differential of about 80 years at this size. Admittedly this example is extreme, as *Arctica islandica* is recognized as one of the longest-lived animals (Jones, 1983). Nevertheless, both examples illustrate how woefully imprecise the assumption that related species have similar size–age relationships can be.

The above discussion clearly demonstrates the need for more infusion of developmental information relating size increase to ontogenetic timing into heterochronic analyses. This is particularly important if studies of heterochrony are to advance beyond the realm of mere pattern description to an exploration of the underlying causes. Where will this size–age data be obtained, especially when dealing with fossils? Fortunately there is hope of obtaining size and age information (even in fossils) by closely examining the skeletal growth records of organisms with "hard parts." Many organisms (and a disproportionately high percentage of fossils) possess hard parts, and the interpretation of their skeletal records along with an analysis of the precision of the data preserved in them constitute the principal subjects of discussion throughout the remainder of this chapter.

3. Sclerochronology

> . . . as if we could not count, in the shells of cockles and snails, the years and months of their life, as we do in the horns of bulls and oxen, and in the branches of plants,—
>
> Leonardo da Vinci

As the quotation from Leonardo da Vinci's notebook (Richter, 1883) demonstrates, the idea that organisms record age and growth rate information in selected accretionary tissues is not new. It remained, however, until the final quarter of the last century for rigorous field studies to be conducted which began documenting the veracity of age markers preserved in the skeletons of vertebrates, invertebrates, and in the woody tissues of plants (Barker, 1970). Of these, the annual rings in tree trunk sections are probably most familiar. Dendrochronology, the study of these tree rings, has provided a high-precision technique for determining the age and growth rate of trees (Fritts, 1976). It has also significantly enhanced our understanding of terrestrial climate change during the last 100 centuries as chronologies are cross-dated back in time (Hitch, 1982).

The counterpart to dendrochronology in organisms possessing mineralized tissues has been termed "sclerochronology," the study of periodic features in the skeletal portions of animals. Though applicable to any taxon with hard parts, the term was originally applied to the study of yearly density banding (the alternating light and dark bands seen in cross sections) in scleractinian corals (Hudson *et al.*, 1976). Since that original designation, the definition has been expanded to include other invertebrates, particularly mollusks (Jones, 1983), as well as vertebrates. To illustrate the type of sclerochronological records that are potentially available, the diversity of information in them, and the methods used to interpret them, I will focus upon the mollusks, where such research has proceeded farthest.

3.1. Growth Increments Preserved in Molluscan Shells

Perhaps the one structural feature of the molluscan shell that has historically attracted the most attention is the banding or growth increment variation associated with so many molluscan species. Such growth patterns are often prominently displayed on the external surfaces of shells and have long been the subject of serious biological and paleobiological research (Clark, 1974; Lutz and Rhoads, 1980). The usefulness of external shell growth patterns for age and growth rate

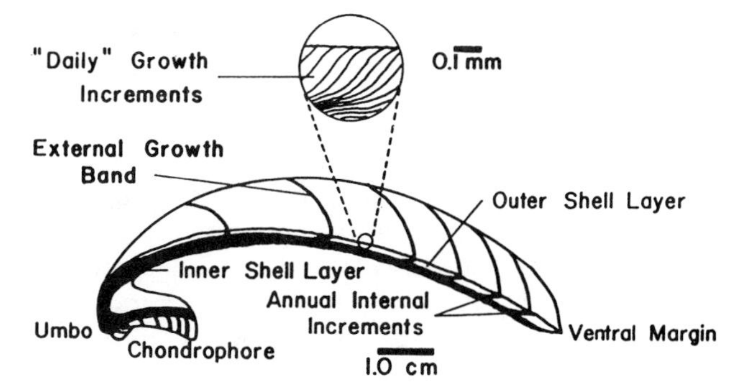

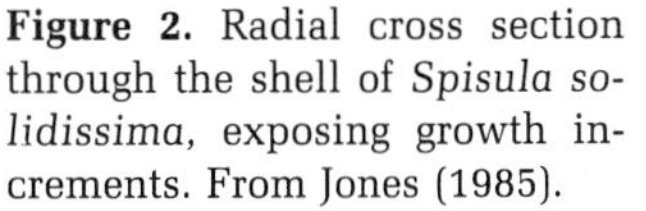
Figure 2. Radial cross section through the shell of *Spisula solidissima*, exposing growth increments. From Jones (1985).

determination is limited, however, by both the inability to distinguish true periodic features from random disturbance marks and by the extreme crowding of growth lines near the margins of mature shells. In the last two decades these problems have been surmounted with the recognition of periodic growth patterns *within* shells. Internal shell growth patterns are known from all classes of mollusks, but those in the Bivalvia have been studied most extensively. This is a result of the relative ease with which a complete ontogenetic growth record can be obtained by sectioning a shell along the axis of maximum growth (Rhoads and Pannella, 1970). Analogous growth increment records are very difficult to obtain from coiled or spiral shells (e.g., gastropods) using current techniques (Lutz and Rhoads, 1980). This section reviews the major types of internal shell growth patterns, referring chiefly to bivalves, and discusses their origin and application to age determination.

Upon cross-sectioning a typical bivalve shell from umbo to ventral margin, the entire growth increment record is revealed. Many species, such as *Spisula solidissima* (Fig. 2), possess prominent growth increments in both the inner and outer shell layers, which are visible to the unaided eye because of their large size (millimeter to centimeter range). The combination of one light and dark band has been documented as representing 1 year of shell growth (Jones, 1983). Each true annual increment can be traced out onto the shell surface, where it corresponds to a concentric growth "band" or "ring"; however, the presence of false rings resulting from serious interruptions in shell formation (e.g., storms) makes it unwise to count external bands in age determination without first consulting the shell interior.

As Fig. 2 indicates, annual features are not the only type of growth increments present in molluscan shells. Smaller increments are also observed when shell cross sections are studied at higher magnifications. In fact, an entire hierarchy of cyclic groupings of growth increments was observed by Barker (1964), who first interpreted these to form in direct response to periodic environmental stimuli on several levels. These ranged from annual increments, corresponding to the yearly cycle of temperature and salinity, down to subdaily increments, thought to reflect subdaily tidal rhythms.

The most recent and thorough literature review of microgrowth increment patterns within the molluscan shell is provided by Lutz and Rhoads (1980), building on earlier compendia by Rhoads and Pannella (1970) and Clark (1974). In their review, which is summarized below, documented microgrowth patterns are assigned to one of five temporal categories (semidiurnal and diurnal, fortnightly, monthly, annual, and semiperiodic or random), not too unlike those of Barker (1964):

1. *Semidiurnal and diurnal.* Pannella and MacClintock (1968) provided convincing evidence that diurnal periodicities were recorded in molluscan shells when they counted between 360 and 370 growth lines and between 720 and 725 lines in specimens of *Mercenaria mercenaria* that had been growing for periods of 1 and 2 years, respectively. Since then, daily growth increments have been reported in many molluscan species. Even subdaily lines thought to form in response to subdaily tidal cycles have been observed (e.g., Evans, 1972). Analogous patterns have been identified in fossil bivalves as well (Barker, 1970).

2. *Fortnightly.* Such patterns are usually expressed as cyclical clusterings of microgrowth increments which thicken and thin at intervals of approximately 15 days. They have been identified primarily in mollusks inhabiting tidal zones (e.g., Lutz and Rhoads, 1977).

3. *Monthly.* Evidence for the clustering of microgrowth increments with a monthly periodicity has been presented by numerous authors. Again, organisms living within the range of tidal influence appear most likely to exhibit monthly growth periodicities [e.g., *Mercenaria mercenaria* (Kennish and Olsson, 1975) and chitons (Jones and Crisp, 1985)].

4. *Annual.* Lutz and Rhoads (1980) state that annual microgrowth increment patterns have been observed in the majority of shells examined in detail. These patterns are generally a result of variable growth rates, which in turn are a function of seasonal changes (particularly in temperature) in the environment.

5. *Random or semiperiodic.* Included in this category are those patterns that form randomly or without well-defined periodicities. Examples include distinct patterns resulting from damage through storms, predation attempts, or physiologically induced patterns, such as those arising from spawning events.

While all of the patterns cited above (with the probable exception of the last) could theoretically be used in age determination, it should not be surprising that annual patterns have found the widest application, given their ubiquity and overall significance. The others can be used if within-year resolution is required and if the periodicities are well documented, as not all repeating fabrics in molluscan shells are necessarily periodic (Jones, D. S., 1981, 1985).

3.2. Chemical Periodicities in Molluscan Shells

Alternating microstructural increments are not the only records of time contained in a shell. Periodic chemical changes may also provide temporal information if the length of the period can be established. Of particular importance in this regard are the changing ratios of stable oxygen ($^{18}O/^{16}O$) and carbon ($^{13}C/^{12}C$) isotopes in shell carbonate. The oxygen and carbon isotopic compositions of fossils have been studied more extensively than any other geochemical property, largely because the theoretical factors involved are explainable in straightforward physical–chemical terms and the effects of physiology and diagenesis are fairly well understood. The isotopic techniques are not new and the principles have been reviewed previously (Dodd and Stanton, 1981).

Several factors govern the isotopic ratios measured in shells, of which salinity and temperature are paramount. Fortunately, physiologic (vital) effects are minimal, making it easier to interpret the environmental influences on the isotopic signatures. In the late 1940s, Harold Urey first suggested that variations in the temperature of the water from which $CaCO_3$ is precipitated should lead to measurable variations in the $^{18}O/^{16}O$ ratio of the carbonate. Thus, mollusks could be expected to record in their shells the water temperature they experienced during growth. Epstein *et al.* (1953) empirically determined the nature of this temperature relationship to be nearly linear between 5 and 30°C, with lower $^{18}O/^{16}O$ ratios at higher temperatures. Therefore, if mollusks live in waters where there is a seasonal temperature cycle, the oxygen isotopic composition of their shells should vary cyclically throughout the year in an annual cycle.

With the modern mass spectrometers of the last decade, which require only an extremely small sample size, it is no longer necessary to grind up an entire shell for one analysis. Instead, it is now possible to sequentially sample across an entire shell, recovering numerous discrete samples in a complete ontogenetic sequence. The annual isotopic cycles that are revealed (Fig. 3) have led directly to the ability to calculate the age, growth rate, and life history of numerous molluscan species, including modern and fossil gastropods and bivalves (e.g., Wefer and Killingley, 1980; Krantz *et al.*, 1984; Donner and Nord, 1986) as well as cephalopods (e.g., Landman *et al.*, 1983; Taylor and Ward, 1983).

Biogenic carbonates also contain an entire spectrum of minor (Mg, Sr) and trace elements (e.g., Ba, Cd, Co, Cu, Fe, Mn, Na, Zn). Abundances of these elements may also vary periodically with growth rate (Rosenberg, 1980) and provide additional chemical age determination tools.

3.3. Combined Approach

Both growth increment analysis and chemical analysis have advantages: the former offers high-resolution results and is fairly inexpensive, whereas the latter can be employed on taxa without growth lines, but is expensive. In both approaches the periodicity of the variation needs to be examined carefully to ensure accuracy. However, when an analysis of shell growth increments is combined with geochemical studies, a powerful interpretive tool is produced, as the techniques complement one another.

Microsampling across major shell increments (Fig. 3) has resulted in detailed, high-resolution isotope records, which can confirm the annual periodicity of the growth increments, document the seasonal temperature range over which shells grew, examine the influence of temperature upon growth, and in general reconcile the two records. A good example of this reconciliation is to be found by comparing the results of Fig. 3 with those in Fig. 4. In the former study the annual cycles in the oxygen isotope record corresponded exactly with the internal growth increments, clearly identifying them as annual and permitting precise age determi-

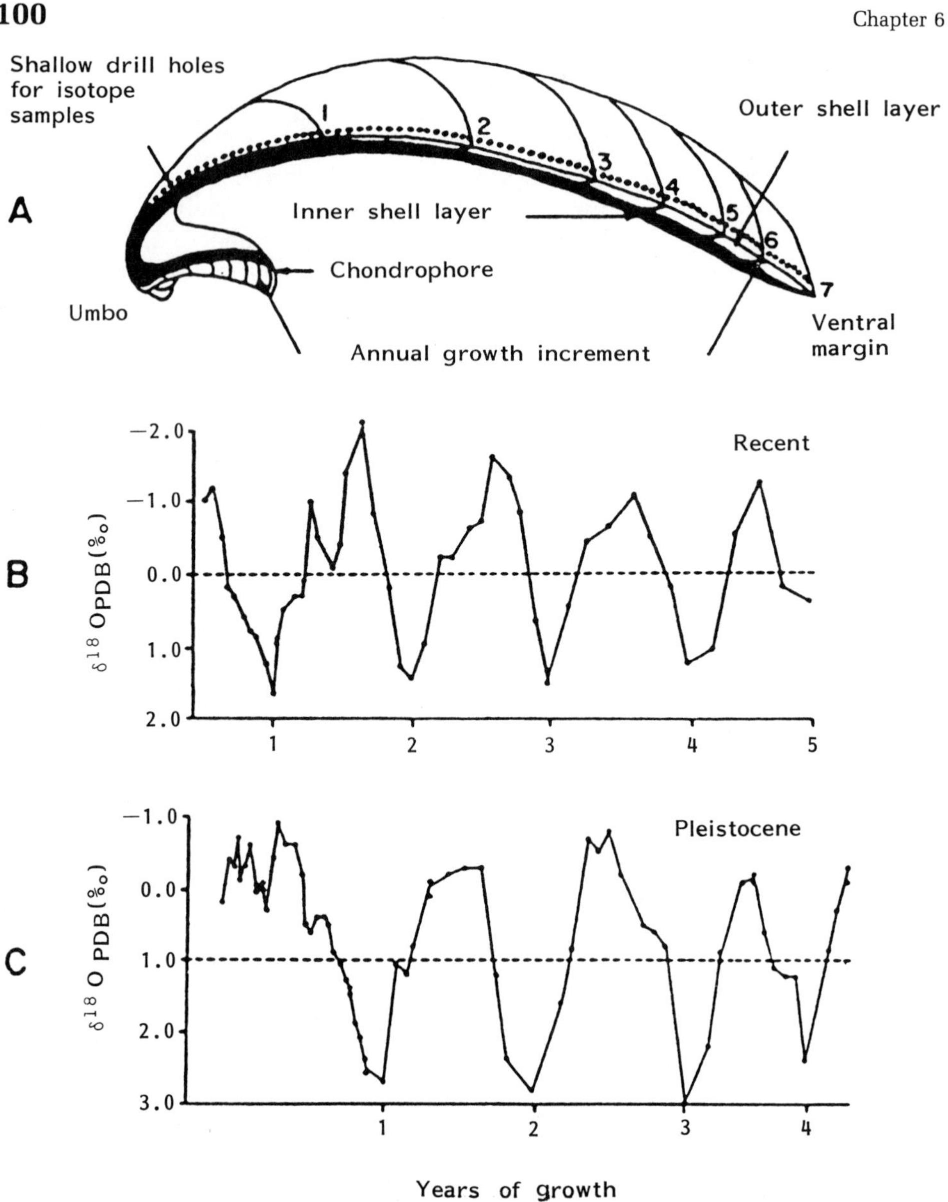

Figure 3. (A) Radial cross section through the shell of *Spisula solidissima* indicating annual shell increments (numbered) and position of tiny drill holes for isotopic sampling. (B, C) Comparison of annual cycles in $^{18}O/^{16}O$ profiles between modern and fossil specimens. From Jones (1985).

nations. In the latter case the annual oxygen isotope cycles did not match the prominent growth rings on the external surface of the scallop shells, indicating that the rings were not truly annual phenomena and should not be used for age determination. Used in tandem, the chemical and microstructural approaches represent an excellent method for interpreting age and growth rate data encoded in skeletons.

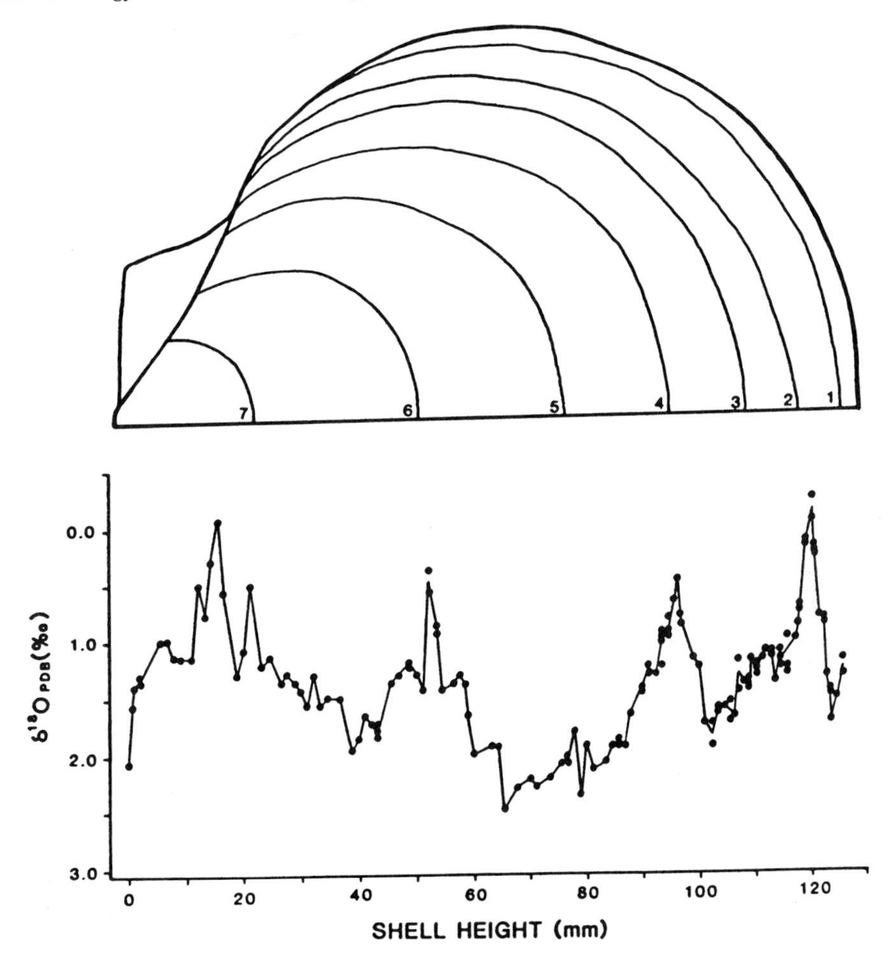

Figure 4. The position of external growth rings on the shell of the sea scallop, *Placopecten magellanicus,* when compared to annual cycles in the oxygen isotope record, indicates that several of the rings are not true yearly markers. From Krantz *et al.* (1984).

4. Sclerochronology and Age–Size Relations in Modern and Fossil Taxa

4.1. Mollusks

Annual internal growth increments have received wide use in molluscan age and growth rate determination, gradually replacing studies using external rings. The result has been an increase in the number of species for which accurate age–size data are now available. In almost every species examined, the number of annual internal shell increments suggests life spans greater than previously imagined, with many common coastal bivalve species known to live for several decades. Zolotarev (1980) reports that half of the mollusks studied from the far eastern seas of the USSR have life spans of greater than 20 years and many live over 50 years. Microgrowth increment analyses have also shown certain species

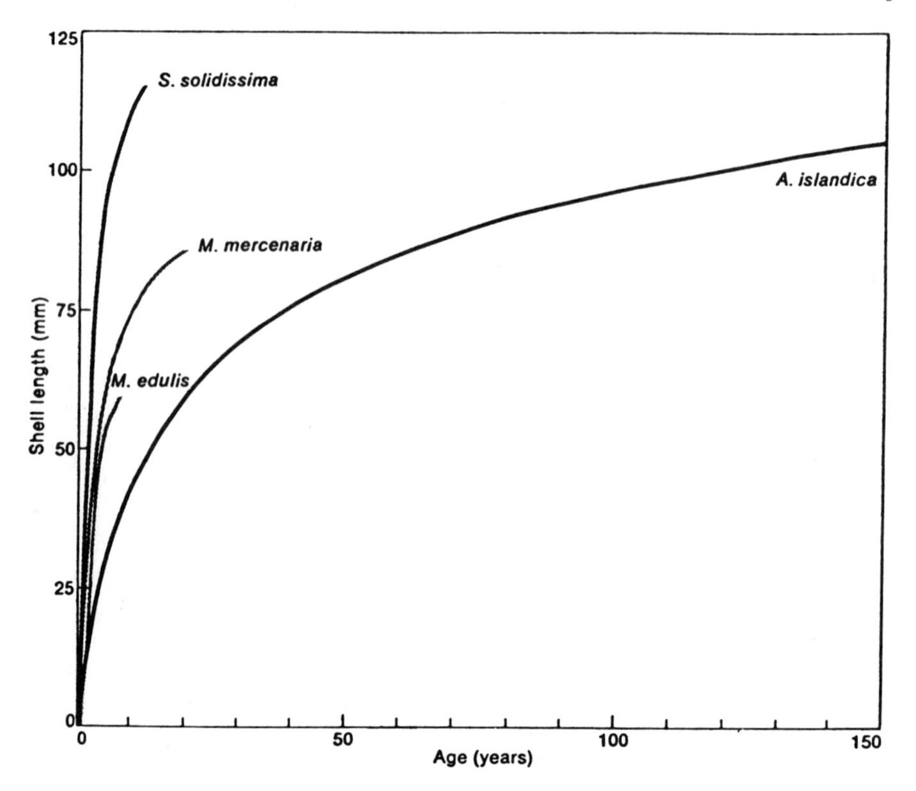

Figure 5. Size–age curves for several common coastal bivalve species constructed on the basis of annual internal growth increments. From Jones (1983).

to be extremely long-lived, surviving for over 100 and in some cases 200 years (Jones, 1983).

By measuring the size of annual increments (or yearly isotopic cycles), growth curves can be constructed (Fig. 5) and growth rates compared. This procedure can work for well-preserved fossil shells as well as modern specimens. Hence, it should be possible to examine molluscan heterochronic changes in the context of time and not just size, by using the techniques outlined previously. In rare cases among modern species, growth increment and/or isotopic patterns have also been used to determine the age at sexual maturity (Rhoads and Pannella, 1970; Jones *et al.*, 1986), another important factor in heterochronic analysis.

It is apparent that a great deal of highly precise life history information of value to the heterochronist is potentially interpretable from the molluscan shell using the sclerochronological methods discussed earlier. A legitimate question may then be posed: To what extent can other, nonmolluscan taxa be expected to have analogous skeletal records? The following sections address this query.

4.2. Other Invertebrates

Many other invertebrate taxa besides mollusks contain periodic microgrowth patterns which could potentially yield age and growth rate information. By and large these taxa have not been investigated with the same intensity as the mollusks and therefore less size–age information is to be found in the literature. Forms with accretive skeletons have received the most attention, but growth lines have a widespread taxonomic distribution.

A partial list to illustrate the taxonomic distribution of growth lines in organisms other than mollusks includes the corals, where both daily (e.g., Barnes, 1970) and annual (e.g., Knutson *et al.*, 1972; Weber *et al.*, 1975; Dodge and Vaisnys, 1980) increments have been solidly documented; brachiopods (e.g., Rudwick, 1968; Paine, 1969; Walker and Parker, 1976; Thayer, 1977); echinoids (e.g., Raup, 1968; Weber, 1969; Pearse and Pearse, 1975); barnacles (Bourget, 1980); and even annelid worm jaws (Olive, 1980). Periodicities have also been recognized in layering within stromatoporoids (Meyer, 1981) and algal stromalites (e.g., Jones, C. B., 1981) as well. The reader should consult Clark (1974) and Rhoads and Lutz (1980) for more details.

4.3. Vertebrates

Space does not permit a comprehensive tabulation of all of the age markers recorded from vertebrate skeletons or accretionary tissues [but the reader is referred to the seminal work of Peabody (1961) for an introduction to the literature]. Therefore, to develop a flavor for the diversity of tissues containing age data and the distribution of taxa involved, an admittedly idiosyncratic list has been assembled. Annual growth patterns in fish otoliths (Pannella, 1980) and scales (Fig. 6) have probably received the most attention as sclerochronological age indicators among vertebrates. Both tissues can yield very precise results. Prominent growth rings also occur in the spines and vertebrae of certain fish and these, too, have been interpreted as age indicators (e.g., Voorhies, 1969).

Peabody (1961) was a strong believer in age determination through counting of annual growth zones developed in ectothermic vertebrates when metabolism and growth are arrested or slowed by seasonal cycles in the environment. He thus argued that fishes, amphibians, and reptiles inhabiting regions characterized by strongly contrasting, cold–hot or wet–dry seasons should exhibit annual growth zones observed in bone cross sections. Additional support for this approach has come from the skeletochronology group in Paris (e.g., Castanet *et al.*, 1977), which argues convincingly that ectothermic vertebrates possess cyclic bone growth, which is annual. Nevertheless, workers since Peabody have been more cautious, not applying this idea indiscriminately, but tending to evaluate ectotherms on a case-by-case basis. In addition to fishes, a number of amphibian and reptilian species reveal a match in age with the number of growth layers, for example, toads (Hemelaar and van Gelder, 1980), snakes (Minakami, 1979), sea turtles (Zug *et al.*, 1986), and crocodiles (Fig. 6) (Hutton, 1986).

The situation is more complicated in endotherms, as fewer skeletal markers are preserved in bone due to the dynamic nature of this tissue. Mountain goats and wild sheep often display annual rings on their horns, separating successive periods of rapid horn growth (Fig. 6). Bears may form annual tooth ridges coincident with hibernation cycles (Fig. 6). Layers in the cementum of mammalian teeth and in the periosteal zone of bones are suggested by Klevezal and Kleinen-

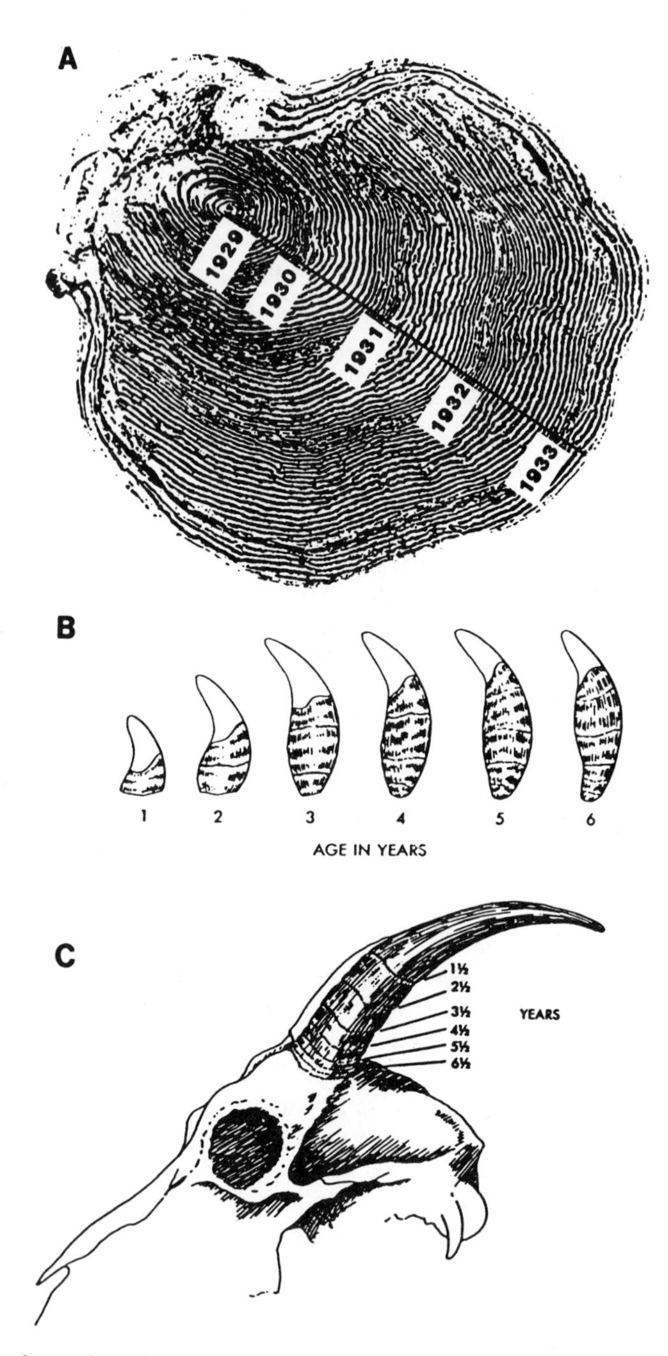

Figure 6. Annual growth markers in selected vertebrate taxa: (A) Band-width variation in fish scales; (B) root annulations on the right upper canine of Alaskan black bear; (C) horn of the mountain goat [see Larson and Taber (1980) for details and original citations]; and (D) laminae in middiaphysis cross sections from long bones of young Nile crocodiles [from Hutton (1986)].

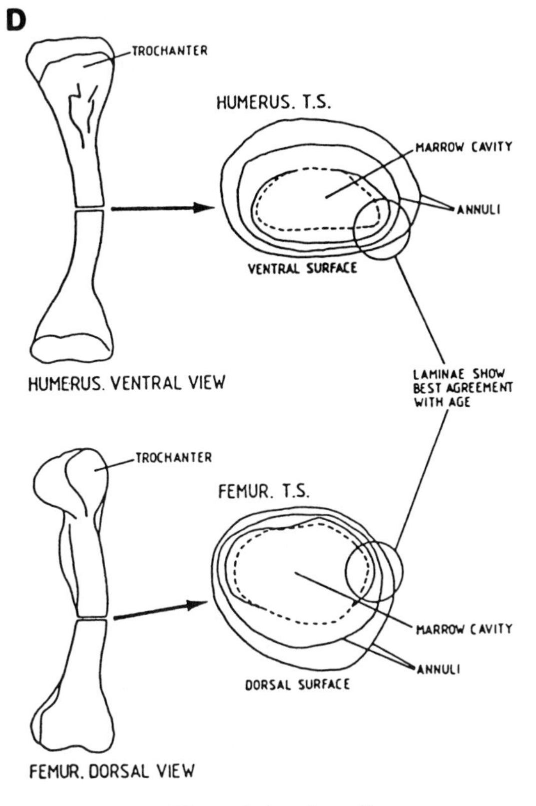

Figure 6. *(continued)*

berg (1967) to be highly accurate indicators of year class. Cementum is deposited on the roots of teeth each year in alternating light (summer) and dark (winter) bands. The exact nature of the depositional process is not known, but the layers occur in virtually all mammals and the aging technique is likely to be effective in any mammal provided the layers can be exposed (Larson and Taber, 1980). Bromage and Dean (1985) have successfully employed incremental growth features in tooth enamel to determine the age at death of Plio-Pleistocene hominids. In a novel application of combined growth increment/stable isotope analysis, Koch and Fisher (1986) have recently reported annual patterns from the tusks of Pleistocene mastodons and mammoths which reveal age and rate of growth.

4.4. Other Age Determination Techniques

There are, of course, many other age determination techniques available to the heterochronist which do not involve sclerochronology. These can be applied in those cases where growth increments are absent or where the material under investigation warrants special consideration. For example, Comfort (1951) considered the intensity of molluscan shell pigmentation to vary inversely with growth rate. Age data for modern and fossil horses have been accurately estimated from patterns of cheek-tooth wear (e.g., Spinage, 1972b), analysis of discrete year classes in large populations, and sequential patterns of dental eruptions (Hulbert, 1982). Similar methods have been applied to primates (Shea, 1983) as well as to other ungulates (Spinage, 1972a). Patterns of crown and root development in teeth of pongids and humans have been used to determine dental ages for *Australopithecus* and early *Homo* (Smith, 1986).

Finally, it should be mentioned that a useful place to search for information on age and growth rate techniques is the literature on wildlife or fisheries management (e.g., Schemnitz, 1980). Management strategies require detailed life history data, which are available for many extant species.

5. Conclusion

In studies of heterochrony, body size is often used as a substitute for age because data on absolute timing of ontogenetic development are rarely available. This assumption of equivalence does not always hold, however, and can lead to erroneous heterochronic inferences, particularly in deciding between rate as opposed to time processes. Age–size data are especially important if heterochronic analyses are to move beyond the realm of pattern description to focus upon the causative mechanisms.

The accretionary skeletal parts of many (most?) fossil and living organisms record detailed life history information, which, if interpreted correctly, can provide important absolute age data. This information is preserved as ordered microstructural and biogeochemical variations that form in response to any of a hierarchy of environmental periodicities. Both invertebrates and vertebrates contain skeletal age indicators, which can be investigated using a variety of sclerochronological methods. Accurate age and growth rate determination, while involving considerably more research effort for the heterochronist, will be rewarded by elimination of the size vs. time problem and an enhanced ability to investigate the ultimate (ecological) causes of heterochrony.

References

Alberch, P., Gould, S. J., Oster, G. F., and Wake, D. B., 1979, Size and shape in ontogeny and phylogeny, *Paleobiology* **5**:296–317.

Barker, R. M., 1964, Microtextural variation in pelecypod shells, *Malacologia* **2**:69–86.

Barker, R. M., 1970, Constituency and origins of cyclic growth layers in pelecypod shells, Ph.D. dissertation, University of California, Berkeley.

Barnes, D. J., 1970, Coral skeletons: An explanation of their growth and structure, *Science* **170**:1305–1308.

Bourget, E., 1980, Barnacle shell growth and its relationship to environmental factors, in: *Skeletal Growth of Aquatic Organisms* (D. C. Rhoads and R. A. Lutz, eds.), pp. 469–491, Plenum Press, New York.

Bromage, T. G., and Dean, M. C., 1985, Re-evaluation of the age at death of immature fossil hominids, *Nature* **317**:525–527.

Castanet, J., Muenier, F. J., and Ricqlès, A., 1977, L'enregistrement de la croissance cyclique par le tissu osseux chez les vertébrés poikilothermes: Données comparatives et essai de synthèse, *Bull. Biol. Fr. Belg.* **111**:183–202.

Clark, G. R., 1974, Growth lines in invertebrate skeletons, *Annu. Rev. Earth Planet. Sci.* **2**:77–99.

Comfort, A., 1951, The pigmentation of molluscan shells, *Biol. Rev.* **26**:285–301.

De Beer, G. R., 1930, *Embryology and Evolution*, Clarendon Press, Oxford.

De Beer, G. R., 1958, *Embryos and Ancestors*, Clarendon Press, Oxford.

Dodd, J. R., and Stanton, R. J., Jr., 1981, *Paleoecology, Concepts and Applications*, Wiley, New York.

Dodge, R. E., and Vaisnys, J. R., 1980, Skeletal growth chronologies of recent and fossil corals, in: *Skeletal Growth of Aquatic Organisms* (D. C. Rhoads and R. A. Lutz, eds.), pp. 493–517, Plenum Press, New York.

Donner, J., and Nord, A. G., 1986, Carbon and oxygen stable isotope values in shells of *Mytilus edulis* and *Modiolus modiolus* from Holocene raised beaches at the outer coast of the Varanger Peninsula, north Norway, *Palaeogeogr. Palaeoclimatol. Palaeoecol.* **56**:35–50.

Emerson, S. B., 1986, Heterochrony and frogs: The relationship of a life history trait to morphologic form, *Am. Nat.* **127**:167–183.

Epstein, S., Buchsbaum, R., Lowenstam, H. A., and Urey, H. C., 1953, Revised carbonate–water isotopic temperature scale, *Bull. Geol. Soc. Am.* **64**:1315–1326.

Evans, J. W., 1972, Tidal growth increments in the cockle *Clinocardium nuttalli*, *Science* **176**:416–417.

Fritts, H. C., 1976, *Tree Rings and Climate*, Academic Press, New York.

Gould, S. J., 1977, *Ontogeny and Phylogeny*, Harvard University Press, Cambridge.

Hemelaar, A. S. M., and van Gelder, J. J., 1980, Annual growth rings in phalanges of *Bufo bufo* (Anura, Amphibia) from the Netherlands and their use for age determination, *Neth. J. Zool.* **30**:129–135.

Hitch, C. J., 1982, Dendrochronology and serendipity, *Am. Sci.* **70**:300–305.

Hudson, J. H., Shinn, E., Halley, R., and Lidz, B., 1976, Sclerochronology: A new tool for interpreting past environments, *Geology* **4**:361–364.

Hulbert, R. C., Jr., 1982, Population dynamics of the three-toed horse *Neohipparion* from the late Miocene of Florida, *Paleobiology* **8**:159–167.

Hutton, J. M., 1986, Age determination of living Nile crocodiles from the cortical stratification of bone, *Copeia* **1986**(2):332–341.

Jones, C. B., 1981, Periodicities in stromatolite lamination from the early Proterozoic Hearne Formation, Great Slave Lake, Canada, *Palaeontology* **24**:231–250.

Jones, D. S., 1981, Repeating layers in the molluscan shell are not always periodic, *J. Paleontol.* **55**:1076–1082.

Jones, D. S., 1983, Sclerochronology: Reading the record of the molluscan shell, *Am. Sci.* **71**:384–391.

Jones, D. S., 1985, Growth increments and geochemical variations in the molluscan shell, in: *Mollusks: Notes for a Short Course* (D. J. Bottjer, C. S. Hickman, and P. D. Ward, eds.), pp. 72–87, Paleontological Society and University of Tennessee, Knoxville.

Jones, D. S., Williams, D. F., and Romanek, C. S., 1986, Life history of symbiont-bearing giant clams from stable isotope profiles, *Science* **231**:46–48.

Jones, P., and Crisp, M., 1985, Microgrowth bands in chitons: Evidence of tidal periodicity, *J. Moll. Stud.* **51**:133–137.

Jungers, W. L. (ed.), 1985, *Size and Scaling in Primate Biology*, Plenum Press, New York.

Kennish, M. J., and Olsson, R. K., 1975, Effects of thermal discharges on the microstructural growth of *Mercenaria mercenaria*, *Environ. Geol.* **1**:41–64.

Klevezal, G. A., and Kleinenberg, S. E., 1967, Age Determination of Mammals from Annual Layers in Teeth and Bones, USSR Academy of Science, Severtsov Institute of Animal Morphology (Translated from Russian), U. S. Department of Commerce, Springfield, Virginia.

Knutson, D. W., Buddemeier, R. W., and Smith, S. V., 1972, Coral chronometers: Seasonal growth bands in reef corals, *Science* **177**:270–272.

Koch, P. L., and Fisher, D. C., 1986, Out of the mouths of mammoths: An isotopic signal of seasons in proboscidean tusks, in: *Geological Society of America Annual Meeting, Abstracts with Program*, Vol. 18, p. 660.

Krantz, D. E., Jones, D. S., and Williams, D. F., 1984, Growth rates of the sea scallop, *Placopecten magellanicus*, determined from the $^{18}O/^{16}O$ record in shell calcite, *Biol. Bull.* **167**:186–199.

Landman, N. H., Rye, D. M., and Shelton, K. L., 1983, Early ontogeny of *Eutrephoceras* compared to Recent *Nautilus* and Mesozoic ammonites: Evidence from shell morphology and light stable isotopes, *Paleobiology* **9**:269–279.

Larson, J. S., and Taber, R. D., 1980, Criteria of sex and age, in: *Wildlife Management Techniques Manual*, 4th ed. (S. D. Schemnitz, ed.), pp. 143–202, Wildlife Society, Washington, D.C.

Lutz, R. A., and Rhoads, D. C., 1977, Anaerobiosis and a theory of growth line formation, *Science* **198**:1222–1227.

Lutz, R. A., and Rhoads, D. C., 1980, Growth patterns within the molluscan shell: An overview, in: *Skeletal Growth of Aquatic Organisms* (D. C. Rhoads and R. A. Lutz, eds.), pp. 203–254, Plenum Press, New York.

McKinney, M. L., 1984, Allometry and heterochrony in an Eocene echinoid lineage: Morphological change as a by-product of size selection, *Paleobiology* **10**:207–219.

McNamara, K. J., 1986, A guide to the nomenclature of heterochrony, *J. Paleontol.* **60**:4–13.

Meyer, F. O., 1981, Stromatoporoid growth rhythms and rates, *Science* **213**:894–895.

Minakami, K., 1979, An estimation of age and life span of the genus *Trimeresurus* (Reptilia, Serpentes, Viperidae) on Amani Oshima Island, Japan, *J. Herpetol.* **13**:147–152.

Olive, P. J. W., 1980, Growth lines in polychaete jaws (teeth), in: *Skeletal Growth of Aquatic Organisms* (D. C. Rhoads and R. A. Lutz, eds.), pp. 561–592, Plenum Press, New York.

Paine, R. T., 1969, Growth and size distribution of the brachiopod *Terebratalia transversa* Sowerby, *Pac. Sci.* **23**:337–343.

Pannella, G., 1980, Growth patterns in fish sagittae, in: *Skeletal Growth of Aquatic Organisms* (D. C. Rhoads and R. A. Lutz, eds.), pp. 519–560, Plenum Press, New York.

Pannella, G., and MacClintock, C., 1968, Biological and environmental rhythms reflected in molluscan shell growth, *J. Paleontol.* **42**:64–80.

Peabody, F. E., 1961, Annual growth zones in living and fossil vertebrates, *J. Morphol.* **108**:11–62.

Pearse, J. S., and Pearse, V. B., 1975, Growth zones in the echinoid skeleton, *Am. Zool.* **15**:731–753.

Raup, D. M., 1968, Theoretical morphology of echinoid growth, *J. Paleontol.* **42**:50–63.

Rhoads, D. C., and Lutz, R. A. (eds.), 1980, *Skeletal Growth of Aquatic Organisms*, Plenum Press, New York.

Rhoads, D. C., and Pannella, G., 1970, The use of molluscan shell growth patterns in ecology and paleoecology, *Lethaia* **4**:413–428.

Richter, J. P., 1883, *Leonardo*, Low, Marston, Searle, and Rivington, London.

Rosenberg, G. D., 1980, An ontogenetic approach to the environmental significance of bivalve shell chemistry, in: *Skeletal Growth of Aquatic Organisms* (D. C. Rhoads and R. A. Lutz, eds.), pp. 133–168, Plenum Press, New York.

Rudwick, M. J. S., 1968, Some analytic methods in the study of ontogeny in fossils with accretionary skeletons, *J. Paleontol.* **42**:35–49.

Schemnitz, S. D. (ed.), 1980, *Wildlife Management Techniques Manual*, 4th ed., The Wildlife Society, Washington, D.C.

Shea, B. T., 1983, Allometry and heterochrony in the African apes, *Am. J. Phys. Anthropol.* **62**:275–289.

Smith, B. H., 1986, Dental development in *Australopithecus* and early *Homo*, *Nature* **323**:327–330.

Spinage, C. A., 1972a, African ungulate life tables, *Ecology* **53**:645–652.

Spinage, C. A., 1972b, Age estimation of zebra, *E. Afr. Wildl. J.* **10**:273–277.

Taylor, B. E., and Ward, P. D., 1983, Stable isotope study of *Nautilus macromphalus* Sowerby (New Caledonia) and *Nautilus pompilius* L. (Fiji), *Palaeogeogr. Palaeoclimatol. Palaeoecol.* **41**:1–16.

Thayer, C. W., 1977, Recruitment, growth, and mortality of a living articulate brachiopod, with implications for the interpretation of survivorship curves, *Paleobiology* **3**:98–109.

Voorhies, M. R., 1969, Taphonomy and Population Dynamics of an Early Pliocene Vertebrate Fauna, Knox County, Nebraska, University of Wyoming Contributions in Geology, Special Paper 1.

Walker, K. R., and Parker, W. C., 1976, Population structure of a pioneer and a late stage species in an Ordovician ecological succession, *Paleobiology* **2**:191–201.

Weber, J. N., 1969, Origin of concentric banding in the spines of the tropical echinoid *Heterocentrotus*, *Pac. Sci.* **23**:452–466.

Weber, J. N., White, E. W., and Weber, P. H., 1975, Correlation of density banding in reef coral skeletons with environmental parameters: The basis for interpretation of chronological records preserved in the coralla of corals, *Paleobiology* **1**:137–149.

Wefer, G., and Killingley, J. S., 1980, Growth histories of strombid snails from Bermuda recorded in their O-18 and C-13 profiles, *Mar. Biol.* **60**:129–135.

Zolotarev, V. N., 1980, The life span of bivalves from the Sea of Japan and Sea of Okhotsk, *Sov. J. Mar. Biol.* **6**:301–308.

Zug, G. R., Wynn, A. H., and Ruckdeschel, C., 1986, Age determination of loggerhead sea turtles, *Caretta caretta*, by incremental growth marks in the skeleton, *Smithson. Contrib. Zool.* **427**:1–34.

II

Heterochrony in Major Groups

Chapter 7

Heterochrony in Plants

The Intersection of Evolution Ecology and Ontogeny

EDWARD O. GUERRANT, JR.

1. Introduction

Phylogenetic changes in ontogenetic rates or timing are termed heterochrony. Evolutionary changes in organismal form necessarily arise from alterations in ontogenies, and so it is hardly surprising that heterochrony has profoundly affected the evolution of plants as well as animals. However, because the life cycles, body plans, and growth of plants and animals are so different, the effects of heterochrony are expressed differently in plants than they are in animals. The indeterminate or open growth habit and modular construction of plants lead to much greater environmentally induced phenotypic variation in form than is found in animals. Consequently, even though zoocentric theory has much to offer botanists, the study of heterochrony in plants must take on a character of its own.

The scope of heterochrony's influence on plants is great, affecting both the haploid gametophytic and diploid sporophytic generations, their interaction, and even the relationship of the generations to each other within the life cycle. Effects of heterochrony have been implicated at a wide variety of levels of structural organization, from the cellular up through shoot systems, to the plant as an integrated whole organism.

EDWARD O. GUERRANT, JR. • Department of Botany and Plant Pathology, Oregon State University, Corvallis, Oregon 97331; and Department of Biology, Lewis and Clark College, Portland, Oregon 97219.

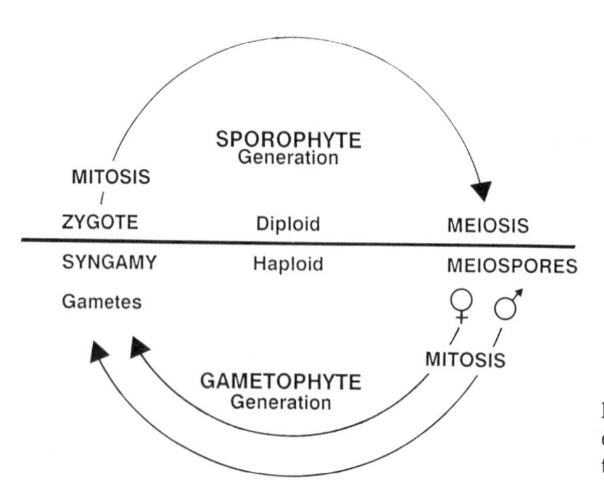

Figure 1. Schematic representation of life cycle of plants, illustrating the alternation of generations.

This chapter surveys various ways in which heterochrony may have affected plants evolutionarily. It is not intended to be a thorough review. The history of botanical thought and work on the genetic and developmental bases of heterochronic change in plants are discussed within the context of the effects of heterochrony at various levels of structural organization. The paucity of paleobotanical literature cited is not intended to reflect negatively on this important source of information. It is simply beyond the scope of this preliminary foray into heterochrony in plants. Finally, this chapter seeks to draw attention to the study of plants as whole organisms within an evolutionary framework.

2. Life Cycle and Growth in Higher Plants

An understanding of the life cycle of plants and how they grow is necessary to appreciate the ways in which heterochrony can affect plant form and function.

An alternation of heteromorphic generations characterizes the life cycle of higher plants (Fig. 1). A full cycle includes at least two genomes operating in distinct organisms; a diploid sporophyte generation alternates with a haploid gametophyte. The gametophyte generation may consist of one haploid plant housing both sexes or the sexes can be on separate plants. The most dramatic difference between the life cycles of plants and animals follows meiosis. Gametes are produced by meiosis in most higher animals, and they are the only haploid cells in their life cycle. In plants, by contrast, the haploid products of meiosis, meiospores, divide mitotically to form the gametophyte generation.

In contrast to the unitary body plan of animals, plants are often characterized as organisms of modular construction (Hallé and Oldeman, 1970; Hallé *et al.*, 1978; Harper and White, 1974; Prévost, 1978; White, 1984). The plant body may be seen as a collection of just three organ types—leaves, stems, and roots—which are iteratively deployed during development. Leaves and even some stems are abscised, or shed from the plant as a normal part of ontogeny. Even though the open or indeterminate growth habit of plants contrasts sharply with the closed or determinate construction of some animals, the gametophytes of many higher

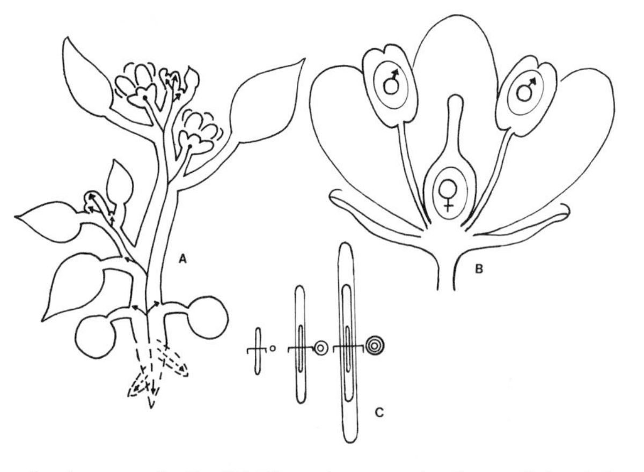

Figure 2. Schematic illustrations: (A) Sporophytic body of a typical dicot. The shoot portion is outlined with a solid line, and the root portion with a dashed line. Internal lines represent root and shoot axes, with meristems indicated by arrowheads. There are two cotyledons (circular in outline), a lateral stem with two leaves of its own has grown from the axil of the first epicotylar leaf, and individual flowers have grown from the axils of the next two. (B) The location of the microgametophyte (male) and megagametophyte (female) in the flower, which otherwise is sporophytic. (C) Three increments of growth in girth and length. Notice how once a particular length has been attained, new growth in length occurs by terminal addition.

plants, for example, the flowering plants, have determinate growth culminating in a definite number and arrangement of cells.

Plant growth differs markedly from that of animals, even in the cellular details. New plant cells arise at the apex, and except for those giving rise to the vascular cambium, each has but limited powers of mitotic division thereafter. Each cell is enclosed in an extracellular cellulosic cell wall, which provides considerable structural support. The vast bulk of organ size increase in plants is due to an extreme hydrostatically driven enlargement of cells. There are no migrations of cells or tissues during development, features so prevalent and important for heterochronic change in animals. Nor is there a Weissmann's Barrier, or embryonic separation of the botanical analogs of germinal cell lines from somatic; meiosis may occur in many places around the plant body in cell lineages that have independently undergone considerable mitotic division since their most recent common parent cell.

The manner in which plants grow and are constructed is extremely diverse, and the description offered here is limited largely to the primary body of the sporophyte of higher vascular plants. It is intended to illustrate how heterochrony may be viewed in an indeterminately growing organism, and is not a thorough description of plant growth (Foster and Gifford, 1974; Wareing and Phillips, 1981).

The flowering plant body is constructed on a remarkably simple theme (Fig. 2). It can be described as an axis that can add cells, and therefore initiate growth in length only at the shoot and root apices. Only the shoot apical meristem produces lateral appendages, the leaves and their axillary buds. Lateral roots arise internally, away from the root apex. The iteration of this pattern of organization is the basis for the open, or indeterminate growth habit of plants, and it results in a modular construction. Members of the organ category "leaf" are all considered serially homologous, and may assume a variety of forms, including bud scales, foliage leaves, spines, tendrils, and even floral appendages. Leaves typically exhibit strictly determinate growth, and give rise to no other structures (but see

Dickinson, 1978). In contrast, buds arise from regions of cells usually positioned in the axils of leaves that retain meristematic activity. These buds may grow into lateral axes and differentiate almost immediately, or remain dormant for definite or indefinite periods of time. A stem and its associated leaves is referred to as a shoot. When a lateral shoot grows, it can either reiterate the activity of the primary axis or have distinctive and often position-dependent properties (Tomlinson, 1978; Hallé *et al.*, 1978). A unit of shoot construction above individual organs that has determinate growth is called a module (Hallé *et al.*, 1978; Tomlinson, 1982; White, 1984). Tomlinson (1982) distinguishes two levels of development in modular organisms, a descriptive group that includes most plants: ontogenetic development refers to the sequence of addition of modules, and primordial development, the process whereby particular modules are elaborated from some meristematic precursor.

This simplistic representation is neither exhaustive nor universally accurate. For example, only passing reference was made to the extremely important phenomenon of secondary growth, whereby conifers and dicot trees grow in girth. Nevertheless, it is a good working model, or basic conceptual framework with which to begin comparing the growth and development of plant forms.

3. Consequences of Open Growth

The extreme environmentally induced phenotypic plasticity inherent in the open growth of plants is such that the concept of homology does not apply to plants in the same way it does to organisms such as vertebrates or insects. These have a one-to-one correspondence of body components between related organisms that is not found in plants. The question of homology that seems to have dominated botanical thought revolves around attempts to determine to which of the three categories of homologous structures (leaves, stems, or roots) a particular structure belongs (Sattler, 1966, 1984; Kaplan, 1984; Stevens, 1984; Tomlinson, 1984a,b). Implicit in this approach is a view of plants as being simply a collection of parts. Unfortunately, it is the parts themselves that have become the primary focus of attention, and the study of plants as integrated whole organisms appears to have suffered as a result. Perhaps this has resulted from too strict an adherence to zoocentric concepts of homology and structural comparability.

If we are to determine how heterochrony affects the form of plants as whole organisms, we must have a concept of homology that permits us to distinguish genetically based aspects of form and simultaneously to recognize variability that is due to phenotypic plasticity. Phenotypic plasticity is not just bothersome noise, it is a phenomenon of considerable evolutionary and ecological interest, which appears to be especially important in plants (Bradshaw, 1965; Jain, 1979; Stearns, 1982; Scheiner and Goodnight, 1984; Sultan, 1987).

Although no consensus has emerged, promising first steps have been made toward comparing plants as whole organisms, and these focus on developmentally homologous components of form generation (Hallé and Oldeman, 1970; Hallé *et al.*, 1978; Sachs, 1982; Guerrant, 1984). The notion of developmental homologs used in this paper is intentionally vague. It is part of an attempt to shift the focus of attention of comparisons in plants from the products of development to their underlying processes. There are many ways in which this goal can be accomplished, so to constrain the notion of what may constitute a developmental homolog may be counterproductive.

4. The Scope of Heterochrony

The effects of heterochrony on plants are pervasive. The causes of heterochrony can act directly on either of the generations, sporophytic or gametophytic. Heterochronic change in either generation may cause an associated change in the other, so there is the interaction between generations to consider as well.

On a macroevolutionary scale, changes in the relationship of the generations to one another within the life cycle as a whole may have been mediated by heterochronic processes. The life cycle of vascular plants is generally considered to be dominated by the diploid sporophyte generation. The ascendency of the sporophyte over the gametophyte appears to be a synapomorphy of the vascular plants (Mishler and Churchill, 1984, 1985). In the cladistic sister group of vascular plants, it is the green, leafy gametophyte that is often thought of simply as a moss.

Within these broad categories, heterochronic processes have been implicated at several levels of structural organization, from the cellular on up through organ complexes.

4.1. Gametophytic Heterochrony

It is fitting to begin a discussion of heterochrony in plants by considering the evolution of gametophytes, especially the extremely reduced representatives of the sexual generation in flowering plants, the pollen and embryo sac. Since 1943, Takhtajan has repeatedly stressed the role of paedomorphosis in the evolution of both micro- and megagametophytes, or the male and female gametophytes, respectively. He maintains that the effects of heterochrony are extremely common across a broad phylogenetic range of plants (Takhtajan, 1943, 1954, 1972, 1976). Among the most exemplary of recent analyses of heterochrony is a study by Mishler (1986) of moss gametophytes.

Takhtajan was not just a lonely champion of a role for isolated cases of heterochrony in the evolution of plants (Stebbins, 1974, p. 119); he also elaborated an entire theoretical overview of how morphological evolution is influenced by changes in development (Takhtajan, 1972). His view encompasses developmental insertions and deletions, in addition to more strictly heterochronic changes of ontogenetic rates and timing. Although Takhtajan's terminology differs from that of Gould (1977) and Alberch *et al.* (1979) (for example, he uses neoteny to mean paedomorphosis), he conceptually distinguishes processes similar to neoteny and progenesis.

Consider the following passage:

> The simplified gametophytes of flowering plants could emerge only by strong acceleration of the processes leading to sexual maturity and a corresponding retardation of other

developmental processes in the ancestral gametophytes. The premature completion of the ontogeny of gametophytes, having started at relatively late developmental stages, gradually shifted to earlier and earlier stages. Consequently, gametogenesis also shifted to ever-earlier phases of development. This led, finally, in the early angiosperms or their immediate ancestors, to the loss of those phases of development at which, in most gymnosperms, the gametangia are formed. Gametogenesis in flowering plants takes place at such an early phase of development that the gametangia cannot even be formed, and the gametes are formed without them. Moreover, the development of the gametes themselves is also cut short, and they become extremely simplified. (Takhtajan, 1976, p. 213.)

In this passage, Takhtajan presages Gould's (1977) emphasis on the importance of relative acceleration and retardation. In addition to unnamed processes, he also appears to invoke both progenetic and neotenic phenomena as well as postdisplacement. For example, he points out that the production of gametes (gametogenesis) occurs at such a young and early stage in flowering plants that an ancestrally antecedent and apparently necessary developmental precursor, the gametangia (gamete-producing structures), are not formed at all. The shifting in relative timing of sequentially related events results in a seemingly necessary precursor being lost. This is a relatively neglected phenomenon, worthy of more study.

The megagametophyte of angiosperms, often having only eight nuclei in seven cells, is still more complex than the microgametophyte—the pollen grain. Angiosperm pollen grains border on the lower limit of multicellularity, with but two mitotic divisions giving rise to only three cells. Takhtajan (1976, p. 214) concluded that there is "every reason to suppose that the male gametophyte of angiosperms originated by way of both neoteny (terminal abbreviation and subsequent structural and functional modification) and basal abbreviation." Following the first mitotic division of the haploid meiotic product, the generative cell divides to form the two sperm that participate in double fertilization, a characteristic of flowering plants. The generative cell performs the ancestral function of antheridia and their spermatogenous cells, but with much greater economy of time, energy, and material.

Very early maturation, with its truncated range and altered sequence of events, appears to preclude complex structures from being formed in these extremely reduced organisms. In micro- and megagametophytes, one or a few cells perform functions that in their ancestors were associated with specialized tissues and organs. Consequently, structural homologies become uncertain, and perhaps they are more likely to be resolved by theoretical advances than by empirical ones.

The microgametophyte is housed within a very durable structure, the pollen wall, which is well represented in the fossil record. Stone *et al.* (1981) showed the extremely reduced pollen wall of *Tapeinochilos*, a banana relative, to be paedomorphic. Its pollen wall goes through only the earliest of the stages of development that its relatives do.

Takhtajan's work preceded the conceptual and terminological advances of Gould (1977) and Alberch *et al.* (1979). Gould's (1977) book and the paper it inspired by Alberch *et al.* (1979) are the cornerstones on which modern thinking about the relationship between ontogeny and phylogeny are built. For the sake of brevity, and to avoid repetitious citation, they may be referred to hereafter simply as the Gould–Alberch models. These advances and the increased adoption of cladistic theory by botanists that has taken place only in the last few years have made it such that the relationship of ontogeny and phylogeny may now be studied in a more rigorous and falsifiable fashion than before.

The work of Mishler (1986) on moss gametophytes is exemplary in this regard. Drawing on his cladistic analyses, comparative developmental studies, and knowledge of the ecology of *Tortula*, he concluded that heterochrony has been an extensive and important influence on the evolution and ecology of this monophyletic group. He identified multiple instances of both paedomorphosis and peramorphosis, often with important ecological correlates. Furthermore, he was able to identify probable instances of the processes of neoteny and hypermorphosis.

For example, the size and form of successive leaves change along the moss stalk from small "juvenile" leaves to the larger "mature" forms (a phenomenon similar to heteroblasty in vascular plants); each leaf form has ecological advantages. "Juvenile" leaves have a greater ability to give rise to new ramets (physiologically independent individuals) if detached from the parent plant, while a variety of characters of "mature" leaves confer advantages to the plant with respect to water relations. Specialized brood leaves, which resemble "juvenile" leaves in a number of morphological and ecological characters, have originated independently in at least five lineages. Since the 19th century, stalks bearing these leaves have been considered to represent a form of asexual reproduction. Stalks with brood leaves appear to be paedomorphic, and in at least one case probably originated by neoteny. In some taxa, structurally elaborate features of "mature" leaves, such as perforations near their bases, and elaborately branched surface papillae, occur relatively late in the ontogeny of individual leaves, and the latter character, at least, is peramorphic. Together these features facilitate capillary movement of water up the stalk, and the tips of the papillae aid in gas exchange in leaves otherwise covered with water.

Mishler (1986) also emphasized how paedomorphosis can complicate the study of phylogenetic reconstruction, a problem not limited to plants (Larson, 1980). Where developmentally late characters are lost in one or more members of a clade, the independently derived paedomorphic taxa may resemble both one another and truly primitive members of the group, if the terminal characters were originated within the group in question.

4.2. Sporophytes

4.2.1. Cellular and Tissue

The effects of heterochrony at cellular and tissue levels of structural organization in the sporophyte are best documented for the woody tissue, xylem. Xylem is the portion of the vascular system that conducts water and mineral nutrients. Along with Takhtajan, the wood anatomist Carlquist was among the few early advocates of a role for heterochrony in the evolution and ecology of higher plants. Carlquist (1962) suggested how paedomorphosis (*sensu* de Beer, 1930) accounts for the anomalous structures of certain woods. He later expanded on this theme, showing how both cellular and tissue aspects of the xylem are affected (Carlquist, 1975). Others have followed Carlquist's lead (Anderson, 1972; Cumbie, 1963, 1967a,b; Gibson, 1973). Previously, Chrysler (1937) had suggested, in Haeckelian terms, that "abbreviated recapitulation" might account for persistently juvenile aspects of the xylem of tuberous cycads.

Wood is accumulated xylem tissue, and like the shells of molluscs, it is a permanent and relatively unchanging record of one aspect of an individual's ontogeny. It is a complex tissue, which originates in two different ways. Primary xylem forms directly from tissues produced by the apical meristem. In herbaceous plants, this may be the only xylem produced. However, in many plants, a concentric layer of cells near the periphery of the axis remains meristematic. This layer becomes the vascular cambium, and is responsible for secondary xylem, which forms the vast bulk of most woody plants. Once produced, xylem remains in place, with new cells being added toward the periphery of the axis. By comparing cells in a radial transect outward from the center of an axis to the cambium, ontogenetic changes from "juvenile" to "mature" structure can be documented. This is also true in fossil woods that have a vascular cambium. Carlquist (1962) used characteristics of the primary xylem to represent the "juvenile" condition. To the degree that characteristics of the primary xylem are protracted into the secondary xylem, paedomorphosis is seen to have occurred.

The usage of the term juvenile in this context is problematic (Richards, J. H., personal communication). The degree to which it is legitimate to have the "juvenile" condition of xylem being the product of a different meristem than the "mature" condition raises serious questions about homologies that need to be resolved. A similarly problematic situation with Meckel's cartilage in vertebrates was discussed by Hall (1984). While there seems to be little controversy over whether Meckel's cartilage is homologous throughout the vertebrates, it is produced by different developmental mechanisms in different groups within the vertebrates. Nevertheless, given that primary xylem always precedes secondary xylem during the ontogeny of an axis, Carlquist's usage is at least descriptively accurate within each axis.

Carlquist uses the term paedomorphosis in a generally descriptive way. He emphasized (Carlquist, 1975) that this is not a widespread phenomenon applicable to all dicot woods, but is limited to herbaceous plants, and plants that have evolved woodiness from herbaceous ancestors, stem succulents, and rosette trees, as well as other similar specialized situations. Paedomorphosis is not portrayed by Carlquist as a single phenomenon, but a common outcome of a variety of processes that lead to various alternatives in herbaceous structure. The functional significance of paedomorphic wood seems often to be associated with a release from mechanical stress.

Additional exampies of paedomorphosis in wood anatomy are mentioned by Takhtajan (1976, citing the work of Melet, 1968), and concern three taxa of cushionlike plants living in high mountainous regions of eastern Pamir. In these instances, paedomorphosis was thought to be related to the extremely harsh habitats in which they grow.

Evidence of permanent "juvenility" at the cellular level can be found in the vessel elements—the cellular conduits of vascular transport (Fig. 3). The beginning of secondary growth in typical woody dicots is accompanied by gradual changes toward a more mature pattern of wood anatomy. Within the primary

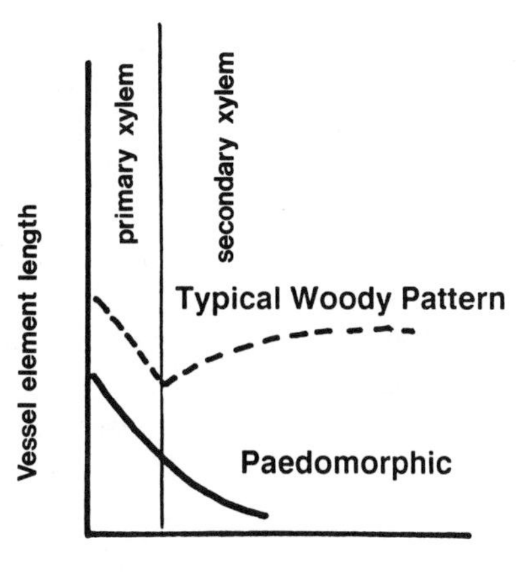

Figure 3. Age on length curves for two dicots: cell lengths of vessel elements as a function of age at which they were produced. Adapted from figures in Carlquist (1962, 1975).

xylem, there is a progressive decrease in cell length from the first formed xylem elements (protoxylem) through those that follow (metaxylem). However, there is generally a rather abrupt reversal once a vascular cambium is formed: successive tracheary elements of woody dicots and gymnosperms usually get progressively longer in the secondary xylem. Carlquist (1962, 1975) interprets a monotonic decrease in vessel element length across the primary–secondary growth boundary as the retention of a juvenile characteristic of growth. The phenomenon is not limited to gross cell length, but is accompanied by several other cellular features, such as changes in cell shape, aspects of cell wall pitting, and features of other cell types in the xylem tissue. For example, in the rays of paedomorphs there is a high proportion of erect cells and few or no procumbent cells. Philipson and Butterfield (1967) presented a contrasting view of the causes of radial variation in the sizes of wood elements. It is based on the details of cell division in the cambial initials that allow it to cover an increasing circumference as the axis grows in girth.

4.2.2. Determinate Organs

Individual determinate organs are well suited to heterochronic study. Determinate growth and relatively simple construction, which borders on two-dimensional in some leaves, do not present the problems of homology found in indeterminate aspects of plant form. Some of the most explicit examples of heterochrony at the level of the organ can be found in studies of floral appendages, and will be treated later in the context of whole flowers.

A notable group of early workers, led by E. W. Sinnott, empirically converged on a series of developmental factors responsible for the generation of plant size and form that are remarkably similar to the parameters of the Gould–Alberch models of developmental relations. For example, Hammond (1941a,b) identified the genetic and developmental bases of leaf shape differences between various relatives of cotton, *Gossypium*. The relatively deeper sinus in the leaf of one variety is produced by a delay in the appearance of lateral lobes in the young primordium, an increased relative growth rate in length, and the maintenance of the increased rate for a longer period of time. Sinnott and his students combined comparative developmental studies and genetic analyses in a way that, curiously, has been attempted by few recent workers (but see Richards and Barrett, 1984).

Sinnott and Kaiser (1934) were quick to adopt Huxley's (1932) suggestion that growth could be studied profitably from an allometric perspective. By viewing the control of shape not as the attainment of a particular ratio at a given stage of life, but as a developmental relationship of relative growth rates in specific dimensions, they were able conceptually to distinguish size and shape, and see that shape could be controlled in a variety of ways. They showed that striking differences in mature fruit shape within inbred lines of squash (*Cucurbita*) and peppers (*Capsicum*) could be attributed to developmental differences described by the intercept and slope, respectively, of allometric plots. In *Cucurbita*, the difference in intercept, which translated to a shape difference in the primordium, was controlled by one gene.

Sinnott (1927, 1931, 1935, 1936) showed that the developmental determinants of size and shape in squash fruits are inherited independently. He studied two inbred lines of *Cucurbita pepo*, characterized either by flat, disk-shaped fruits or smaller, more spherical ones. Shape was under simple one-gene, two-allele Mendelian control, with flat the dominant allele. Genes governing shape apparently act by controlling relative rates of growth in different dimensions. Size, on the other hand, was under polygenic control (see also, Sinnott and Dunn, 1935; Sinnott, 1963). A similar situation was described for the fruits of *Capsicum* (Kaiser, 1935), although in one of the crosses between different lines of peppers, factors controlling size and shape appeared to be linked. In a review of the literature on the genetic control of morphology in plants, Gottlieb (1984) concluded that qualitative features are often under simple genetic control, while size is polygenic [see Bachmann (1983) and Hilu (1983) and also Coyne and Lande (1985) and reply by Gottlieb (1985)].

In a study of proximal causes underlying differences in organ sizes in plants of different body sizes, Sinnott (1921) concluded that the size of the apical meristem was of critical importance. In effect, he identified the parameter of initial apex size as being a significant developmental determinant of mature organ size in plants. Houghtaling (1935) discovered that initial size of the ovary primordium before flowering was an important developmental determinant of size differences seen among mature fruits in a number of inbred lines of tomato. Whaley (1939) found a direct relationship between the size of the apical meristem and flower and fruit sizes, and suggested the relationship was cause and effect.

Although a positive relationship between shoot apex size and mature leaf size has been found in some studies (Abbe *et al.*, 1941; Guerrant, 1984)—indeed some even refer to it as "Sinnott's law" (Grafius, 1978)—this relationship does not describe the growth of all plants. For example, von Maltzahn (1957) showed that a large size difference in the leaves of two inbred strains of *Cucurbita pepo* was due to differences in duration of leaf growth—again a character of considerable importance to contemporary views of heterochronic change. The leaves of pumpkins and Chinese gourds are borne on meristems of similar size, but those of the pumpkin become larger because they grow for a longer period of time than do those of the Chinese gourd.

4.2.3. Shoots

A shoot is a composite structure: a stem and its leaves produced by an apical meristem. Heterochrony therefore has more to work on in shoots than it does with individual organs. Shoots can exhibit indeterminate or determinate patterns of growth, and they can be reproductive or vegetative.

Flowers were the subjects of the earliest heterochronic investigations on plants in which the Alberch *et al.* (1979) methodology was used (Guerrant, 1982; Lord, 1982). They are convenient units to study by this zoologically inspired methodology, in part because floral form is relatively stable within and among individual plants. As a consequence of having a discrete number and arrangement of parts, the nature of the homologous bases of comparisons involving individual flowers is conceptually similar to that of whole animals. That is, there is a one-to-one correspondence of homologous parts between flowers within and among individual plants.

No such stable relationship is available at the level of the plant body as an integrated whole organism. Therefore, the way in which the structure of plants, as integrated whole organisms, is to be compared between individuals within or between taxa must be on some basis other than simply the matching of homologous parts.

4.2.3.1. Reproductive Shoots: Flowers. Guerrant (1982) observed that the flowers ("adults") of the evolutionarily derived hummingbird-pollinated larkspur, *Delphinium nudicaule,* resemble the buds ("juveniles") of more generalized bumblebee-pollinated species, and not their flowers. Comparative developmental studies showed that the external appearance of *D. nudicaule* flowers are paedomorphic by the process of neoteny.

The directionality of evolutionary change in this case was argued on the basis of cladistic logic using the outgroup substitution approach to outgroup analysis. The neotenic features, including the budlike, forward-pointing sepal orientation, contribute to a generally tubular floral form, which along with a reddish color are often associated with hummingbird-pollinated flowers. Hummingbird-pollinated flowers typically produce more nectar than do bumblebee-pollinated flowers, and the nectariferous petals of *D. nudicaule* are larger and peramorphic in form. It appears that selection did not act in a single unified way on the flowers. Rather, the ecological necessities of attracting and rewarding a new pollinator elicited a variety of heterochronic responses in different parts of the flower.

Other work on heterochrony in flowers has focused on cleistogamous flowers. The phenomenon of cleistogamy is one in which individual plants produce dimorphic flowers. It has evolved repeatedly in a number of unrelated families, and has been studied extensively [see reviews by Lord (1981, 1984) and Clay (1982)]. The apparent ecological significance of cleistogamy lies in the guarantee of seed set by the cleistogamous (CL) flowers, which may pollinate themselves in the bud and never open, combined with the opportunity for outcrossing that is afforded by the more generalized, open chasmogamous (CH) flowers. The CL floral form is generally considered to have been evolutionarily derived from CH flowers (Lord, 1981, 1984). Although this supposition has not been tested by explicit phylogenetic studies, it seems reasonable given that relatives of these species usually bear only open flowers.

Lord and her co-workers have studied two species of annuals, *Lamium amplexicaule* and *Collomia grandiflora*, and a perennial, *Viola odorata* [reviewed in Lord (1984)]. Common denominators among these and other cleistogamous species are a reduced androecium and lack of corolla enlargement at anthesis. Many CL flowers mature precociously, so the ability to set seed quickly may be an important adaptive aspect as well. These three species admirably demonstrate the considerable range of variation that exists among lineages in the extent to which CL flowers may differ developmentally from their CH counterparts. *Lamium* is the least divergent, and its cleistogamous flowers have been described as progenetic dwarfs (Lord, 1982). The CL flowers of *Collomia* are more divergent: aspects of their structure can be attributed either to neoteny or predisplacement (Minter and Lord, 1983; Lord, 1984). Still other characteristics, such as the reduced number of pollen-producing locules in an anther, are unique to the CL floral form (Minter and Lord, 1983). Comparative developmental studies demonstrated that in these two species the floral primordia of the CL and CH flowers are indistinguishable at inception. However, the CL flowers of *Viola*, which are most divergent at sexual maturity, differ from CH flowers even at inception, at least in size (Mayers and Lord, 1983).

Lord's work is noteworthy for a variety of reasons, including its place in the history of ideas. She first interpreted her data from a more traditional perspective of plant morphology, but in later studies has adopted a more modern evolutionary outlook. Her earlier work on this subject (Lord, 1979, 1980) does not mention the Gould–Alberch models. Rather, she interpreted her data from the perspective of Goebel's (1900) concept of "arrest" and Lindman's (1908) notion of precocity. Ultimately she found that neither of these theoretical frameworks, which preceded the modern synthesis, was adequate to explain her data. Until relatively recently even mainstream plant morphologists, like their zoological counterparts (Ghiselin, 1980), were not interpreting their data within the context of contemporary evolutionary ideas. For another example, compare Kaplan's (1973, 1980) treatment of Goebel's concept of arrest before and after the influence of Gould (1977).

Based on the success of these initial attempts to interpret comparative developmental information *a posteriori*, the Alberch *et al.* (1979) methodology seems to be a useful tool for addressing questions of floral evolution. Because the process of progenesis involves a reduction in the age at sexual maturity, an important life-history character, it may be possible to use the theory to predict some aspects of the developmental relationships between taxa having appropriate phylogenetic and ecological relationships.

The evolution of autogamy in some temperate zone herbaceous plants is a suitable situation in which to seek a case of progenesis in flowering plants (Guerrant, 1984). Flowers of plants that habitually self-pollinate are often conspicuously smaller than those of the outbreeding taxa from which they evolved (Ornduff, 1969; Wyatt, 1983). In addition, the ability to flower and mature seeds earlier in the spring than their outbreeding ancestors is considered to have been of primary selective importance for the evolution of autogamy in species of at least three genera of annual wildflowers. Evolutionarily derived, inbreeding species of *Leavenworthia* (Lloyd, 1965; Solbrig and Rollins, 1977), *Clarkia* (Moore and Lewis, 1965), and *Limnanthes* (Arroyo, 1973; Ritland and Jain, 1984) all live in habitats in which the summer drought arrives earlier in the spring than it does in those of their outbreeding ancestors. Arroyo concluded that the main selective force leading to the evolution of inbreeding in *Limnanthes floccosa* was for the ability to flower and set seed earlier than *L. alba*, which she argued was its immediate ancestor. The ability to flower earlier in *L. floccosa* is genetically determined and is an important factor that allows this species to persist and flourish in the habitats it now occupies, which are edaphically and climatically poorer than those of *L. alba* (Arroyo, 1973; Ritland and Jain, 1984).

With respect to adaptation, small flower size is often thought of, at least implicitly, as a consequence of inbreeding. For example, Solbrig and Rollins (1977) wrote, "once a plant has acquired the ability to self habitually, changes that decrease the cost of pollinator attraction may be favored." It seems from this statement that the change in breeding system is assumed to precede, and then in a sense to drive, the change in size. This energy optimization argument postulates that energy saved may then be reallocated to features enhancing competitive ability, reproductive output, or both. The progenesis hypothesis provides an alternative, though not mutually exclusive view. It suggests that smaller flower size and concomitant shape changes in some inbreeders such as *L. floccosa* are manifestations of a broader pattern of paedomorphosis, resulting from selection to mature at a younger age. In other words, the flowers of the derived, inbreeding *L. floccosa* may be smaller and have a predictably different shape because they mature at a younger age than their immediate ancestor.

Comparative developmental studies showed that flowers of *L. floccosa* do mature into flowers at a younger age than do those of *L. alba*, and that they have considerable overlap in their size–shape growth trajectories during ontogeny consistent with a progenetic origin (Guerrant, 1984) (Fig. 4). Even so, a major developmental difference between them is that floral tissues of *L. floccosa* have a considerably higher relative growth rate than do those of *L. alba*—a difference easily described by the parameters of the Gould–Alberch models, but not by the term progenesis. Perhaps we should be cautious, and not let our preconceptions of what can happen be overly influenced by a rigid terminology. Beyond this developmental difference, a strict and simple interpretation of the hypothesis is confounded for two reasons. One is that the flowers originate on vegetative shoots of different sizes, so the initial conditions are not identical. A second and more profound problem concerns the phylogenetic premise of *L. floccosa* being derived from *L. alba*. Recent electrophoretic and interfertility data do not support the hypothesis that *L. floccosa* is a derivative of *L. alba* (McNeill and Jain, 1983). Nevertheless, despite these confounding factors, the developmental relationship between the two taxa is such that relatively few developmental differences distinguish them, and that these are described by the parameters of the Alberch *et al.* (1979) methodology.

4.2.3.2. Vegetative Shoots. A variety of characteristics of vegetative shoots seem particularly well suited to heterochronic study. One is the phenomenon of heteroblasty, where the size and shape of successive leaves change in relation to one another along an axis. Another shoot level attribute is the rate of leaf formation.

Heteroblasty has been studied extensively from a traditional morphological point of view. Although these studies do not use the terminology of heterochrony, or even have an explicitly evolutionary perspective, many identify developmental

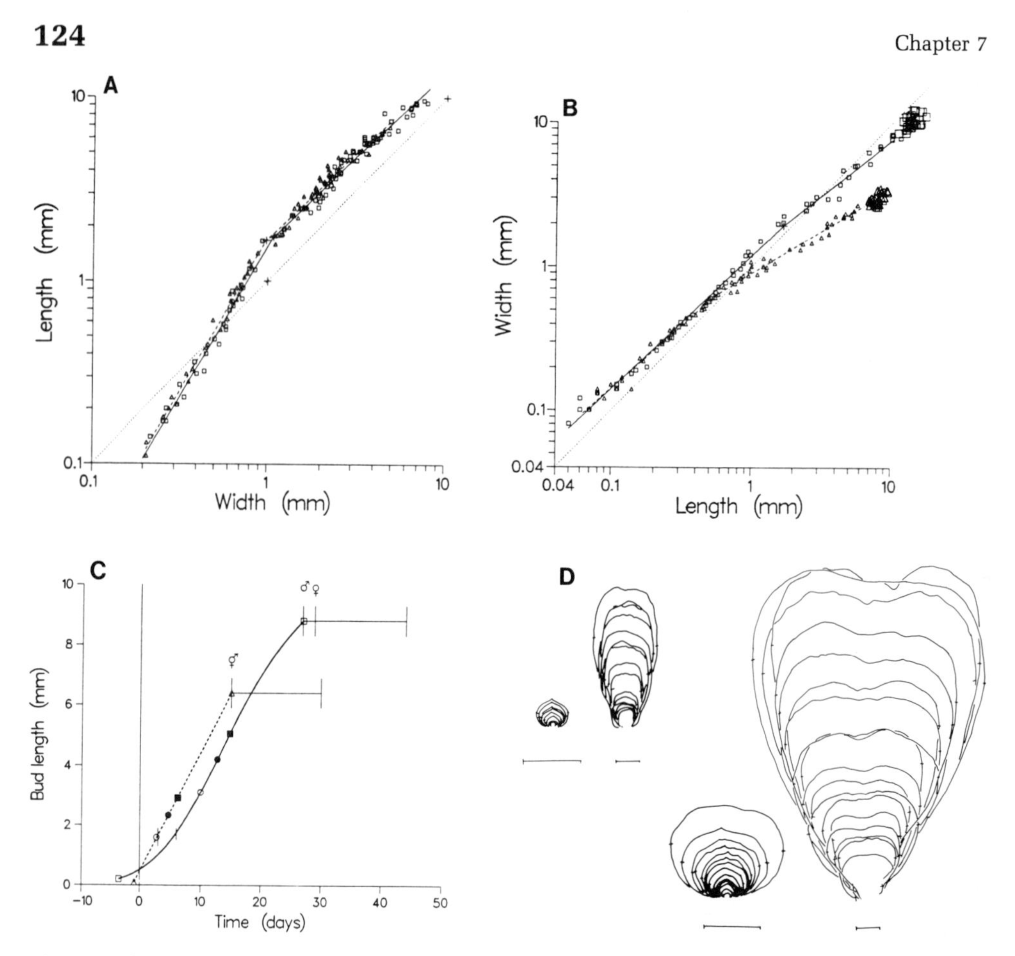

Figure 4. Flower growth of (□) *Limnanthes alba* and (△) *L. floccosa.* (A) Allometric plot of bud growth in length and width; the species cannot be distinguished statistically in these features. (B) Allometric plot of petal growth in length and width; slopes and intercepts are statistically indistinguishable during the initial portion of growth, but both slopes and intercepts differ after the slopes change. (C) Growth in length over time, and timing of ontogenetic events during flower growth. (D) Tracings of some of the petals whose dimensions are indicated in B, showing the progression of growth in size and shape. The two figures on the left are *L. floccosa* before and after the slope changed, and the two on the right are *L. alba.* All scale bars are 1 mm, but note that the scale is different before and after the slope changed. The smaller illustration in each pair fits into the open space of the larger.

factors similar to the parameters of the Gould–Alberch models. For example, Kaplan (1973) showed that divergence in leaf form along the axis of *Acorus* could be attributed in part to differences in the timing of the onset of maturation. This result is similar to other studies in which the reduced size and simplified form of "juvenile" leaves are associated with earlier maturation (Crotty, 1955; Cutter, 1955), though see Bruck and Kaplan (1980) for a counterexample. Size differences in the apical meristem at the time of leaf inception may also affect forms of leaves in a heteroblastic series (Crotty, 1955; Kaplan, 1973; Mueller, 1982). Wulff (1985) showed that the transition from "juvenile" to "adult" leaves in seedlings of the legume *Desmodium paniculatum* was correlated with seed weight. Larger seeds gave rise to plants that produced "adult" leaves at earlier nodes. It appears

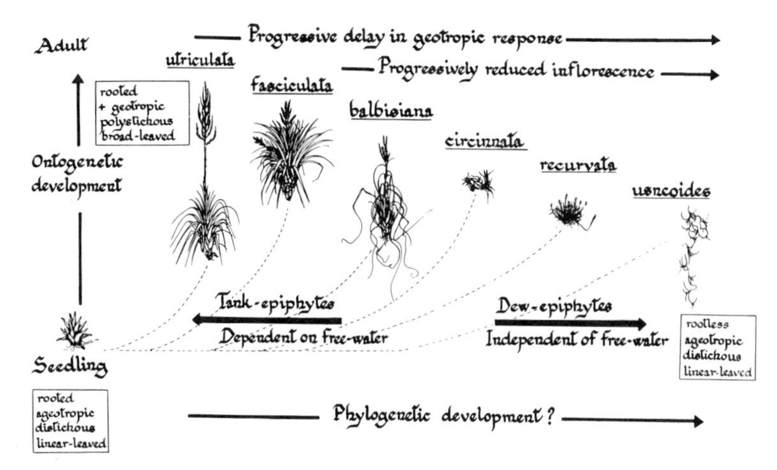

Figure 5. Diagrammatic representation of relationships between ontogeny and potential phylogeny in several species of *Tillandsia*. Figure reprinted from Tomlinson (1970) with permission.

that leaf form at specific nodes is not inalterably predetermined, but that initial size and duration of growth are important parameters affecting the form of leaves in a heteroblastic series.

In *Tillandsia*, a genus in the pineapple family that includes the so-called Spanish moss, Tomlinson (1970) described a situation where both heteroblasty and phyllotaxy are involved in an apparent paedomorphocline. Although the phylogenetic relationships among these taxa are explicitly conjectural, the ecological correlates of morphological juvenilization are apparent (Fig. 5). The seedlings of all taxa show no geotropic growth response and have narrow, linear leaves in a distichous phyllotaxis. Subsequent leaves of the tank epiphytes become broader, the phyllotaxy polystichous, and the shoots become strongly negatively geotropic such that the leaves are oriented in the proper direction to be able to accumulate water used for growth. In contrast, the leaves of Spanish moss, *T. usneoides*, remain linear throughout life, the phyllotaxy does not transform from distichous, and the shoots do not develop a geotropic response. Tomlinson suggested that all of these characters can be interpreted as paedomorphic. It is important to note, however, that *T. usneoides* is not simply a neotenic form. It also has new features, such as continuous branching, internodal elongation, and inflorescence abbreviation, of which some are potentially peramorphic (Tomlinson, P. B., personal communication). Indeed, even though individual flowers and other modules of shoot construction are relatively small in this species, the number of these that a plant can produce is a separate issue. Consequently, total reproductive output of a plant is not necessarily related to the size of any of its modules. All of this reflects the distinctive impacts ontogenetic development and primordial development have on the form and function of whole plants and their component parts.

This example admirably illustrates how plant structure may be compared among individuals at the level of the whole organism, which is a major problem when studying plants, given their extreme phenotypic variability. The bases of comparison are not the structures themselves, but homologous aspects of devel-

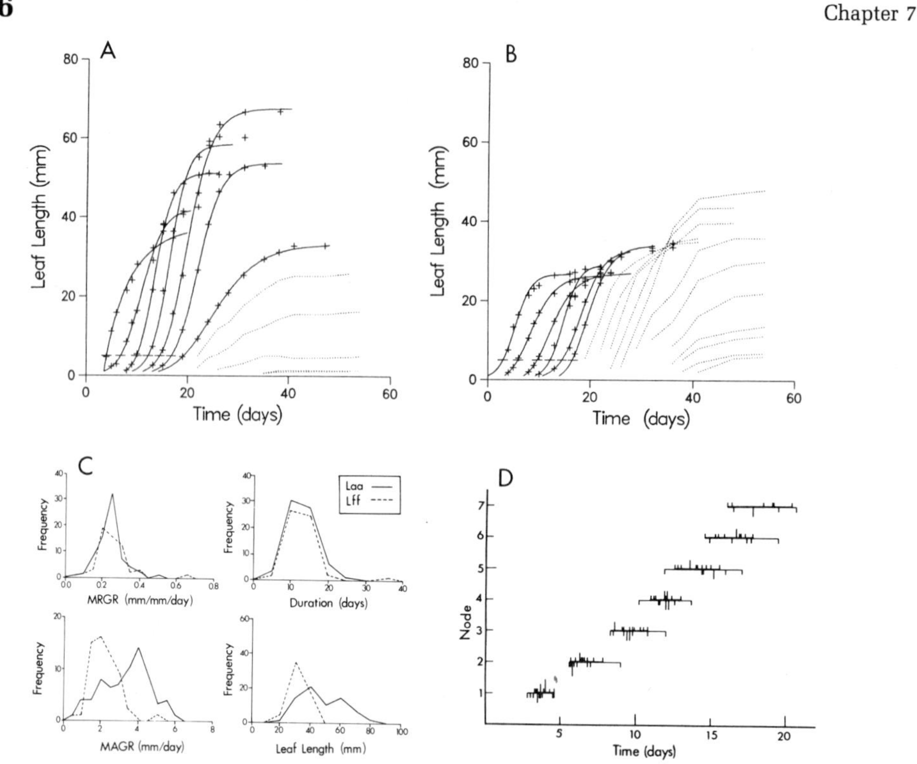

Figure 6. Shoot growth in *Limnanthes alba* and *L. floccosa*. (A,B) Each line represents the growth of a single leaf, and each figure represents all of the leaves of the primary axis of a single plant. (C) Relative frequency polygons for leaves of the two taxa indicating the mean relative growth rate (NS), duration of leaf growth (NS), mean absolute growth rate (Laa > Lff, $p < 0.001$), and maximum leaf length (Laa > Lff, $p < 0.001$). (D) Leaf production rate, indicating the times at which leaves became 5 mm long, as indicated by the horizontal lines on A and B.

opmental relationships that give rise to the structures. The notion of developmental homologs is not limited to the study of plants (van Valen, 1982; Roth, 1984), but it may be especially useful in comparisons involving these and other indeterminately growing organisms (Sachs, 1982; Guerrant, 1984; Groff, P., personal communication). The bases of comparison among these *Tillandsia* species, such as phyllotaxy and geotropic response, are qualitative aspects or elements of construction, and they serve to illustrate the usefulness of developmental homologs in comparing the indeterminate aspects of plant form. The expressly nonevolutionary classification of tree architecture of Hallé *et al.* (1978; Hallé and Oldeman, 1970) provides a rich source of descriptive variables of shoot construction.

Developmental bases of shoot growth of a more quantitative nature may also be used to compare the structures of plants. In a comparative developmental study of shoot growth between the same two species of *Limnanthes* mentioned previously, Guerrant (1984) showed that seemingly minor differences in the initial size of the shoot apical meristem of the mature embryo in the seed appear to be sufficient to account for dramatic differences in mature plant form and ecological functioning (Fig. 6). Differences in mature size and parallel differences in absolute growth rates are important factors in the ecology of these taxa (Ritland and Jain,

1984). By means of an analogy with the growth of funds in interest-bearing bank accounts, differences in size and absolute growth rates can been seen as consequences of what appear to be underlying developmental determinants. Leaves of the two taxa grow in length for the same amount of time and at the same mean relative growth rate (a measure equivalent to economic compound interest rate) as one another. Leaves of the self-pollinating *Limnanthes floccosa*, however, grow more slowly in absolute terms and mature at a smaller size than do those of *L. alba*. These relationships suggest that a difference only in the botanical analog of initial capital (shoot apical meristem and leaf primordium sizes) would be sufficient to account for the macroscopic differences in growth and form observed. Subsequent analyses have shown the shoot apical meristem in the seed is smaller in *L. floccosa* than it is in *L. alba* (Guerrant, E. O., Jr., unpublished results).

Sometimes attributes that do not change are as interesting as those that do. The rate that new leaves are produced is a case in point (Fig. 6D). Nothing at present suggests that similar rates of leaf production in the two *Limnanthes* species were not inherited in a form unchanged from their most recent common ancestor. If this is so, then leaf production rate is independent of apex size. Ecological studies have shown that within a number of genera, such as *Floerkea* (Smith, 1983), *Triglochin* (Jefferies, 1984), and *Desmodium* (Wulff, 1985), leaf production rate is insensitive to a variety of environmental perturbations.

The dissection of whole plant form into the combined action of developmental variables, which can be related to one another in ways described by the Gould–Alberch models, results in a comparative view of plants as whole organisms that can filter out the confounding effects of indeterminate growth on the outcome of plant form. By shifting the emphasis of comparison away from the structures per se, which are the products of development, to components of the ontogenetic processes, there is then no need to rely on the pattern of one-to-one correspondence of structures that characterizes homology in animals as whole organisms. If the bases of comparison are homologous elements of construction, then plants may be compared with one another morphologically as whole organisms. An advantage of this approach is that if proper developmental–genetic studies are done, then intrinsic aspects of form generation and also the way in which they respond to particular environmental variables can be identified. The range of what constitutes a developmental homolog is great and might be expected to operate at a variety of levels of structural organization.

4.3. Interaction between Generations

The intimate physical and physiological juxtaposition of the alternating haploid and diploid generations within the plant life cycle means that genetically determined heterochronic change in either may affect the other. Such interactions easily could have cascading phenotypic effects.

Consider a situation where, for whatever reason, natural selection favored small shoot apex size of the vegetative body in a flowering plant. It could easily have the incidental consequence of reducing the size (but not necessarily the number) of floral primordia, which may have happened in *Limnanthes floccosa*. In turn, there would be less sporogenous tissue in the anthers of individual flowers to undergo meiosis, and fewer pollen grains would result. In this way, selection on a nonreproductive aspect of the plant body of the sporophyte may incidentally reduce the population size of microgametophytes within a flower. The consequences would not end here, because pollen is a source of gibberellin, and this growth hormone can affect the expansion of sporophytic portions of the developing flower within which it is produced (Letham *et al.*, 1978; Lord, 1982, 1984). Lord (1984) suggested that smaller anther size in some CL flowers may result in gibberellin levels too low to induce floral opening. In this hypothetical example, selection to reduce the size of the vegetative plant body would lower the population of microgametophytes within individual flowers, which in turn would amplify the original effect.

Although I am not aware of work that has explicitly considered the effects of genetically determined changes in one generation on the other from the perspective of heterochrony, such effects surely exist.

4.4. Relationships between Generations

For plant taxa to persist, both generations must survive to continue the cycle. The relationship between generations within the life cycle and the manner in which each encounters the environment clearly have considerable ecological and evolutionary significance. Different ploidy levels suggest different abilities to withstand environmental change: diploids have more genetic variety on which to draw than do haploids. How each generation encounters the external environment should affect the sorts of habitats that can be colonized successfully. Evolutionary changes in the relationship between the generations is within the domain of heterochronic change.

Cladistic analyses (Mishler and Churchill, 1985) suggest that the life cycle of the original land plants was dominated by the gametophyte, and that the sporophyte was but a simple sessile sporangium. Indeed, they suggest that the sporophyte generation itself may have had a heterochronic origin, mediated by delaying meiosis in the zygote. Their analyses also indicate that within the land plants, the sporophyte or diploid generation became dominant only once and as such is a synapomorphy of the tracheophytes, or vascular plants. The gametophyte generation is dominant within the life cycle of mosses (sister group of the tracheophytes), and the sporophyte is physiologically dependent on it. Although the sporophyte generation is considered dominant in all vascular plants, the relationship between generations varies tremendously. In the cladistically more basal members such as ferns, both generations are photosynthetic and free-living. Within the tracheophyte clade, however, there has been a progressive reduction in the gametophyte generation. Flowering plants represent one extreme in reduced gametophytes. Hegemony of the genetically more diverse diploid sporophyte over the haploid gametophyte generations may be explored profitably from the perspective of heterochrony.

5. Conclusions

The central role of ontogeny is epitomized by van Valen's (1974) felicitous phrase, "Evolution is the control of development by ecology." Evolutionary theory is in general more highly advanced than ecological theory, and both are considerably more mature disciplines than the study of development, at least as far as the latter contributes to a broad understanding of the evolution of form and function, and especially the origins of adaptation. The study of heterochrony was jolted back to life by Gould (1977) and Alberch *et al.* (1979) and is becoming a powerful tool bridging evolutionary and ecological theory. For example, the age at which an organism reaches maturity has long been of considerable evolutionary and ecological interest, and now in light of its developmental significance takes on even greater meaning.

With the exception of Darwin, many of the major advances in our conceptual understanding of biological evolution have been articulated by those who have studied animals exclusively. Such foundational notions as species and homology concepts reflect the peculiarities of animals, and do not describe plants as well. For example, the concept of homology—as a part-for-part correspondence—was inspired by studies of insects and vertebrates by Owen (1843, 1848) (who was, ironically, in search of the essentialistic archetype). And so it was with the study of the relationship between ontogeny and phylogeny. The major theoreticians, von Baer, Haeckel, de Beer, and Gould, are all students of animal life and have greatly influenced the study of heterochrony in plants.

Because individuals in plants are constructed so differently than they are in animals, botanists have a different world to consider. The Gould–Alberch models go a long way toward connecting the empirical study of ontogeny with ecological and evolutionary theory. Applying it to plants as whole organisms requires a concept of homology that can accommodate a plant's open growth and its extreme phenotypic plasticity. Although many botanists have not used the terminology, or even have incorporated heterochrony into their world view, phenomena identified by the parameters of the Gould–Alberch models have been identified empirically in plants and their genetic and developmental bases identified. Initial size, relative growth rates, and duration of growth all have been shown to have genetic and developmental bases in plants.

Theory helps inspire new questions of relationships which had not been considered or found meaningful before. By viewing plant growth, especially between relatives of known cladistic relationship in appropriate ecological circumstances, we can decompose plant growth into its underlying parameters and compare their expression in specified environmental conditions. Evolutionary and ecological theory are being met by recent advances in phylogenetic reconstruction and theory concerning the relationship of ontogeny and phylogeny. This combination augers well for the study of evolution.

ACKNOWLEDGMENTS. I would like to extend my sincere appreciation to the many people who have assisted me during the course of this project. Drs. Ken Chambers, Peggy Fiedler, Paul Groff, Mike McKinney, Linda Newstrom, Chuck Quibell, Jenny Richards, and Barry Tomlinson all kindly gave of their time to help me improve the manuscript in its various forms. Despite their unselfish efforts, any remaining errors or fuzzy thinking are entirely my own. I would also like to thank Dr. Tomlinson and Academic Press for their kind permission to reprint Fig. 5, and Columbia University Press for allowing me to quote from Takhtajan (1976). Margie Gardner typed the manuscript and Linda Sawaya assisted with the illustrations. Finally, I would like to extend my thanks to Lewis and Clark College for financial assistance in making it possible for me to attend the Symposium on Heterochrony in San Antonio, Texas.

References

Abbe, E. C., Randolph, L. F., and Eisnet, J., 1941, The developmental relationship between shoot apex and growth pattern of leaf blade in diploid maize, *Am. J. Bot.* **28**:778–784.

Alberch, P., Gould, S. J., Oster, G., and Wake, D. B., 1979, Size and shape in ontogeny and phylogeny. *Paleobiology* **5**:296–317.

Anderson, L. C., 1972, Studies in *Bigelowia* (Asteraceae), II. Xylary comparisons, woodiness, and paedomorphosis, *J. Arnold Arb.* **53**:499–514.

Arroyo, M. T. Kalin, 1973, Chiasma frequency evidence on the evolution of autogamy in *Limnanthes floccosa* (Limnanthaceae), *Evolution* **27**:679–688.

Bachmann, K., 1983, Evolutionary genetics and the genetic control of morphogenesis in flowering plants, in: *Evolutionary Biology*, Vol. 16 (M. K. Hecht, B. Wallace, and G. T. Prance, eds.), pp. 157–208, Plenum Press, New York.

Bradshaw, A. D., 1965, Evolutionary significance of phenotypic plasticity in plants, *Adv. Genet.* **13**:115–155.

Bruck, D. K., and Kaplan, D. R., 1980, Heterophyllic development in *Muehlenbeckia* (Polygonaceae), *Am. J. Bot.* **67**(3):337–346.

Carlquist, S., 1962, A theory of paedomorphosis in dicotyledonous woods, *Phytomorphology* **12**:30–45.

Carlquist, S., 1975, *Ecological Strategies of Xylem Evolution*, University of California Press, Berkeley.

Chrysler, M. A., 1937, Persistent juveniles among the cycads, *Bot. Gaz.* **98**:696–710.

Clay, K., 1982, Environmental and genetic determinates of cleistogamy in a natural population of the grass *Danthonia spicata*, *Evolution* **36**(4):734–741.

Coyne, J. A., and Lande, R., 1985, The genetic basis of species differences in plants, *Am. Nat.* **126**:141–145.

Crotty, W. J., 1955, Trends in the pattern of primordial development with age in the fern *Acrostichum danaefolium*, *Am. J. Bot.* **42**:627–636.

Cumbie, G. B., 1963, The vascular cambium and xylem development in *Hisbiscus lasiocarpus*, *Am. J. Bot.* **50**:944–951.

Cumbie, G. B., 1967a, Developmental changes in the vascular cambium in *Leitneria floridana*, *Am. J. Bot.* **54**:414–424.

Cumbie, G. B., 1967b, Development and structure of the xylem in *Canavalia* (Leguminosae), *Bull. Torrey Bot. Club* **94**:162–175.

Cutter, E. G., 1955, Experimental and analytical studies of Pteridophytes XXIX. The effect of progressive starvation on the growth and organization of the shoot apex of *Dryopteris aristata* Druce, *Ann. Bot.* **19**:485–499.

De Beer, G. R., 1930, *Embryology and Evolution*, Clarendon Press, Oxford.

Dickinson, T. A., 1978, Epiphylly in angiosperms, *Bot. Rev.* **44**:181–232.

Foster, A. S., and Gifford, E. M., Jr., 1974, *Comparative Morphology of Vascular Plants*, Freeman, San Francisco.

Ghiselin, M. T., 1980, The failure of morphology to assimilate Darwinism, in: *The Evolutionary Synthesis, Perspectives on the Unification of Biology* (E. Mayr and W. B. Provine, eds.), pp. 180–193, Harvard University Press, Cambridge.

Gibson, A. C., 1973, Wood anatomy of Cactoideae (Cactaceae), *Biotropica* **5**:29–65.

Goebel, K., 1900, *Organography of Plants Especially of the Archegoniatae and Spermaphyta. Part 1. General Organography*, Clarendon Press, Oxford.

Gottlieb, L. D., 1984, Genetics and morphological evolution in plants, *Am. Nat.* **123**:681–709.

Gottlieb, L. D., 1985, Reply to Coyne and Lande, *Am. Nat.* **126**:146–150.

Gould, S. J., 1977, *Ontogeny and Phylogeny*, Harvard University Press, Cambridge.

Grafius, J. E., 1978, Multiple characters and correlated response, *Crop Sci.* **18**:931–934.

Guerrant, E. O., Jr., 1982, Neotenic evolution of *Delphinium nudicaule* (Ranunculaceae): A hummingbird-pollinated larkspur, *Evolution* **36**(4):699–712.

Guerrant, E. O., Jr., 1984, The role of ontogeny in the evolution and ecology of selected species of *Delphinium* and *Limnanthes*, Ph. D. dissertation, University of California, Berkeley.

Hall, B. K., 1984, Developmental processes underlying heterochrony as an evolutionary mechanism, *Can. J. Zool.* **62**:1–7.

Hallé, E., and Oldeman, R. A. A., 1970, *Essai sur l'architecture et la dynamique de croissance des arbes tropicaux*, Masson, Paris.

Hallé, F., Oldeman, R. A. A., and Tomlinson, P. B., 1978, *Tropical Trees and Forests—An Architectural Analysis*, Springer-Verlag, Berlin.

Hammond, D., 1941a, The expression of genes for leaf shape in *Gossypium hirsutum* L. and *Gossypium arboreum* L. I. The expression of genes for leaf shape in *Gossypium hirsutum* L., *Am J. Bot.* **28**(1):124–138.

Hammond, D., 1941b, The expression of genes for leaf shape in *Gossypium hirsutum* L. and *Gossypium arboreum* L. II. The expression of genes for leaf shape in *Gossypium arboreum* L., *Am. J. Bot.* **28**(1):138–150.

Harper, J. L., and White, J., 1974, The demography of plants, *Annu. Rev. Ecol. Syst.* **5**:419–463.

Hilu, K. W., 1983, The role of single-gene mutations in the evolution of flowering plants, in: *Evolutionary Biology*, Vol. 16 (M. K. Hecht, B. Wallace, and G. T. Prance, eds.), pp. 97–128, Plenum Press, New York.

Houghtaling, H. B., 1935, A developmental analysis of size and shape in tomato fruits, *Bull. Torrey Bot. Club* **62**(5):243–251.

Huxley, J. S., 1932, *Problems of Relative Growth*, MacVeagh, London.

Jain, S. K., 1979, Adaptive strategies: Polymorphism, plasticity and homeostasis, in: *Topics in Plant Population Biology* (O. T. Solbrig, S. K. Jain, G. B. Johnson, and P. H. Raven, eds.), pp. 160–187, Columbia University Press, New York.

Jefferies, R. L., 1984, The phenotype: Its development, physiological constraints, and environmental signals, in: *Perspectives on Plant Population Ecology* (R. Dirzo and J. Sarukhan, eds.), pp. 347–358, Sinauer, Sunderland, Massachusetts.

Kaiser, S., 1935, The factors governing shape and size in *Capsicum* fruits: A genetic and developmental analysis, *Bull. Torrey Bot. Club* **62**:433–454.

Kaplan, D. R., 1973, Comparative developmental analysis of the heteroblastic leaf series of axillary shoots of *Acorus calamus* L. (Araceae), *La Cellule* **69**(3):253–290.

Kaplan, D. R., 1980, Heteroblastic leaf development in *Acacia* Morphological and morphogenetic implications, *La Cellule* **73**(2):137–203.

Kaplan, D. R., 1984, The concept of homology and its central role in the elucidation of plant systematic relationships, in: *Cladistics: Perspective on the Reconstruction of Evolutionary History* (T. Steussy and T. Duncan, eds.), pp. 51–70, Columbia University Press, New York.

Larson, A., 1980, Paedomorphosis in relation to rates of morphological evolution in the salamander *Aneides flavipunctatus* (Amphibia, Plethodontidae), *Evolution* **34**(1):1–17.

Letham, D. S., Goodwin, P. B., and Higgins, T. J. V., 1978, *Phytohormones and Related Compounds: A Comprehensive Treatise*, Vol. II, Elsevier/North-Holland, Oxford.

Lindman, C. A. M., 1908, Ueber das Bluten von *Lamium amplexicaule* L., *Ark. Bot.* **8**:1–25.

Lloyd, D. G., 1965, Evolution of self compatibility and racial differentiation in *Leavenworthia* (Cruciferae), *Contrib. Gray Herb.* **195**:3–195.

Lord, E. M., 1979, The development of cleistogamous and chasmogamous flowers in *Lamium amplexicaule* (Labiatae): An example of heteroblastic infloresence development, *Bot. Gaz.* **140**:39–50.

Lord, E. M., 1980, An anatomical basis for the divergent floral forms in the cleistogamous species, *Lamium amplexicaule* L. (Labiatae), *Am. J. Bot.* **67**:1430–1441.

Lord, E. M., 1981, Cleistogamy: A tool for the study of floral morphogenesis, function and evolution, *Bot. Rev.* **47**:421–449.

Lord, E. M., 1982, Floral morphogenesis in *Lamium amplexicause* L. (Labiatae) with a model for the evolution of the cleistogamous flower, *Bot. Gaz.* **143**:63–72.

Lord, E. M., 1984, Cleistogamy: A comparative study of intraspecific floral variation, in: *Contemporary Problems in Plant Anatomy* (R. A. White and W. C. Dickison, eds.), pp. 451–494, Academic Press, New York.

Mayers, A. M., and Lord, E. M., 1983, Comparative flower development in the cleistogamous species *Viola odorata*. II. An organographic study, *Am. J. Bot.* **70**(10):1556–1563.
McNeill, C. I., and Jain, S. K., 1983, Genetic differentiation studies and phylogenetic inference in the plant genus *Limnanthes* (section Inflexae), *Theor. Appl. Genet.* **66**:257–269.
Melet, L. S., 1968, The phenomenon of paedomorphosis in the secondary wood of some cushion-plants of the eastern Pamir [in Russian], Izv. *Div. Biol. Sci. Takjikistan Acad. Sci.* **2**:19–22.
Minter, T. C., and Lord, E. M., 1983, A comparison of cleistogamous and chasmogamous floral development in *Collomia grandiflora* Dougl. ex Lindl. (Polemoniaceae), *Am. J. Bot.* **70**(10):1499–1508.
Mishler, B. D., 1986, Ontogeny and phylogeny in *Tortula* (Musci: Pottiaceae), *Syst. Bot.* **11**(1):189–208.
Mishler, B. D., and Churchill, S. P., 1984, A cladistic approach to the phylogeny of the "bryophytes," *Brittania* **36**:406–424.
Mishler, B. D., and Churchill, S. P., 1985, Transition to a land flora: Phylogenetic relationships of the green algae and bryophytes, *Cladistics* **1**(4):305–328.
Moore, D. M., and Lewis, H., 1965, The evolution of self-pollination in *Clarkia xantiana*, *Evolution* **19**:104–114.
Mueller, R. J., 1982, Shoot ontogeny and the comparative development of the heteroblastic leaf series in *Lygodium japonicum* (Thunb.) SW, *Bot. Gaz.* **143**(4):424–438.
Ornduff, R., 1969, Reproductive biology in relation to systematics, *Taxon* **18**:121–133.
Owen, R., 1843, Lectures on the comparative anatomy and physiology of the invertebrate animals, delivered at the Royal College of Surgeons in 1843, Longmans, Brown, Green, and Longmans, London.
Owen, R., 1848, *On the Archetype and Homologies of the Vertebrate Skeleton*, R. and J. E. Taylor, London.
Philipson, W. R., and Butterfield, B. G., 1967, A theory of the causes of size variation in wood elements, *Phytomorphology* **17**:155–159.
Prévost, M. F., 1978, Modular construction and its distribution in tropical woody plants, in: *Tropical Trees As Living Systems* (P. B. Tomlinson and M. H. Zimmerman, eds.), pp. 223–231, Cambridge University Press, Cambridge.
Richards, J. H., and Barrett, S. C. H., 1984, The developmental basis of tistyly in *Eichornia paniculata* (Pontederiaceae), *Am. J. Bot.* **71**(10):1347–1363.
Ritland, K., and Jain, S., 1984, The comparative life histories of two annual *Limnanthes* species in a temporally variable environment, *Am. Nat.* **124**(5):656–679.
Roth, V. L., 1984, On homology, *J. Linn. Soc. Biol.* **22**:13–29.
Sachs, T., 1982, A morphogenetic basis for plant morphology, in: *Axioms and Principles of Plant Construction* (R. Sattler, ed.), pp. 118–131, Nijhoff/Junk, The Hague.
Sattler, R., 1966, Towards a more adequate approach to comparative morphology, *Phytomorphology* **16**(4):417–429.
Sattler, R., 1984, Homology—A continuing challenge, *Syst. Bot.* **9**(4):382–394.
Scheiner, S. M., and Goodnight, C. J., 1984, The comparison of phenotypic plasticity and genetic variation in populations of the grass *Danthonia spicata*, *Evolution* **38**(4):845–855.
Sinnott, E. W., 1921, The relation between body size and organ size in plants, *Am. Nat.* **55**:385–403.
Sinnott, E. W., 1927, A factorial analysis of certain shape characters in squash fruits, *Am. Nat.* **61**:333–334.
Sinnott, E. W., 1931, The independence of genetic factors governing size and shape, *J. Hered.* **22**:381–387.
Sinnott, E. W., 1935, Evidence for the existence of genes controlling shape, *Genetics* **20**:12–21.
Sinnott, E. W., 1936, A developmental analysis of inherited shape differences in Cucurbit fruits, *Am. Nat.* **70**:245–254.
Sinnott, E. W., 1963, *The Problem of Organic Form*, Yale University Press, New Haven.
Sinnott, E. W., and Dunn, L. C., 1935, The effect of genes on the development of size and form, *Rev. Camb. Philos. Soc.* **10**:123–151.
Sinnott, E. W., and Kaiser, S., 1934, Two types of genetic control over the development of shape, *Bull. Torrey Bot. Club* **61**(1):1–7.
Smith, B. H., 1983, Demography of *Floerkea proserpinacoides*, a forest floor annual. I Density-dependent growth and mortality, *J. Ecol.* **71**:391–404.

Solbrig, O. T., and Rollins, R. C., 1977, The evolution of autogamy in species of the mustard genus *Leavenworthia*, *Evolution* **31**:265–281.
Stearns, S. C., 1982, The role of development in the evolution of life histories, in: *Evolution and Development* (J. T. Bonner, ed.), pp. 237–258, Springer-Verlag, Berlin.
Stebbins, G. L., 1974, *Flowering Plants. Evolution above the Species Level*, Harvard University Press, Cambridge.
Stevens, P. F., 1984, Homology and phylogeny: Morphology and systematics, *Syst. Bot.* **9**(4):395–409.
Stone, D. E., Sellers, S. C., and Kress, W. J., 1981, Ontogenetic and evolutionary implications of a neotenous exine in *Tapeinochilos* (Zingiberales: Costaceae) pollen, *Am. J. Bot.* **68**(1):49–63.
Sultan, S. E., 1987, Evolutionary implications of phenotypic plasticity in plants, in: *Evolutionary Biology*, Vol. 21 (M. K. Hecht, B. Wallace, and G. T. Prance, eds.), pp. 127–178, Plenum Press, New York.
Takhtajan, A., 1943, Correlations of ontogenesis and phylogenesis in higher plants [in Russian with English summary], *Tr. Erevansk. Gos. Univ.* **22**:71–176.
Takhtajan, A., 1954, *Essays on the Evolutionary Morphology of Plants* 1959. American Institute of Biological Sciences, Arlington, Virginia.
Takhtajan, A., 1972, Patterns of ontogenetic alterations in the evolution of higher plants, *Phytomorphology* **22**(2):164–171.
Takhtajan, A., 1976, Neoteny and the origin of flowering plants, in: *Origin and Early Evolution of Angiosperms* (C. B. Beck, ed.), pp. 207–219, Columbia University Press, New York.
Tomlinson, P. B., 1970, Monocotyledons—Toward an understanding of their morphology and anatomy, in: *Advances in Botanical Research*, Vol. 3 (R. D. Preston, ed.), pp. 207–292, Academic Press, New York.
Tomlinson, P. B., 1978, Branching and axis differentiation in tropical trees, in: *Tropical Trees As Living Systems* (P. B. Tomlinson and M. H. Zimmerman, eds.), pp. 187–207, Cambridge University Press, Cambridge.
Tomlinson, P. B., 1982, Chance and design in the construction of plants, in: *Axioms and Principles of Plant Construction* (R. Sattler, ed.), pp. 162–183, Nijhoff/Junk, The Hague.
Tomlinson, P. B., 1984a, Homology in modular organisms—concepts and consequences, Introduction, *Syst. Bot.* **9**(4):373.
Tomlinson, P. B., 1984b, Homology: An empirical view, *Syst. Bot.* **9**(4):374–381.
Van Valen, L. M., 1974, A natural model for the origin of some higher taxa, *J. Herpetol.* **8**:109–121.
Van Valen, L. M., 1982, Homology and causes, *J. Morphol.* **173**:305–312.
Von Maltzahn, K. E., 1957, A study of size differences in two strains of *Cucurbita pepo* L., *Can. J. Bot.* **35**:809–832.
Wareing, P. F., and Phillips, I. D. J., 1981, *Growth and Differentiation in Plants*, 3rd ed., Pergamon Press, Oxford.
Whaley, W. G., 1939, The relation of organ size to meristem size in the tomato, *Am. Soc. Horticult. Sci.* **37**:910–912.
White, J., 1984, Plant metamerism, in: *Perspectives on Plant Population Ecology* (R. Dirzo and J. Sarukhan, eds.), pp. 15–47, Sinauer, Sunderland, Massachusetts.
Wulff, R. D., 1985, Effect of seed size on heteroblastic development in seedings of *Desmodium paniculatum*, *Am. J. Bot.* **72**(11):1684–1686.
Wyatt, R., 1983, Pollinator–plant interactions and the evolution of breeding systems, in: *Pollination Biology* (L. Real, ed.), pp. 51–95, Academic Press, New York.

Chapter 8

Heterochrony in Colonial Marine Animals

JOHN M. PANDOLFI

1. Introduction

Since the publication of *Ontogeny and Phylogeny* (Gould, 1977), many workers studying fossil organisms have used heterochrony as a working hypothesis to account for phylogenetic changes in evolving lineages. Previous studies have dealt largely with solitary, aclonal organisms, but heterochrony has also been identified in clonal colonial organisms: graptolites (Rickards, 1977), bryozoans (Schopf, 1977; Anstey, 1987), and tabulate corals (Pandolfi, 1984a, 1987).

The modular construction and the preservation by accretionary growth of the history of development within many clonal colonial organisms allow meaningful developmental comparisons between primitive and derived character states in skeletonized fossil marine colonial animals. Colonial animals are unique for studying the application of heterochrony to changes in phylogeny because developmental shifts occur at two developmental levels. Heterochrony in clonal *colonial* animals may involve either or both developmental processes within the

JOHN M. PANDOLFI • Department of Geology, University of California, Davis, California 95616. *Present address:* Australian Institute of Marine Science, Townsville M.C., Queensland 4810, Australia.

organism: (1) the developmental history or *ontogeny* of each module (e.g., zooids, thecae) composing the colony and (2) the developmental history or *astogeny* of the colony as a whole. Therefore, traditional heterochronic concepts conceived at one developmental level for aclonal animals must be evaluated at two developmental levels when dealing with colonial clones. A new conceptual framework for evaluating apparent heterochronic patterns in clonal colonial animals must be defined before discussion of the implications of heterochrony for the evolutionary history of clonal colonial species can be attempted. In this chapter, I discuss the developmental criteria for examining heterochrony in clonal colonial animals. I develop a conceptual framework for the analysis of heterochrony in these organisms and use it in studies of heterochronic change in fossils. I then discuss several evolutionary consequences of heterochronic processes in clonal colonial animals.

McNamara (1986) used de Beer's (1930) definition of heterochrony as "the phenomenon of changes through time in the appearance or rate of development of ancestral characters." It is assumed explicitly here, as it is implicit in all studies of heterochrony, that ancestor–descendant relationships can be identified with sufficient accuracy for meaningful discussion. In the cases of heterochrony reviewed in this chapter the ways in which ancestor–descendant hypotheses have been determined probably vary among workers. Identifying correct ancestor–descendant relationships from the fossil record is extremely difficult (Fink, this volume) and most of the phylogenies where heterochrony has been identified in the literature in colonial animals are not based on rigorous phylogenetic analyses.

Several developmental differences exist between astogeny and ontogeny in clonal colonial animals. In clonal organisms, ontogeny refers to both growth and reproduction of the module. The module, throughout its ontogeny, may or may not have the potential to reproduce sexually. Because astogeny involves the development of the whole organism, the colony also may be treated as an individual entity: growth occurs through generations of modules; thus, growth of the colony is not constrained by the increase in size of the module. The potential for the colony to reproduce sexually depends on at least one of the modules having this capacity.

2. Growth and Astogenetic Stages in Clonal Colonial Animals

Most clonal organisms exhibit indeterminate growth, but growth may be determinate for colonies that raise their tissue up above the substratum before overgrowth by other organisms can occur (Jackson, 1979). Graptolites may also exhibit determinate growth (Finney, 1986). Growth in clonal colonial animals not only involves an increase in the area and volume of the organism, but also the addition of new modules. I therefore consider asexual multiplication by budding as a form of colony growth (Ryland, 1981).

The most fundamental change in growth is the advent of sexual maturity, and this change is stressed more than any other in discussions of heterochrony because it provides an unambiguous standard against which to measure morphological development. Identifying specific heterochronic processes is contingent upon recognition of a time standard, and onset of sexual maturity is usually preferred. It is important, therefore, to look for the onset of sexual maturity in any organism for which heterochrony is postulated. Of course, where onset of sexual maturity cannot be ascertained, rate changes and onset timing of characters can often be used to detect heterochronic processes.

In the analysis of heterochrony in clonal colonial animals, the optimal goal, at least at the astogenetic level, would be to determine the onset of sexual maturity of the colony. The next best alternative would be to determine real growth time within the colonial skeleton, for example, growth lines. The third alternative is to determine reliable astogenetic stages in the hope that at least they will allow cross-comparison within lineages, if not across them.

2.1. Growth Rates and Timing of Sexual Maturity

Are there hard-part morphological criteria to estimate the onset of sexual maturity in colonial animals? Distinct differences in morphological characters may provide clues to mark the onset of sexual maturity, as long as these differences occur in varied environments, allowing environmental influences to be segregated from genetic ones. Modern analogs may be helpful where data are available on growth rates and onset of sexual maturity. Seasonal growth bands occur in many fossil and recent corals and estimates of growth rate are easily obtained using standard X-radiographic techniques (e.g., Buddemeier *et al.*, 1974).

To the extent that sexual reproduction and growth are exclusive processes, onset of sexual maturity is negatively correlated with growth rates. Growth patterns in clonal colonial animals may involve alternating periods of high asexual growth rates, and low growth rates that correspond to episodes of sexual reproduction. This occurs in ascidians and hydroids, for example (Haven, 1971; Sugimoto and Nakuachi, 1974; Yamaguchi, 1975; Stebbing, 1980). Jackson (1979) reasoned that colonies that grow very fast early in astogeny will delay sexual reproduction until they are well established, whereas colonies without early fast growth will have earlier sexual reproduction. Jackson and Hughes (1985) later confirmed this in bryozoan species in Discovery Bay, Jamaica. It is important to determine growth rates in fossil colonial animals not only to identify the possible onset of sexual maturity, but also because differential growth rates can increase or decrease the length of ontogenetic and astogenetic developmental stages. Such data are invaluable in the analysis of heterochronic processes in colonies. Unfortunately, when colonies undergo asexual fragmentation, age and size can be decoupled, hence growth rates can be difficult to determine for the entire clone [see Wahle (1983) for an example involving gorgonians]. In Paleozoic tabulate and rugose corals there is no evidence for asexual fragmentation, and age and size often can be detected by growth lines on the colony exterior. Finally, timing of sexual maturity may be phenotypically plastic (Harvell and Grosberg, in press).

Scrutton and Powell (1980) documented astogenetic variability in the extinct favositid corals. They noticed abundant polymorphic corallites occurring during periods of peak colony growth alternating with mostly monomorphic corallites during periods of low colony growth. Jackson (1979) reasoned that because sexual reproduction may often depend on the activities of a few isolated modules, polymorphism decreases the probability that a given surviving module would be capable of sexual reproduction. In favositids, then, the onset of sexual reproduction may occur at the first astogenetic stage where most corallites are monomorphic. Careful serial sectioning through all stages of astogeny is needed to test this hypothesis, since abundance of polymorphic corallites could be determined at each serial section. If true, once phylogenetic patterns or polarity of transformation of character states are reliably determined, heterochronic processes could be analyzed in favositids. Neoteny could be distinguished from progenesis, and acceleration from hypermorphosis, because the timing of emplacement of a strictly reproductive character (or suite of characters) could be determined.

2.2. Growth Rates and Colony Growth Form

Variations in growth rate of the colony or of the modules of the colony in different stages of astogeny may be reflected in the overall growth form (Jackson, 1979) of the colony. For example, tabulate corals that expand rapidly horizontally along the substratum early in astogeny produce colonies with a sheet morphotype, whereas colonies that expand horizontally and vertically at similar rates produce a mound morphotype (Pandolfi, 1984b). Shifts in sedimentation rates may cause temporary changes in growth rates, which allow shifts in colony morphotype from sheets to mounds and vice versa. The potential morphological flexibility of clonal colonial animals, which enables them to change morphology through astogeny, provides the selective basis for developmental changes to operate in their evolution, provided that morphogenetic changes are heritable. The evolutionary biologist must be careful to distinguish morphological changes resulting from genetically programmed alterations in developmental timing from nonheritable phenotypic responses to environmental changes that occur during the life of the colony. This may be very difficult to sort out because as an organism grows, the nature of its relationship to its habitat changes, even if the habitat does not change (Grosberg, R.K., *personal communication*).

2.3. Astogenetic Stages

If we are to recognize heterochrony in the fossil record of clonal colonial animals, we must be able to determine relative developmental stages of these organisms using only their hard-part morphology. Astogenetic stages in colonial organisms may be useful in deciphering relative colony ages of hypothesized ancestral and descendant colonies. Astogenetic stages can be observed in clonal colonial animals by two methods. The first is by delineating the astogenetic units in a developing colony and the second is through colonial regeneration events. *Astogenetic units* are groups of modules within a colony within which development is distinct from other such groups. For example, in bryozoans astogenetic units may occur as zones of astogenetic change and zones of astogenetic repetition (Boardman *et al.*, 1970). An astogenetic sequence can repeat itself and so a given astogenetic unit may occur several times or places in astogeny. Examples include the branching points of arborescent bryozoans (Boardman *et al.*, 1970), fragments

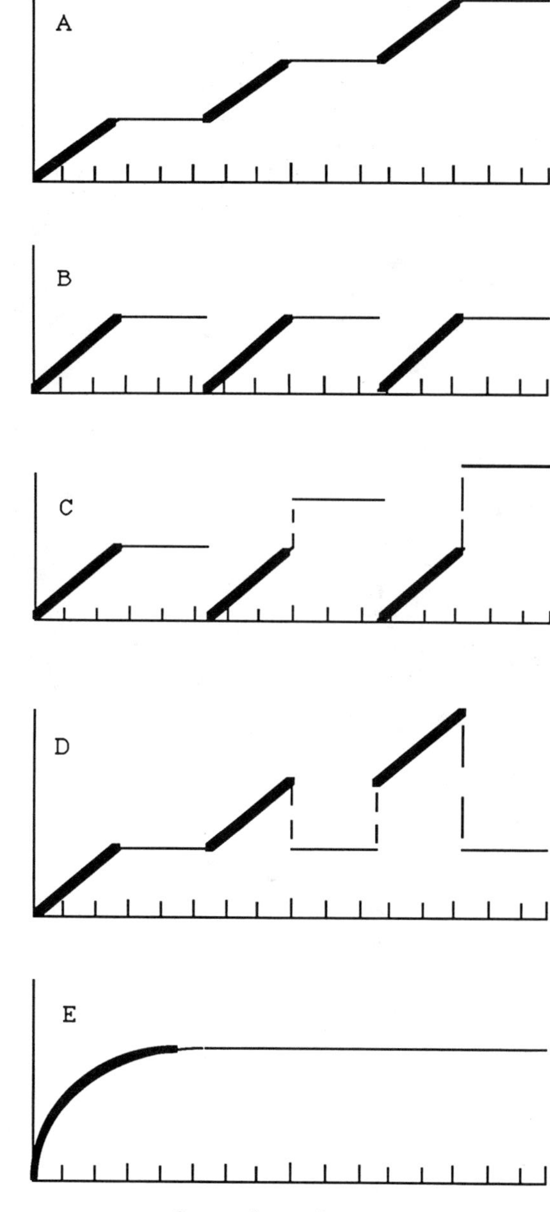

Figure 1. Five models to illustrate the development of astogenetic units in a colonial skeleton. Heavy lines denote zones of astogenetic change and light lines denote zones of astogenetic repetition. (A) Both zones of astogenetic change and repetition change through astogeny. Initial and subsequent zones of change and repetition are uniquely different and these differences are linear throughout astogeny. (B) Neither zones of astogenetic change nor repetition change through astogeny. Initial and subsequent zones are all similar. (C) Only initial and subsequent zones of astogenetic repetition are different. Usually, the primary zone of astogenetic change is different from all subsequent zones of astogenetic change because of the presence of the first-formed or founder individual of the colony. Therefore, cases B and C may be rare. (D) Only initial and subsequent zones of astogenetic change are different. (E) Original model of astogenetic change and astogenetic repetition of Taylor and Furness (1978).

formed by asexual fission in scleractinian corals (Highsmith, 1982), and regeneration points in Paleozoic Tabulata (see below).

Morphometric plots, such as the ones shown in Fig. 1, may be helpful in identifying shifting developmental stages when they are compared between hypothesized ancestors and descendants. Taylor and Furness (1978) constructed a model to illustrate astogenetic variation throughout the development of bryozoan colonies. Figure 1 is an extension of their model to include several zones of as-

Figure 2. *Favosites* sp. from the Middle Devonian Hamilton Group of New York State showing regeneration event. This specimen was subject to a sedimentation event in which most of the coral surface was suffocated. A few corallites did survive and these mimicked the early astogeny of the coral when there were high asexual growth rates, and initial corallites expanded horizontally across the sea floor to provide a broad base for subsequent development of the colony. Note that the coral later died of another sediment suffocation event in which none of the corallites survived. Scale bar = 0.5 cm.

togenetic change and repetition. Five models are shown in Fig. 1 to illustrate the various situations that may occur.

Regeneration occurs when the colony is reduced in size before increasing again. Physical and/or biological disturbances may be responsible for such events. Astogenetic stages during regeneration may mimic earlier astogenetic stages within the colony. Regeneration events can be viewed as reiterations of astogenetic units and therefore may give rare clues to astogenetic development if early astogeny is unavailable or if developmental stages are otherwise difficult to differentiate (e.g., in massive colonies). Figure 2 illustrates the use of regeneration events in deciphering developmental stages in clonal corals. A colony of *Favosites* from the Middle Devonian Hamilton Group of New York State has been almost completely buried by sediment suffocation. Only two or three corallites survived and these subsequently mimicked the earlier astogenetic stages, when the coral was characterized by high asexual budding rates and horizontal expansion of the corallites, which provided a broad base for the new colony to develop.

Table I. Heterochronic Processes in Colonial Animals[a]

General	Specific
Acceleration	Acceleration Progenesis Predisplacement
Retardation	Neoteny Hypermorphosis Postdisplacement

[a] Heterochronic processes may occur in any combination at the astogenetic and/or ontogenetic levels.

3. Pathways of Heterochronic Evolution

3.1. Conceptual Framework for General Heterochronic Processes and Results

Table I shows the general and specific heterochronic processes outlined by Alberch *et al.* (1979) which may occur in any combination at either the astogenetic and/or ontogenetic developmental levels [see McNamara (1986) for cogent discussion of general and specific heterochronic processes]. The general heterochronic processes are acceleration and retardation (Gould, 1977) and they may affect either reproductive or somatic structures. Retardation of reproductive structures may result in a delay of the onset of timing of maturation in the descendant taxon, whereas acceleration of reproductive structures may result in an earlier onset in timing of maturation. Retardation may yield a paedomorphic result due to neoteny or postdisplacement or a peramorphic result due to hypermorphosis. Acceleration (*sensu* Gould, 1977) may yield a peramorphic result due to acceleration (*sensu* Alberch *et al.*, 1979) or predisplacement or a paedomorphic result due to progenesis. Figure 3 provides a conceptual framework for the morphological results of heterochronic processes in clonal colonial animals. In all cases, either of the general heterochronic processes of retardation or acceleration may be acting.

Clonal animals in which developmental changes through phylogeny have been reported are graptolites, bryozoans, and tabulate corals. Table II gives all the instances I have been able to obtain from the literature in which heterochronic patterns have been identified in the evolution of fossil clonal colonial animals. I have interpreted the examples in Table II in terms of the conceptual framework advanced in this chapter. It is usually not clear either when sexual maturation occurred in the fossil colonies or what the relative ages of inferred ancestral and descendant taxa were. Therefore, it may not always be possible to differentiate among acceleration, hypermorphosis, and predisplacement in peramorphic examples and neoteny, progenesis, and postdisplacement in paedomorphic examples. In some of the examples given in Table II, however, I have inferred specific processes on the basis of the loss or addition of developmental stages and/or the

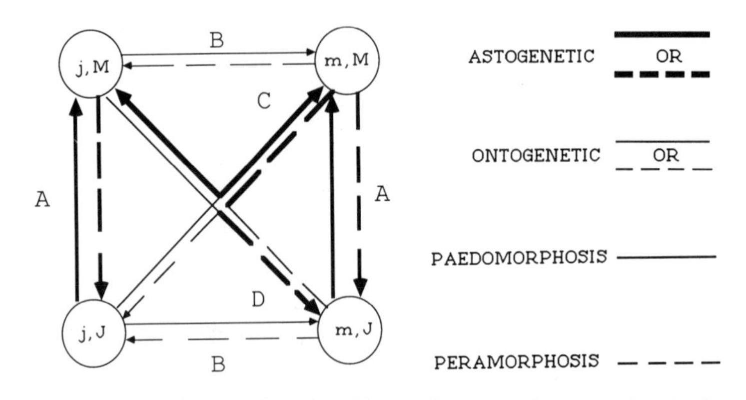

Figure 3. Model showing morphological results of heterochronic evolution in clonal colonial animals. Four starting points are recognized: (1) juvenile module, juvenile colony; (2) juvenile module, mature colony; (3) mature module, juvenile colony; and (4) mature module, mature colony. Arrows point from ancestor to descendant or from parent to offspring. Four general outcomes of heterochrony are possible: (A) Heterochrony at only the astogenetic developmental level resulting in either astogenetic paedomorphosis or astogenetic peramorphosis. (B) Heterochrony at only the ontogenetic developmental level resulting in either ontogenetic paedomorphosis or ontogenetic peramorphosis. (C) Heterochrony at both developmental levels and the same morphological result occurs. This results in either ontogenetic and astogenetic paedomorphosis or ontogenetic and astogenetic peramorphosis. (D) Heterochrony at both developmental levels, but where paedomorphosis occurs at one level, peramorphosis occurs at the other. This results in either astogenetic paedomorphosis and ontogenetic peramorphosis or astogenetic peramorphosis and ontogenetic paedomorphosis. Of course variations on this general scheme will result when multiple heterochronic processes produce multiple heterochronic results at either or both developmental levels. j, Juvenile module; J, juvenile colony; m, mature module; M, mature colony.

basis of significant size changes within developmental stages that could be calibrated against colony age.

In general, four situations with respect to general heterochronic processes may arise: (1) only astogenetic retardation and/or astogenetic acceleration; (2) only ontogenetic retardation and/or ontogenetic acceleration; (3) both astogenetic and ontogenetic retardation and/or both astogenetic and ontogenetic acceleration; and (4) both astogenetic retardation and ontogenetic acceleration and/or both astogenetic acceleration and ontogenetic retardation. For example, in two of the taxa in Table II, heterochronic processes have occurred at both developmental levels and in each instance only one heterochronic process was acting at each level. An example from my work on tabulate corals involves the hypothetical ancestor *Lichenaria simplex* and the inferred descendant *Eofletcheria utahia* (Rigby and Hintze, 1977). Rounded corallites with thick walls are characteristic of juvenile corallites of *L. simplex*, whereas polygonal corallites with thin walls characterize adult corallites. In *E. utahia* proximal portions of the colony are cerioid, with thin-walled polygonal corallites, and distal portions are fasciculate, with thick-

Table II. Reported Occurrences of Heterochrony in Fossil Marine Colonial Animals

Taxon	Character	Product[a]	Process	Reference
Monograptus sedgwickii–M. halli	Thecae	A—Peramorphosis	Acceleration	Elles (1922)
		O—Paedomorphosis	?	Rickards (1977)
Monograptus pseudoplanus–M. planus	Thecal hooks	A—Peramorphosis	?	Rickards *et al.* (1976); Sudbury (1958)
Petalograptus–Cephalograptus cometa extrema	Early thecae	A—Paedomorphosis	?	Bulman (1963), Rickards (1977)
Didymograptus–Tetragraptus	Stipe reduction	A—Paedomorphosis	?	Elles (1922), Bulman (1963)
Tetragraptus–Dichograptus	Stipe reduction	A—Paedomorphosis	?	Elles (1922), Bulman (1963)
Climacograptus normalis–C. transgrediens	Median septum	A—Paedomorphosis	Postdisplacement	Waern (1948); this chapter
Climacograptus transgrediens–C. medius	Median septum	A—Paedomorphosis	Postdisplacement	Waern (1948); this chapter
Dicellograptus–Leptograptus	Thecal apertures	O—?	?	Finney (1986)
Glossograptus–Nanograptus	Early astogeny	A—Paedomorphosis	Progenesis	Finney (1986)
Glossograptus ciliatus–Apoglossograptus lyra	Thecal curvature	O—Peramorphosis	Acceleration	Finney (1986)
Nemagraptus subtilis–N. gracilis	Thecae	O—Peramorphosis	?	Finney (1986)
Glossograptus–Corynoides	Early astogeny	A—Paedomorphosis	Progenesis	Finney (1986)
Glyptograptus euglyphus–Climacograptus brevis	Early astogeny	A—Paedomorphosis	Progenesis	Finney (1986)
Climacograptus typicalis–C. pygmaeus	Thecae growth rate	?—Peramorphosis	Acceleration	Mitchell (1986)

(continued)

Table II. (*Continued*)

Taxon	Character	Product[a]	Process	Reference
Monograptus cyphus–M. priodon	Hooked thecae	A—Paedomorphosis	?	Elles (1923)
Monograptus priodon–M. tumescens	Unhooked thecae	A—Peramorphosis	?	Elles (1923)
Atavograptus atavus–A. Praestrachani	Geniculate thecae	A—Paedomorphosis	?	This chapter
Atavograptus praestrachani–A. strachani	Sicular length	A—Paedomorphosis	Neoteny	This chapter
Cyclostomata–Cryptostomata	Zooecia	A—Peramorphosis	?	Cummings (1910)
Fenestellids	?	A—Peramorphosis	?	Cummings (1910)
Proboscina	Branching	A—Peramorphosis	Acceleration	Lang (1904)
Heteropora	Zooid walls	A—Paedomorphosis	?	Larwood and Taylor (1979), Anstey (1987)
Ceriocava	Zooid walls	A—Paedomorphosis	?	Larwood and Taylor (1979), Anstey (1987)
Stenolaemates	Zooid walls	A—Peramorphosis	?	Larwood and Taylor (1979), Anstey (1987)
Sarsiflustra japonica	Ancestrula	A—Peramorphosis	?	Dzik (1975)
Phylactolaemata–Gymnolaemata	Larval vs. adult growth	A—Peramorphosis	?	Jebram (1973)
Gymnolaemata	Position of buds	A—Peramorphosis	?	Jebram (1973)
Stenolaemata	Varied	A—18 Peramorphosis, 22 Paedomorphosis	Varied	Anstey (1987)[b]
Prasopora	Monticules	A—Paedomorphosis	?	Anstey (1981)
Lichenaria simplex–Eofletcheria utahia	Corallite shape and wall thickness;	O—Paedomorphosis	?	This chapter
	Late astogeny	A—Peramorphosis	Acceleration	This chapter
Nyctopora goldfussi–Calapoecia	Corallite ontogeny	A—Paedomorphosis	Retardation	Scrutton (1984); this chapter

[a] A, Astogenetic; O, ontogenetic.
[b] Anstey (1987) reported 40 occurrences of astogenetic heterochrony in Paleozoic Stenolaemates.

walled circular corallites. Therefore ontogenetic paedomorphosis likely occurred from an ontogenetic heterochronic process at a late astogenetic stage. Astogenetic peramorphosis has also resulted, because closely packed, thin-walled corallites characteristic of adult *L. simplex* occur only early in the astogeny of *E. utahia*. The critical morphological developmental relationship is that between colony form (cerioid vs. fasciculate) and corallite morphology (rounded with thick walls vs. polygonal with thin walls). Of course, close packing of circular forms results in polygonal corallites (Thompson, 1917), but such a simple physical explanation may be insufficient, because thick-walled, rounded juvenile corallites in *L. simplex* are cerioid.

The developmental change at the astogenetic level seems to have occurred along with that at the module level. Developmental stages of the descendant colony transcend those in the ancestral colony, at which time an ontogenetic character was retarded. Changes in developmental timing resulted in a large morphological discontinuity between ancestor and descendant and were dependent on the interplay of both developmental levels. Since colonies function as individuals and are therefore selected as individuals (Cowen and Rider, 1973), the characters of the colony are significant in evolution. These characters may also include the hierarchical arrangement of developmental levels. For example, even though modules are subsumed and can be sacrificed or changed for the interests of the colony, each developmental level is subject to selection. There is feedback between levels; developmental changes at one level may affect the other, just as organ-level changes in an aclonal organism may affect the behavior of the entire individual. Because developmental changes are selected for, they are probably (necessarily?) functional, and in the clonal organism must be harmonious between levels. The relationship between the two developmental levels in the *Lichenaria–Eofletcheria* example indicates that heterochronic changes in the colony may have been selected for at both developmental levels.

Another example involves the evolution of the graptolite *Monograptus halli* from *M. sedgwickii* (Rickards, 1977). *Monograptus sedgwickii* has thecae that are similar throughout astogeny. *Monograptus halli* has proximal thecae that are similar to the thecae of *M. sedgwickii*, and distal thecae in which full growth of the metathecal hook is retarded, resulting in ventrally rather than proximally opening tubes (Rickards, 1977). Thus, ontogenetic paedomorphosis occurred in the distal thecae. In this example astogenetic peramorphosis again was the result of changes in developmental timing in modules at a particular astogenetic stage.

Developmental changes at the astogenetic level may also occur as a result of colony-wide changes in ontogenetic features or as a result of a morphogenetic gradient (Urbanek, 1973; Anstey, 1987). Such heterochronic evolution is common in graptolites and bryozoans [e.g., *Peronopora*, (Hickey, 1987, 1988)].

3.2. Ontogenetic versus Astogenetic Heterochrony

Ontogenetic heterochrony is different in clonal organisms than in aclonal ones, first because timing of sexual maturity of the colony can be controlled either by the modules or the astogenetic units of the colony. Colony maturity may be reached by the asexual budding of a critical number of monomorphic modules or by the introduction of one or more reproductive polymorphs, which initiate sexual reproduction within a particular astogenetic stage. These polymorphs may persist or replicate themselves, or may appear and disappear periodically. Ontogenetic neoteny or hypermorphosis of certain modules may result in delay of maturation for the colony, and ontogenetic acceleration or progenesis may result in an early onset of maturation for the colony. Second, when ontogenetic heterochrony occurs, size changes detectable in aclonal organisms with determinate growth (McNamara, 1986) may not be detectable in the modules of colonies that display indeterminate growth.

In clonal animals, standardization by developmental stage may be unsatisfactory because it may not be possible to determine the interaction of the four relevant properties: size, shape, age, and developmental stage. Therefore a specific heterochronic process may not always be determinable. In astogenetic heterochrony, size differences between ancestor and descendants may not be directly related to colony age, because indeterminate growth, colony fragmentation, and regeneration may result in extension or truncation of dimensions of heterochronic characters. Therefore it may not be possible to compare correctly the size of ancestor and descendant. In clonal animals with indeterminate growth, adult ancestors could mistakenly be compared with juvenile descendants when only size standardization is utilized [see Gould (1977) for an example involving *Gryphaea*].

It is often not possible to tell the exact age of clonal colonial animals. Gould's (1977) clock model and the ontogenetic trajectory model (Alberch *et al.*, 1979) for determining specific heterochronic processes are based on the acquisition of age data in ancestor and descendant, although less desirable methods exist when all life history data are not available (Gould, 1977). An example involves the bryozoan *Leptotrypella pellucida* from the Late Devonian Threeforks Formation of Montana (Prezbindowski and Anstey, 1978). Based on width of the exozone, axial ratio, and zooecial size, it was concluded that the fossil specimens actually represented a form with earlier astogenetic stages of the type specimens of *L. pellucida* from the Middle Devonian Traverse Group of Michigan, but expressed in the adult astogenetic stages. In this example the heterochronic result was astogenetic paedomorphosis. Prezbindowski and Anstey ascribed the evolution of the large, underdeveloped colonies from Montana to the process of neoteny, but because they did not report size changes relative to colony age, it is unclear whether the heterochronic pattern was produced by accelerated development of reproductive organs (astogenetic progenesis) or the retarded development of somatic features (astogenetic neoteny). In addition, because of the potential for indeterminate growth, even if the heterochronic colonies actually had undergone astogenetic progenesis, a smaller size of the colony would not necessarily have resulted. If we are to recognize specific heterochronic processes in clonal organisms with indeterminate growth we must be able to determine the relative ages of developmental stages within heterochronic taxa. By inferring the age of the heterochronic taxa based on other skeletal characters (in this case, number of diaphragms), Anstey (1987) later reported this example to be the result not of the process of astogenetic neoteny, but of astogenetic progenesis.

Because graptolite form is determinate among rhabdosomes of the same or nearly related species (Mitchell, 1986; Finney, 1986), it may sometimes be possible to reckon size changes of morphological characters between hypothesized ancestors and inferred descendants. An example of an astogenetic morphological feature that shows a size change between ancestor and descendant involves the hypothesized lineage *Climacograptus miserabilis* to *C. normalis* to *C. medius* (Davies, 1929). Throughout the lineage more and more thecae are formed before the median septum is developed. Because the formation of the median septum is progressively delayed in descendant species, astogenetic paedomorphosis has resulted. The heterochronic pattern may have been due to astogenetic postdisplacement in which, by comparison with the ancestor, the median septum has initiated development at a later astogenetic stage with respect to other parts of the organism. The median septum is thus smaller in the descendants of the series than in the ancestors. The median septum is clearly an astogenetic feature of the colony, because its formation depends on the development of a group of thecae. Packham (1962) has identified a similar pattern of delay in the formation of the median septum in three lineages involving *Glyptograptus tamariscus*.

In other graptolite examples (Table II), it is possible to determine specific heterochronic processes because the size of the whole colony in the descendant is changed. Finney (1986) reported three examples in which specific heterochronic processes could be identified. *Glossograptus* may have given rise to both *Nanograptus* and *Corynoides* (Finney, 1978). The early astogeny of the ancestor was repeated in both descendants, after which time growth was truncated as sexual maturity may have been attained at an early astogenetic stage (Finney, 1986). Both descendants are very much smaller than the ancestor, suggesting astogenetic progenesis as the heterochronic process. A similar example involves the evolution of *Climacograptus brevis* from *Glyptograptus euglyphus* (Table II).

3.3. Specific Heterochronic Processes

Table III shows how the specific heterochronic processes occur when they act alone at the astogenetic level of development in colonies. Astogenetic neoteny occurs when (1) *astogenetic units* of a given astogenetic stage asexually bud at a slower rate than in the ancestor; (2) the constituent *modules* of astogenetic units of a given astogenetic stage simply take longer to develop; or (3) astogenetic morphological characters develop at a slower rate in the descendant (Table III). In the former two cases astogenetic stages are "drawn out" in the descendant. Onset of sexual maturity may be delayed in the first instance because growth of each individual module was preferred at the expense of asexual multiplication, and in the second instance because an astogenetic unit is composed of modules that reach maturity later in time than in the ancestor. In addition, reproductive polymorphs may appear later in astogeny in the descendant. In astogenetic neoteny, later ancestral astogenetic stages may be lost in the descendant.

Astogenetic progenesis occurs either because sexual maturity occurs earlier in astogeny or because astogenetic characters have a shorter growth period [limiting signal of Alberch *et al.* (1979)]. In the former case, astogenetic units present early in development contain sexually mature modules (Table III); such astogenetic units present in the ancestor may not have had sexually mature modules, or, if they were mature, they were not utilized in a reproductive capacity. Size

Table III. Specific Heterochronic Processes at the Astogenetic Developmental Level

Neoteny: Reduced rate or cessation of somatic development in astogentic units or astogenetic characters by:
1. Slower rate of asexual multiplication within astogenetic units
2. Longer time to develop of astogenetic units or astogenetic characters
3. Later appearance of reproductive polymorphs in astogeny

Progenesis: Early onset of sexual maturity of colony so somatic and/or size changes are stopped or severely retarded by:
1. Presence of sexually mature modules in astogenetic units early in astogeny
2. Reduction of growth period in astogenetic characters

Postdisplacement: One or more astogenetic units and/or astogenetic characters commence development at a later astogenetic stage with respect to other astogenetic units and/or astogenetic characters

Acceleration: Increased rate of somatic development in astogenetic units and/or astogenetic characters by:
1. Faster rate of asexual multiplication within astogenetic units
2. Rapid development of astogenetic units or astogenetic characters
3. Possible appearance of reproductive polymorphs early in astogeny

Hypermorphosis: Delay in onset of sexual maturity causes extension of the early stages of the colony by:
1. Prolongation of development by astogenetic units in early astogenetic stages
2. Possible increase in growth period of astogenetic characters

Predisplacement: One or more astogenetic units and/or astogenetic characters commence development at an earlier astogenetic stage with respect to other astogenetic units and/or astogenetic characters

changes between ancestor and heterochronic descendant may or may not be detectable within astogenetic stages.

Astogenetic acceleration occurs either when (1) astogenetic units of a given astogenetic stage asexually bud at a faster rate than in the ancestor; (2) the astogenetic units of a given astogenetic stage simply do not take as long to develop; or (3) astogenetic morphological characters develop at a faster rate in the descendant (Table III). In the first two instances astogenetic stages are completed earlier in the descendant (possibly resulting in earlier onset of sexual maturity) and the descendant colony passes through the adult astogenetic stage of the ancestor. Onset of sexual maturity of the colony may be accelerated in the first instance because asexual multiplication was preferred at the expense of module growth, and in the second instance because an astogenetic unit is composed of modules that reach maturity earlier than in the ancestor. In addition, reproductive polymorphs may appear earlier in astogeny. With astogenetic acceleration, new, late astogenetic stages may appear in the descendant.

During astogenetic hypermorphosis delay in onset of maturation in the colony causes prolongation in development of astogenetic units of early astogenetic stages. The descendant colony may transcend the ancestral colony by adding on astogenetic stages or by increasing the growth period of astogenetic characters relative to those in the ancestor. As in astogenetic progenesis, heterochronic size changes may not occur or be detectable.

3.4. Timing of Sexual Maturation in Heterochronic Colonies

Heterochronic processes are usually identified using the developmental stage of adulthood or attainment of sexual maturity as the criterion of standardization.

If Gould (1977) is correct in asserting that most effects of heterochrony result from a change in the timing of sexual maturation, then it is essential to understand how these effects would be distributed throughout the two developmental levels within a clonal colonial animal.

Table IV summarizes the influence of the interaction of neoteny, progenesis, acceleration, and hypermorphosis acting simultaneously at the two developmental levels within the colony on the timing of sexual maturity of the descendant colony with respect to the ancestral colony. It is constructed to show the ontogenetic effects on the sexual maturity of the descendant colony. Where neoteny and/or acceleration occur, sexual maturity is assumed to be delayed and/or accelerated, respectively. Predisplacement and postdisplacement are left out of Table IV because they have virtually no effect on onset of sexual maturity within the colony. Table IV illustrates only one specific heterochronic process acting at each level, but resulting in 16 combinations. Extrapolations of several heterochronic processes occurring at one or both levels are left to the reader.

The *enhance* situation arises when the same general heterochronic process is acting at both developmental levels (Table IV). Timing of sexual maturity of the module is affected in the same direction as the colony as a whole and so the ontogenetic level enhances the heterochronic result of either delay or acceleration of timing of sexual maturity of the colony. Enhancement of timing of sexual maturity can result in either paedomorphosis or peramorphosis at either developmental level, depending upon the two specific heterochronic processes acting.

The *impede* situation arises when the general heterochronic processes acting at both developmental levels are different in direction (Table IV). Timing of sexual maturity of the modules is affected in the opposite way as the colony as a whole and so the ontogenetic level impedes the heterochronic result of either delay or acceleration of timing of sexual maturity of the colony. When timing of sexual maturity is impeded, either paedomorphosis or peramorphosis can result at either developmental level.

The impede situation may result in modules having the capacity to reverse the polarity of timing of sexual maturity of the heterochronic colony (Table IV). The situation arises when astogenetic neoteny occurs with either ontogenetic progenesis or acceleration and when astogenetic acceleration occurs with either ontogenetic neoteny or hypermorphosis. For example, it is conceivable that sexual maturity could be delayed by hypermorphosis in modules and therefore astogenetic acceleration would not result in earlier onset of sexual maturity. The heterochronic process at the ontogenetic level would result in the reversal of the relative timing of sexual maturity of the colony.

Astogeny can also influence ontogeny with respect to timing of onset of sexual maturation; modules may not be able to reach sexual maturity, because of their placement in astogeny. For example, in some cheilostomes the earliest zooids may not be as morphologically developed as the zooids formed in later astogenetic stages (Boardman *et al.*, 1970).

4. Paleobiological and Evolutionary Implications

Heterochrony has not been generally cited as an evolutionary phenomenon in clonal colonial animals, in part because it has not been investigated in these

Table IV. Timing of Sexual Maturity When Heterochrony Acts at Both Developmental Levels

	Astogenetic			
Ontogenetic	Neoteny	Progenesis	Acceleration	Hypermorphosis
Neoteny	Modules may enhance later sexual maturity of colony	Modules may impede earlier sexual maturity of colony	Modules may impede or enhance earlier sexual maturity of colony	Modules may enhance earlier sexual maturity of colony
Progenesis	Modules impede or reverse later sexual maturity of colony	Modules may enhance earlier sexual maturity of colony	Modules may enhance earlier sexual maturity of colony	Modules may impede later sexual maturity of colony
Acceleration	Modules impede or reverse later sexual maturity of colony	Modules may enhance earlier sexual maturity of colony	Modules may enhance earlier sexual maturity of colony	Modules may impede later sexual maturity of colony
Hypermorphosis	Modules may enhance later sexual maturity of colony	Modules may impede earlier sexual maturity of colony	Modules may impede or reverse earlier sexual maturity of colony	Modules may enhance later sexual maturity of colony

organisms. The presence of heterochronic patterns, however, in many colonial groups suggests that developmental changes may be an important evolutionary process in these organisms, with significant implications.

4.1. Cladogenesis

Cladogenesis may be the result of heterochronic processes in graptolites. Elles (1922) noted stipe reduction in evolutionary lineages among graptolites in which she identified three astogenetic stages: (1) a *Didymograptus* stage, with a nema and a single stipe extending in both directions away from the nema; (2) a *Tetragraptus* stage, with a further generation of dichotomous branching at one or both ends of the initial stipe; and (3) a *Dichograptus* stage, with two generations of dichotomous branching in which dichotomous branching occurs at any one or all ends of the *Tetragraptus*-stage stipes. The evolutionary history of the graptolites suggested to Elles (1922) that stipe reduction proceeded from the *Dichograptus* to the *Tetragraptus* to the *Didymograptus* stage. Although the *Didymograptus–Tetratgraptus–Dichograptus* series are grades, not clades, stipe reduction series exist within the plexus [see Cooper and Fortey (1982) for a recent discussion]. Therefore, in graptolite evolution, heterochronic loss of astogenetic stages resulted in astogenetic paedomorphosis and thus heterochronic processes may have been responsible for cladogenesis between major groups of graptolite taxa.

In the bryozoans, Larwood and Taylor (1979) argued that the free-walled stenolaemates (Trepostomata, Cryptostomata, and Cystoporata) were probably derived from fixed-walled stenolaemates (Cyclostomata) through a relative retardation in the growth of interzooecial walls through phylogeny. They argued that because ontogenetic paedomorphosis resulted from the lengthening of interzooecial walls throughout the life of the colony, the stenolaemates were able to develop larger colonies with more varied growth forms, producing major bryozoan clades. [Anstey (1987), however, interpreted free vs. fixed interzooecial walls as astogenetic features which varied according to placement of zooid through astogeny. Therefore, an ancestral distal feature occurs in descendant forms proximally, resulting in *astogenetic peramorphosis*.]

I have suggested that the evolution of tabulate corals involved changes in developmental timing of a primary zone of astogenetic change that may have been brought about by ontogenetic and astogenetic heterochronic processes (Pandolfi, 1984a). Although the early phylogeny of Tabulata is poorly understood, cladogenesis involving the Lichenariida and the Auloporida may have resulted from heterochronic processes. In addition, the evolution of coenenchymate taxa from *Nyctopora goldfussi* may have resulted from colony-wide retardation of ontogenetic development of juveniles (Pandolfi, 1987; Scrutton, 1984).

4.2. Heterochrony and Colony Integration

4.2.1. Predicted Relationships

Colony integration, the interdependence of modules comprising a colonial animal, has been reviewed by Coates and Oliver (1973; see also Coates and Jack-

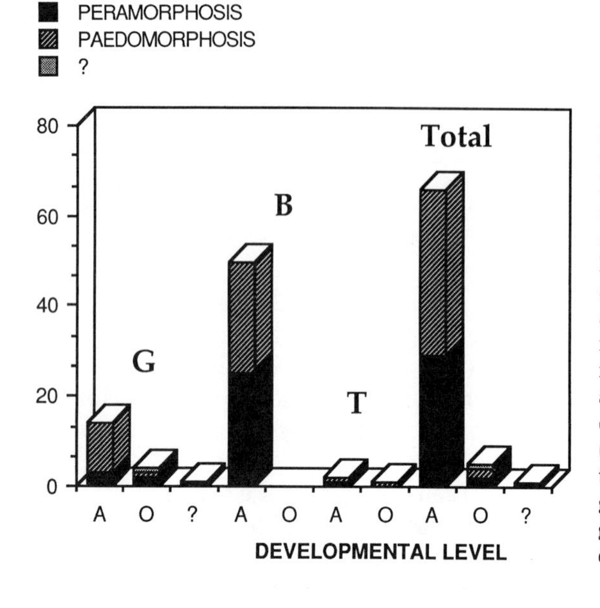

Figure 4. Frequency of heterochrony in colonial marine animals throughout the Phanerozoic era. G, Graptolithina; B, Bryozoa; and T, Tabulata. A, Astogenetic heterochrony; O, ontogenetic heterochrony. Heterochrony is common in bryozoans and may be common in graptolite lineages; in both groups it has led to the formation of higher taxa. Heterochrony is not known from colonial rugose (or scleractinian) corals. Only one example of heterochrony is known from tabulate corals (Pandolfi, 1984a). Note the high proportion of astogenetic as opposed to ontogenetic heterochrony in bryozoans and graptolites, testifying to the individuality of the colony.

son, 1985). Degrees of skeletal and soft-tissue fusion, module communication, and amount of tissue and skeleton between modules are all different aspects of colony integration. Colonies with high integration have modules whose functions are communal, and a tight developmental link might be expected to exist between ontogeny and astogeny. Poorly integrated colonies have modules that are almost or completely individualized with independent functions and, in some taxa, no soft part connections in adult stages. One might expect astogeny and ontogeny to be decoupled in poorly integrated colonies.

The degree to which ontogeny of modules is linked to colony astogeny might control heterochronic changes, including timing of sexual maturation of the colony. Colonies with tight ontogenetic–astogenetic linkage might be expected to undergo astogenetic heterochrony regardless of the ontogenetic stages of the modules or ontogenetic heterochrony with large effects on the development of the colony. Where astogeny and ontogeny are completely decoupled, heterochrony might be expected to result in astogenetic changes with little effect on modules, or in ontogenetic changes during any stage in astogeny.

4.2.2. Examples from the Fossil Record

Graptolites are highly integrated clonal colonies. The morphology of the module is tightly controlled by its position in astogeny. In fact, developmental organization in graptolites is very precise, as a newly formed stipe will bear thecae with the exact morphology of the thecae produced on the original stipe at the same time (Urbanek, 1960). Mitchell (1986) suggested that low levels of intraspecific morphological variation exhibited by graptoloid rhabdosomes in comparison with other colonial organisms was indicative of their high integration.

Figure 4 is a histogram of frequency of heterochrony reported in the literature in three groups of clonal colonial animals studied, both separately and combined.

While the occurrences of heterochrony thus far reported may not reflect the true distribution of heterochrony at the two developmental levels, Fig. 4 shows that the highly integrated graptolite colony displays an astogenetic to ontogenetic heterochrony ratio of greater than 3:1. Ontogenetic heterochrony occurs in only four examples. In two of these, small developmental changes in the ontogeny of a few modules have resulted in large morphological discontinuities between hypothesized ancestor and inferred descendant (*Dicellograptus–Leptograptus* and *Nemograptus subtilis–N. gracilis*).

In the *Nemagraptus subtilis* to *N. gracilis* lineage, Finney (1986) suggested recapitulation (peramorphosis) in the ontogeny of thecae on the main stipe. The rhabdosomes of the two species are very different because in *N. gracilis* lateral branches are developed, whereas in *N. subtilis* they are not. Finney interpreted the evolution of *N. gracilis* from *N. subtilis* as involving the addition of a new stage to the ontogeny of thecae on the main stipe, enabling them to give rise to a second theca and a lateral branch. Each individual theca undergoes the same heterochronic change, and thus developmental changes at the ontogenetic level result in major morphological changes in the colony. Finney (1986) attributed the small developmental changes with large morphological effects to relatively small genetic changes.

Bryozoa and corals display varying degrees of colony integration. Bryozoans displaying high levels of colony integration tend to share the biological functions among the modules of the colony, and consequently, the zooids are often polymorphic. Lidgard (1986) investigated trends in modes of growth in bryozoan zooids and colonies and found that, in contrast to intrazooidal budding, where asexual replication is a discontinuous process where each zooid essentially stops growing prior to the initiation of the next bud, zooidal budding, in which buds expand continuously, was capable of decoupling the outward growth of the colony as a whole from the ontogeny of individual zooids.

Coates and Jackson (1985) traced the history of the degree of colony integration throughout the Phanerozoic for corals and some bryozoans. They noted that cheilostomes showed a more uniform, progressive attainment of higher levels of integration than the corals, perhaps due to the fact that long-term trends toward higher integration in corals have been repeatedly interrupted or reversed by changes in the growth and development of reefs. Dimorphic corallites have been found in only a few genera of tabulate corals (Oliver, 1975), rarely in scleractinians [e.g., *Acropora* (Rosen, 1986)], and never in rugose corals. Rugose corals have a characteristically low level of colony integration, whereas tabulate and scleractinian corals throughout their history have undergone periods of both low and high colony integration (Coates and Jackson, 1985).

On the basis of colony integration, bryozoans, where colony integration may be very high, and tabulate and scleractinian corals should undergo astogenetic heterochrony more often than ontogenetic heterochrony; but ontogenetic heterochrony should also occur. In addition, ontogenetic heterochrony should be dominant in rugose corals. Of the 50 reported examples of heterochrony reviewed here in Bryozoa, all involve the astogenetic level of development (Fig. 4). Of course, the reported occurrences may be skewed toward astogenetic heterochrony because workers have failed to find ontogenetic heterochrony in their taxa. Anstey (1987), however, found no cases in bryozoan heterochrony where ontogenetic changes were independent of astogeny, but found a large number of heterochronic examples where zooid-level changes were involved. Changes in development at the ontogenetic level seem to be regulated by, or at least a manifestation of, selection for changes in development at the colony-wide level. In the tabulate corals, only two lineages have been interpreted to arise as a result of heterochrony. In one example (*Lichenaria simplex–Eofletcheria utahia*) heterochrony occurred at both developmental levels, whereas in the other (*Nyctopora goldfussi–Calapoecia*), astogenetic heterochrony occurred as colony-wide retardation in the development of juvenile modules. There are no reported cases of heterochrony in scleractinian or rugose corals.

Clonal colonial animals might also be expected to display a gradient involving astogenetic to ontogenetic heterochrony as colony integration decreases as a result of growth form. Coates and Jackson (1985) noted that tabulate and scleractinian corals tend to show periods of high integration during times of reef development, and periods of low integration when reefs were poorly developed or absent. Although there is little information on heterochrony in fossil corals, periods of reef growth should prove to be times where astogenetic heterochrony is dominant, whereas during periods of low reef growth ontogenetic heterochrony should be dominant.

4.3. Colonies As Individuals

Figure 4 shows the frequency of heterochrony reported in clonal colonial animals. Paedomorphosis occurs at about the same frequency as peramorphosis, indicating no preferential heterochronic result in colonial animals (see McNamara, this volume, for similar results). Astogenetic heterochrony occurs in 66 of the 72 reports of heterochrony. I subjected these data to a χ^2 goodness-of-fit test with the extrinsic hypothesis that heterochrony is equally likely to occur at the ontogenetic vs. the astogenetic developmental level. However, the χ^2 test shows that heterochronic processes appear to operate at the astogenetic as opposed to the ontogenetic developmental level ($\chi^2 = 55.889$, d.f. $= 2$, $p < 0.005$). This result lends strong support to the treatment of clonal colonial animals as individuals and not as a grouping of autonomous subunits. Studies of changes in development of modules within colonies, however, must still be undertaken in order to understand their influence upon astogenetic heterochrony.

If colonies are not treated as individuals in phylogenetic analyses, heterochrony may be incorrectly interpreted. An example involves the graptolites. Elles (1922, 1923) noted two heterochronic trends in the monograptids. In *progressive* forms, new thecal elaborations are first developed in the proximal portion of the rhabdosome. Progressive evolution occurs when these thecal types spread distally through phylogeny. The morphological result of progressive evolution is paedomorphosis. In *regressive* forms, the proximal portion of the colony is the only place where former thecal elaborations are retained. Regressive evolution occurs when thecae become simplified distally and spread this alteration proximally. The morphological result of regressive evolution is peramorphosis.

Gould (1977) interpreted regressive and progressive evolution in the graptolites incorrectly when he equated progressive evolution with recapitulation (peramorphosis) and regressive evolution with paedomorphosis (see also Anstey, 1987). He worked under the tenet that clonal colonial animals are not to be regarded as individuals. For the progressive sequence, Gould reasoned that since proximal thecae are the oldest, features of their late ontogeny are at first not attained by the younger distal thecae. Then, "by a progressive acceleration in development, even the youngest thecae of descendant species go through the entire potential ontogeny." Gould confounded astogeny with ontogeny when he stated that "youngest thecae are the first to lose adult characters and to retain permanently their juvenile form. . . . [This] tendency spreads to older thecae until all are paedomorphic." In the latter case, youngest refers to an astogenetic stage, which in reality is a *later* stage. Gould (1977) acknowledged this difference in a footnote (pp. 433–434), but denied the existence of colonies as entities in and of themselves, i.e., that a colony is also an individual that goes through developmental stages composed of modules or groups of modules.

In astogenetic development, the oldest modules are the earliest formed modules and represent early astogenetic stages. In addition, the very presence of an intracolony trend involving progression of thecal morphotype suggest an astogenetic overprint that is independent of ontogeny. Mitchell (1986) recently gave developmental evidence for the uniqueness of astogenetic characters in phylogenetic analyses and Finney (1986) treated graptolite colonies as individuals. Cooper and Fortey (1982) emphasized the importance of astogenetic characters, particularly those of the proximal end of the rhabdosome, to graptolite phylogeny. Boardman *et al.* (1970), Cumings (1904, 1910), and Taylor and Furness (1978) treated bryozoan colonies as individuals. In a recent review in fragmentation of coral colonies as a reproductive mechanism, Highsmith (1982) gave the following reasons for treating coral colonies as individuals: (1) they recruit as individuals; (2) they do not reproduce sexually until the colony has reached a certain size or age; (3) they often have characteristic growth forms for each species and in some cases even repair damage so as to restore the original colony shape [as graptolites indeed do; see Urbanek (1973)], suggesting integration of activities; (4) they translocate materials within the colony; and (5) the ecologically relevant mortality rate is that for entire colonies.

5. Summary

Clonal colonial animals have two distinct developmental levels: the module and the colony. Astogenetic stages in clonal colonial animals can be compared and utilized in development analyses just as ontogenetic stages are utilized in aclonal animals. The astogenetic level of development in a colony imparts an individuality to the colony such that the colony may be viewed as an individual. Heterochronic processes may operate at the ontogenetic or astogenetic level or both, and any specific and general heterochronic process or combination of processes or morphological results may occur. In the examples of heterochrony thus far reported, astogenetic heterochrony occurs much more often than ontogenetic heterochrony in the fossil record of clonal colonial animals. The apparent predominance of astogenetic vs. ontogenetic heterochrony may reflect the individuality of the colony as a whole.

The interaction of two heterochronic processes at the two developmental levels may provide insight into the developmental relationship that exists at the two levels within a colony. For example, astogenetic heterochrony may occur as a result of colony-wide modifications in the development of ontogenetic characters. This interaction may also result in the alteration of the colony's timing of sexual maturity. Colony integration may dictate the developmental level at which heterochrony affects a clonal animal or the degree of morphological effects from those processes. Small developmental changes in ontogeny or astogeny may bring about large morphological changes.

Heterochrony in clonal animals may have important implications for the macroevolution of these organisms. Examples of heterochrony from the fossil record of Tabulata, Graptolithina, and Bryozoa show that evolutionary changes in development can result in cladogenesis in clonal colonial animals. Analysis of the relation connecting degree of interdependence of modules of the colony, colony integration, and the distribution of evolutionary changes in development at the two developmental levels within a colony may provide evidence for the degree of colony-wide vs. module control over development in clonal colonial animals. Large morphological differences may occur when highly integrated colonies display ontogenetic heterochrony. If there exists a relationship between colony integration and level of developmental control, then it is predicted that highly integrated colonies display astogenetic heterochrony or ontogenetic heterochrony with large morphological affects, and that poorly integrated colonies display ontogenetic heterochrony. In addition, because times of reef-building correspond to times of high integration in colonial marine animals (Coates and Jackson, 1985), it is predicted that astogenetic heterochrony occurs more often during times of reef-building than when reefs were absent. While the fossil record of heterochronic events is poorly understood, future work on heterochrony in colonial animals should provide a test for such predictions.

ACKNOWLEDGMENTS. I am grateful to J. R. Beerbower, Richard Cowen, R. K. Grosberg, P. W. Signor, and J. E. Sorauf for discussions of the ideas presented in this paper. S. J. Carlson, Richard Cowen, R. K. Grosberg, and P. W. Signor critically read previous drafts of the paper. Patricia Almeida and Mark Lambert photographed Fig. 2. This research was supported in part by grants from the Geology Department, University of California, Davis; the Graduate School, University of California, Davis; the Geological Society of America; Sigma Xi, the Scientific Research Society; and the Theodore Roosevelt Memorial Fund, American Museum of Natural History.

References

Alberch, P., Gould, S. J., Oster, G., and Wake, D. B., 1979, Size and shape in ontogeny and phylogeny, *Paleobiology* **5**:296–317.

Anstey, R. L., 1981, Zooid orientation structures and water flow patterns in Paleozoic bryozoan colonies, *Lethaia* **14**:287–302.

Anstey, R. L., 1987, Astogeny and phylogeny: Evolutionary heterochrony in Paleozoic bryozoans, *Paleobiology* **13**:20–43.

Boardman, R. S., Cheetham, A. H., and Cook, P. L., 1970, Intracolony variation and the genus concept in Bryozoa, in: *Proceedings North American Paleontological Convention*, pp. 294–320.

Buddemeier, R. W., Maragos, J. E., and Knutson, D. W., 1974, Radiographic studies of reef coral exoskeletons: Rates and patterns of corals growth, *J. Exp. Mar. Biol. Ecol.* **14**:179–200.

Bulman, O. M. B., 1963, The evolution and classification of the Graptoloidea, *Q. J. Geol. Soc. Lond.* **119**:401–418.

Coates, A. G., and Jackson, J. B. C., 1985, Morphological themes in the evolution of clonal and aclonal marine invertebrates, in: *Population Biology and Evolution of Clonal Organisms* (J. B. C. Jackson, L. W. Buss, and R. E. Cook, eds.), pp. 67–106, Yale University Press, New Haven.

Coates, A. G., and Oliver, W. A., Jr., 1973, Coloniality in zoantharian corals, in: *Animal Colonies: Development and Structure Through Time* (R. S. Boardman, A. H. Cheetham, and W. A. Oliver, Jr., eds.), pp. 3–27, Dowden, Hutchison & Ross, Stroudsburg, Pennsylvania.

Cooper, R. A., and Fortey, R. A., 1982, The Ordovician graptolites of Spitsbergen, *Bull. Br. Mus. Nat. Hist. (Geol.)* **36**(3):157–302.

Cowen, R., and Rider, J., 1973, Functional analysis of fenestellid bryozoan colonies, *Lethaia* **5**:145–164.

Cumings, E. R., 1904, The development of some Paleozoic Bryozoa, *Am. J. Sci.* **17**:49–78.

Cumings, E. R., 1910, Paleontology and the recapitulation theory, *Proc. Indiana Acad. Sci.* **1909**:305–340.

Davies, K. A., 1929, Notes on the graptolite faunas of the Upper Ordovician and Lower Silurian, *Geol. Mag.* **66**:1–27.

de Beer, G. R., 1930, *Embryology and Evolution*, Clarendon, Oxford.

Dzik, J., 1975, The origin and early phylogeny of the cheilostomatous Bryozoa, *Acta Palaeontol. Pol.* **20**:395–423.

Elles, G. L., 1922, The graptolite fauna of the British Isles, *Proc. Geol. Assoc.* **33**:168–200.

Elles, G. L., 1923, Evolutional palaeontology in relation to the Lower Palaeozoic rocks, *Rep. Br. Assoc. Adv. Sci.* **91**:83–107.

Finney, S., 1978, The affinities of *Osograptus, Glossograptus, Cryptograptus, Corynoides,* and allied graptolites, *Acta Palaeontol. Pol.* **23**:481–495.

Finney, S. C., 1986, Heterochrony, punctuated equilibrium, and graptolite zonal boundaries, in: *Palaeoecology and Biostratigraphy of Graptolites* (C. P. Hughes and R. B. Rickards, eds.), pp. 103–113, Geological Society Special Publication No. 20.

Gould, S. J., 1977, *Ontogeny and Phylogeny*, Harvard University Press, Cambridge.

Harvell, C. D., and Grosberg, R. K., in press, The timing of sexual maturity in clonal organisms, *Am. Nat.* (in press).

Haven, N. D., 1971, Temporal patterns of sexual and asexual reproduction in the colonial ascidian *Metandrocarpa taylori* Huntsman, *Biol. Bull.* **140**:400–415.

Hickey, D. R., 1987, Skeletal structure, development and elemental composition of the Trepostome bryozoan *Peronopora Palaeontology* **30**(4):691–716.

Hickey, D. R., 1988, The role of astogeny in the evolutionary origins and systematics of Paleozoic Bryozoa: An example from the Trepostome bryozoan *Peronopora, J. Paleontol.* **62**(2):180–203.

Highsmith, R. C., 1982, Reproduction by fragmentation in corals, *Mar. Ecol. Progr. Ser.* **7**:207–226.

Jackson, J. B. C., 1979, Morphological strategies of sessile organisms, in: *Biology and Systematics of Colonial Organisms* (G. Larwood and B. R. Rosen, eds.), pp. 499–555, Systematics Association Special Volume 11.

Jackson, J. B. C., and Hughes, T. P., 1985, Adaptive strategies of coral-reef invertebrates, *Am. Sci.* **73**:265–274.

Jebram, D., 1973, The importance of different growth directions in the Phylactolaemata and Gymnolaemata for reconstructing the phylogeny of the Bryozoa, in: *Living and Fossil Bryozoa* (G. Larwood, ed.), pp. 565–576, Academic Press, London.

Lang, W. D., 1904, The Jurassic forms of the 'genera' *Stomatopora* and *Proboscina, Geol. Mag.* **1**:315–322.

Larwood, G. P., and Taylor, P. D., 1979, Early structural and ecological diversification in the bryozoa, in: *The Origin of Major Invertebrate Groups:* (M. R. House, ed.), pp. 209–234, Systematics Association Special Volume 12.

Lidgard, S., 1986, Ontogeny in animal colonies: A persistent trend in the bryozoan fossil record, *Science* **232**:230–232.

McNamara, K. J., 1986, A guide to the nomenclature of heterochrony, *J. Paleontol.* **60**(1):4–13.

Mitchell, C. E., 1986, Morphometric studies of *Climacograptus* (Hall) and the phylogenetic significance of astogeny, in: *Palaeoecology and Biostratigraphy of Graptolites* (C. P. Hughes and R. B. Rickards, eds.) pp. 119–129, Geological Society Special Publication No. 20.

Oliver, W. A., Jr., 1975, Dimorphism in Two New Genera of Devonian Tabulate Corals, U. S. Geological Survey Professional Paper 743-D.

Packham, G. H., 1962, Some diplograptids from the British Lower Silurian, *Palaeontology* **5**:498–526.

Pandolfi, J. M., 1984a, Evidence of heterochrony in early tabulate corals, in: *Geological Society of America, Abstracts with Programs,* Vol. 16, No. 6, p. 617.

Pandolfi, J. M., 1984b, Environmental influence on growth form in some massive tabulate corals from the Hamilton Group (Middle Devonian) of New York State, *Palaeontogr. Am.* **54**:538–542.

Pandolfi, J. M., 1987, Paleobiological studies of colonial marine animals, Ph.D. dissertation, University of California, Davis.

Prezbindowski, D. R., and Anstey, R. L., 1978, A Fourier-numerical study of a bryozoan fauna from the Threeforks Formation (Late Devonian) of Montana, *J. Paleontol.* **52**(2):353–369.

Rickards, R. B., 1977, Patterns of evolution in the graptolites, in: *Patterns of Evolution, As Illustrated by the Fossil Record* (A. Hallam, ed.), pp. 333–358, Elsevier, New York.

Rickards, R. B., Hutt, J. E., and Berry, W. B. N., 1976, The evolution of the Silurian and Devonian graptoloids, *Bull. Br. Mus. Nat. Hist.* **28**(1):1–120.

Rigby, J. K., and Hintze, L. F., 1977, Early Middle Ordovician corals from western Utah, *Utah Geol.* **4**(2):105–111.

Rosen, B. R., 1986, Modular growth and form of corals: A matter of metamers?, *Phil. Trans. R. Soc. Lond. B* **113**:115–142.

Ryland, J. S., 1981, Colonies, growth and reproduction, in: *Recent and Fossil Bryozoa* (G. P. Larwood and C. Nielsen, eds.), pp. 221–226.

Schopf, T. J. M., 1977, Patterns and themes of evolution among the Bryozoa, in: *Patterns of Evolution, As Illustrated by the Fossil Record* (A. Hallam, ed.), pp. 159–207, Elsevier, New York.

Scrutton, C. T., 1984, Origin and early evolution of tabulate corals, *Palaeontogr. Am.* **54**:110–118.

Scrutton, C. T., and Powell, J.H., 1980, Periodic development of dimetrism in some favositid corals, *Acta Palaeontol. Pol.* **25**(3–4):477–491.

Stebbing, A. R. D., 1980, Increase in gonozooid frequency as an adaptive response to stress in *Campanularia flexuosa,* in: *Developmental and Cellular Biology of Coelenterates* (P. Tardent and R. Tardent, eds.), pp. 27–32, Elsevier, New York.

Sudbury, M., 1958, Triangulate monograptids from the *Monograptus gregarius* zone of the Rheidol Gorge, *R. Soc. Lond. Philos. Trans. B* **241**:485–555.

Sugimoto, K., and Nakuachi, M., 1974, Budding, asexual reproduction, and regeneration in the colonial ascidian, *Symplegama reptans, Biol. Bull.* **147**:213–226.

Taylor, P. D., and Furness, R. W., 1978, Astogenetic and environmental variation of zooid size within colonies of Jurassic *Stomatopora* (Bryozoa, Cyclostomata), *J. Paleontol.* **52**(5):1093–1102.

Thompson, D'A. W., 1917, *On Growth and Form,* Cambridge University Press, London.

Urbanek, A., 1960, An attempt at biological interpretation of evolutionary changes in graptolite colonies, *Acta Palaeontol. Pol.* **5**(2):127–234.

Urbanek, A., 1973, Organization and evolution of Graptolite colonies, in: *Animal Colonies: Development and Function Through Time* (R. S. Boardman, A. H. Cheetham, and W. A. Oliver, Jr., eds.), pp. 441–514, Dowden, Hutchison and Ross, Stroudsburg, Pennsylvania.

Waern, B., 1948, in: Waern, B., Thorslund, P., Henningsmoen, G., and Save-Soderbergh, G., Deep boring through Ordovician and Silurian strata at Kinnekulle, Vestergotland, *Bull. Geol. Inst. Univ. Uppsala* **32**:337–474.

Wahle, C. M., 1983, The role of age, size and injury in sexual reproduction among Jamaican gorgonians, *Am. Zool.* **23**:961.

Yamaguchi, M., 1975, Growth and reproductive cycles of the marine fouling ascidians *Ciona intestinalis, Styela plicata, Botrylloides violaceous,* and *Leptoclinum mitsukurii* at Abuatsubo-Moroiso Inlet (Central Japan), *Mar. Biol.* **29**:253–259.

Chapter 9

Heterochrony in Ammonites

NEIL H. LANDMAN

1. Introduction

Ammonites are externally shelled cephalopods and range in geologic age from the Devonian to the Late Cretaceous. They comprise nine orders that may be informally grouped into the paleo-, meso-, and neoammonoidea. The paleoammonoidea consists of the Devonian–Permian goniatites, anarcestids, clymeniids, and prolecanitids. The mesoammonoidea or ceratites range from the Permian to the Triassic. The neoammonoidea consists of the phylloceratids, lytoceratids, ammonitids, and ancyloceratids and ranges from the Jurassic to the Cretaceous.

All these ammonites preserve in their shells a record of their growth and exhibit a number of complex morphological characters, for example, sutures, whose development may be followed through ontogeny. At the same time, the fossil record of ammonites is superb. Therefore, it is not surprising that ammonites and their sister group, the nautiloids, have played a major role in the study of heterochrony, especially recapitulation. At the turn of the century, they were widely cited as prime examples of Haeckel's biogenetic law. Although the results of many of these early studies have been invalidated (Donovan, 1973; Kennedy 1977), much of present-day ammonite taxonomy stems from a strong belief that ontogeny recapitulates phylogeny (Wiedmann and Kullmann, 1980). In this chapter, I review the early studies of recapitulation by examining the ideas of several late 19th–early 20th century workers, after which I describe the ontogeny of ammonites as we understand it today. I spend extra time on the sutural patterns

NEIL H. LANDMAN • Department of Invertebrates, American Museum of Natural History, New York, New York 10024.

because they figure so prominently in most of these recapitulatory schemes. Finally, I examine several examples of recapitulation and other kinds of heterochrony to determine their causes and the role they played in the evolution of ammonites.

2. Historical Background

One of the most enthusiastic students of recapitulation in ammonites and nautiloids was Alpheus Hyatt. In a series of works (Hyatt, 1866, 1883, 1889, 1894) he explained his ideas on the relationship betwen ontogeny and phylogeny. He believed that the ontogeny of a descendant recapitulated the sequence of ancestral adults and therefore provided a literal reading of phylogeny. He postulated a developmental law of acceleration according to which ancestral adult stages appeared earlier in the ontogeny of descendants:

> All modifications and variations in progressive series tend to appear first in the adolescent or adult stages of growth and then to be inherited in successive descendants at earlier and earlier stages according to the law of acceleration, until they either become embryonic, or crowded out of the organization, and replaced in the development by characteristics of later origin. (Hyatt, 1889, p. 1X, item 11)

As a neo-Lamarkian, Hyatt believed that modifications in ontogeny occurred by the terminal addition of acquired characters. The example commonly cited to illustrate his ideas is the development of the "impressed zone" in the shells of nautiloids (Hyatt, 1894). This zone refers to the furrow on the dorsum (inside) of the whorl where it contacts the venter (outside) of the preceding whorl. According to Hyatt, the early whorls of primitive nautiloids were loosely coiled and became more tightly coiled in later ontogeny. Hyatt speculated that the impressed zone arose in later ontogeny due to the pressure of contact between adjoining whorls. This character was inherited and subsequently accelerated so that it first appears in the ontogeny of descendants at a stage when the whorls are not yet touching.

Hyatt's novel contribution to recapitalutory thinking was his so-called old-age theory in which he postulated that an evolutionary sequence of species itself had a life cycle that progressed from youth through maturity to old age. He believed that the life cycle of the individual and the evolutionary history of the group to which it belonged obeyed the same fundamental laws. Hyatt formulated his ideas on the basis of a general conception of cephalopod evolution as a progression from straight-shelled through arcuate to tightly coiled and back to straight-shelled forms. For example, he inferred that early in the evolution of the group, its members exhibited smooth, straight shells, characteristic of early ontogeny. Through acceleration, terminal addition, and condensation, descendant species became progressively more tightly coiled, ornamented, and complex. Late in the evolution of the group, its members regressed to uncoiled, smooth shells suggestive of a second childhood. This occurred by the rapid acceleration of senile features into the earlier ontogeny of descendants and the loss of intermediate features between the morphologically simplified juvenile and the gerontic adult. This retrogressive evolution or racial senescense produced pathological, degraded, and distorted forms and was ultimately followed by extinction.

The theory of recapitulation was accepted by many ammonite workers, who later modified it in the light of their own data. Smith (1898, 1914) reconstructed the phylogeny of a number of groups on the basis of recapitulation, although he assimilated Cope's ideas of retardation as well as Hyatt's law of acceleration. Buckman (1887–1907, 1909, 1918), like Hyatt, inferred a phylogenetic cycle of progressive and retrogressive change, especially in the complexity of ornament. However, he claimed that recapitulation was imperfect because some ancestral characters were lost in the ontogeny of descendants. Trueman (1919, 1922), following Buckman's lead, also inferred a loss of ancestral characters in his study of the evolution of lower Jurassic liparoceratids. Within this group, Trueman recognized three morphotypes: (1) coiled forms in which the whorls barely touch (capricorns), (2) coiled forms in which the early whorls barely touch and the later whorls completely overlap (varicostate forms), and (3) coiled forms in which the whorls completely overlap throughout ontogeny (sphaerocones). According to Trueman, capricorns evolved into varicostate forms by terminal addition, acceleration, and condensation. Varicostate forms evolved, in turn, to sphaerocones by the suppression of the early, loosely coiled stage. [Later studies of these ammonites based on more extensive data have, in general, suggested the opposite line of descent (Spath, 1938; Dommergues, 1982, 1986, 1987) (but see Calloman, 1963, 1980, for another interpretation): varicostate forms are presumed to have descended from sphaerocones by the development of looser coiling in their early whorls; capricorns, in turn, are presumed to have descended from varicostate forms via neoteny.]

Outside North America and Britain, Karpinski (1889, 1890) documented recapitulation in an evolutionary sequence of prolecanitids. Slightly later, Pavlov (1901) demonstrated an example of "reverse recapitulation" (paedomorphosis) in a sequence of Jurassic ammonites. In Germany, Schindewolf (1929, 1942, 1950) incorporated recapitulation and neoteny in a theory of evolutionary history called typostrophism. He believed that the history of a group parallels the life cycle of an individual. This cycle consists of an explosive origin, an intermediate phase marked by increasing specialization, and a final decline. During the initial phase, new characters were introduced into early ontogeny and later appear in descendants at an adult stage of development (neoteny). In contrast, during the next phase, new characters were added at the end of ontogeny and were subsequently accelerated into earlier stages in descendants (recapitulation). Abnormalities and degenerations occur in the last phase before extinction.

Modern workers are curiously divided on the importance of recapitulation in ammonite evolution. For example, Wiedmann and Kullmann (1980, p. 215) believe that sutural recapitulation provides the basis for a higher classification of ammonites: "Many examples show that suture phylogeny is related to suture ontogeny, for the genetic principle of additive typogenesis is mostly involved and the law of recapitulation therefore applies." Similarly, Ruzhentsev (1962) believes that recapitulation is the most common pattern in ammonite evolution and occurs via rapid acceleration concentrating and even deleting stages. Retardation may also occur in which adult descendants resemble juvenile ancestors. However, these are minor exceptions to the general rule. On the other hand, most British workers are suspicious of recapitulation and its usefulness in reconstructing phylogeny (Spath, 1924; Arkell, 1957; Donovan, 1973; Kennedy, 1977; Kennedy and

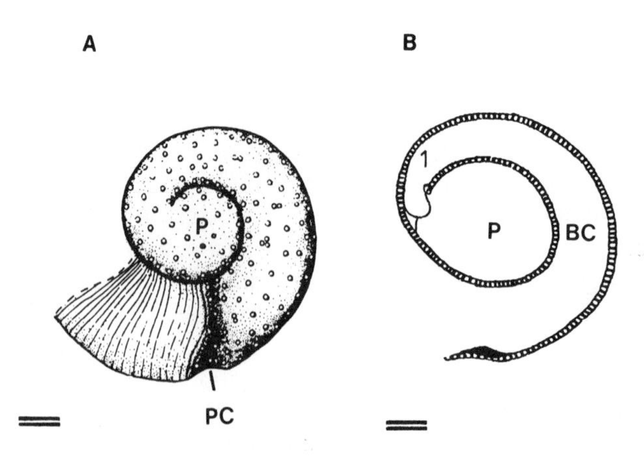

Figure 1. (A) The embryonic shell of ammonites, called the ammonitella, consists of the protoconch (P) followed by approximately one planispiral whorl ending at the primary constriction (PC). (B) Median cross section of an ammonitella reveals the protoconch (P), body chamber (BC), and at least one septum, the proseptum (1). Scale bars = 0.1 mm.

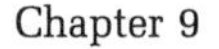

Wright, 1985). According to Kennedy (1977), "the names of Hyatt, Buckman, Perrin Smith, Trueman and others delineate a school which elaborated the theory [of recapitulation] with disastrous consequences. Perhaps the greatest criticism . . . was lack of support in terms of stratigraphic control." As a result, most British workers eschew discussions of recapitulation (Kennedy and Wright, 1985).

3. Ammonite Ontogeny

Hyatt, Smith, and others subdivided ammonite ontogeny into numerous stages, although the criteria used to define these stages were not always the same. These criteria were largely based on the belief that each developmental stage recapitulated a stage in evolutionary history. In addition, the boundary between most stages was transitional. Modern workers are more inclined to describe the ontogenetic trajectory in terms of whorl number or shell diameter and avoid assigning names. They tend to view ontogenetic development as the expression of the life history of the individual and not merely as a manifestation of phylogeny.

3.1. Embryonic Development

The early whorls of all ammonites, including uncoiled species, consist of a spheroidal to ellipsoidal initial chamber (the protoconch) followed by approximately one planispiral whorl terminating in a depression (the primary constriction; Fig. 1A). This applies to all ammonites, from their first appearance in the Devonian to their extinction in the Late Cretaceous (Kulicki, 1979; Bandel, 1986). These early whorls are known as the ammonitella and range from 0.5 to 1.6 mm in diameter (House, 1965; Bogoslovsky, 1976).

The ammonitella is marked by a number of characters that distinguish it from the later whorls. In many Mesozoic ammonites, the ammonitella bears a distinctive microornamentation consisting of uniformly distributed tubercles several microns in diameter (Bandel *et al.*, 1982). In other ammonites, for example, the Devonian genus *Tornoceras*, the ammonitella exhibits another kind of mi-

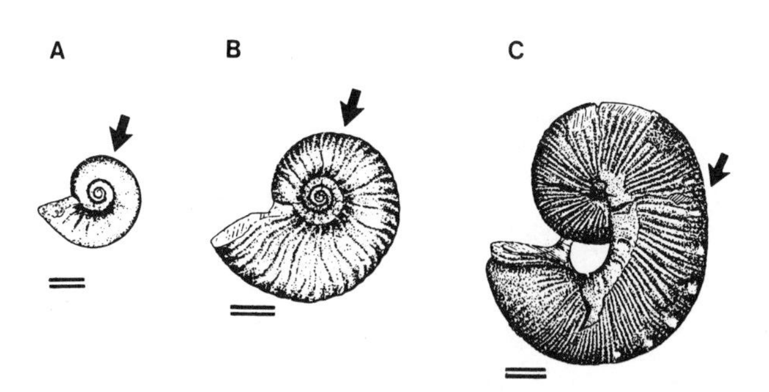

Figure 2. Ontogenetic development within two species of scaphitid ammonites. (A) Ornament is initially absent, but (B) gradually appears after several whorls in *Discoscaphites* sp. Scale bar = (A) 2, (B) 5 mm. (C) Adult shell of *Scaphites cobbani* with a hooklike body chamber. Scale bar = 1 cm. The arrows indicate the base of the body chamber. After Birkelund (1965).

croornamentation (House, 1965). All such ornamentation ends abruptly at the primary constriction and growth lines immediately appear thereafter. This boundary also coincides with a change in the microstructure of the shell wall. Before the constriction, the shell wall consists of several layers of prismatic aragonite. However, at the constriction, a thick pad of nacre develops (Kulicki, 1979; Birkelund, 1980). In addition, the shape of the shell may change at this point, as, for example, in the uncoiled ammonite *Baculites*, which develops into a long shaft (Bandel *et al.*, 1982, Fig. 1C).

On the basis of these morphological observations, the ammonitella has been interpreted as the ammonite embryonic shell (Druschits *et al.*, 1977; Kulicki, 1979; Birkelund, 1980; Bandel, 1982; Landman, 1987a,b). It probably formed by direct development; the uniform surface of the ammonitella provides no evidence for an intervening larval phase. The absence of a larval phase is in accordance with all modern cephalopods whose development is known (Arnold and Carlson, 1986; Bandel and Boletzky, 1979). The embryonic development of ammonites has also been analogized to that of archaeogastropods, in which an originally organic shell is rapidly mineralized to form an initial layer of uniform thickness that preserves the original ornamentation of the organic shell (Bandel, 1986).

Close inspection of preserved ammonitellas indicates that a number of internal features were present at hatching (Fig. 1B) (Landman, 1985; Bandel, 1986). These internal features included the initial portion of the siphuncle, which was attached by strands to the interior of the protoconch, and at least one septum, the prismatic proseptum (Landman, 1985; Bandel, 1986). This septum would have permitted the animal to function with at least one buoyancy chamber, the protoconch. The newly hatched animal may have followed a planktonic life similar to that of the young of many modern coleoids (Boletzky, 1974). However, the diversity in ammonitella size among ammonites suggests variation in the length of such a planktonic stage.

3.2. Postembryonic Growth

Most of the morphological changes in postembryonic growth develop gradually. In planispirally coiled ammonites, the shell grows by adding whorls in a logarithmic spiral (Thompson, 1917). As a result, the shell maintains the same shape throughout ontogeny, although changes may occur (Fig. 2A). Differences

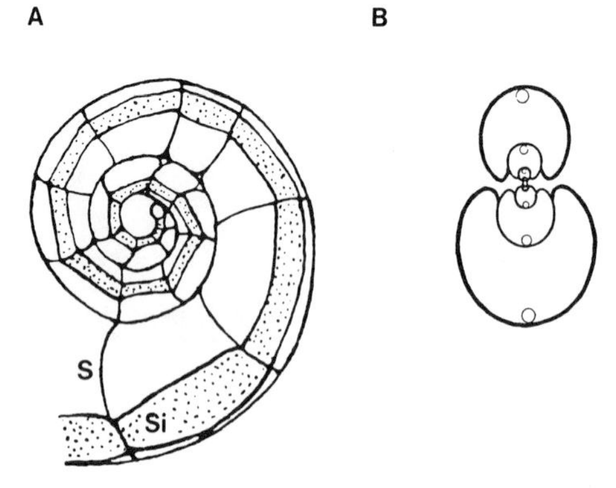

Figure 3. (A) Median and (B) dorsoventral cross sections of an ammonite, showing the septa (S) and siphuncle (Si). The siphuncle may migrate from a dorsal (inner) to a ventral (outer) position during ontogeny. The diagram of the median cross section is modified from Arkell (1957).

in shell shape among species are expressed as differences in the shape of the whorl section, the rate of whorl expansion, and the degree of whorl overlap (Raup, 1967). Species with rapidly expanding whorls and compressed whorl sections tend to exhibit more whorl overlap than those with more slowly expanding whorls and inflated whorl sections (Westermann, 1971).

Ornament is usually absent on the early whorls of the shell, but gradually develops (Fig. 2B). It consists of ridges, tubercles, projecting spines, and keels. According to Westermann (1966), shell shape and the degree of ornamentation covary within species; individuals with inflated whorls and little whorl overlap bear stronger ornament than those with compressed whorls and high whorl overlap. The shell wall itself is initially composed of a prismatic and a nacreous layer. An additional, inner prismatic layer develops after several more whorls (Birkelund, 1980). The thickness of the shell wall increases isometrically or displays a change from negative to positive allometry through ontogeny (Westermann, 1971; Tanabe, 1977).

Internally, the shell is subdivided into chambers that provide buoyancy for continued growth (Fig. 3). Removal of cameral liquid from newly formed chambers was effected by the siphuncle, the tube that runs through all of the chambers. In early ontogeny, the siphuncle may assume a median or dorsal (inner) position (Fig. 3A). Subsequently it may migrate to a ventral (outer) position, although in clymeniids it retains a dorsal position throughout ontogeny. The increase in the thickness of the siphuncular wall is negatively allometric (Westermann, 1971; Tanabe, 1977). The siphuncle passes through the septa via the septal necks, which are tubular extensions of the septa. During ontogeny the septal neck may change from backward-bending (retrochoanitic) to forward-bending (prochoanitic), although this transition may occur over several whorls (Fig. 4). (Doguzhayeva and Mutvei, 1986).

The complex folding of ammonite septa may have served many functions, including providing mechanical support against implosion and facilitating the removal of cameral liquid (Westermann, 1971; Kulicki, 1979; Hewitt and Westermann, 1986; see also Kennedy and Cobban, 1976). Septal thickness increases isometrically or displays a change from slightly negative to slightly positive allometry through ontogeny (Westermann, 1971; Tanabe, 1977). The first septum (proseptum) is prismatic; all subsequent septa are nacreous (Kulicki, 1979; Land-

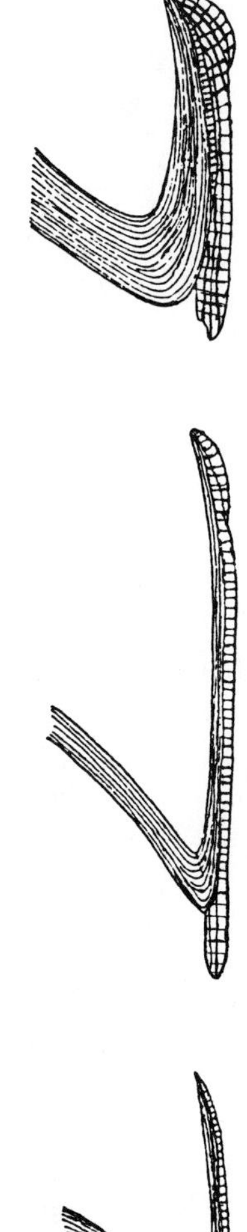

Figure 4. The change in the direction of bending of the septal neck (in cross section) from backward-bending (retrochoanitic) to forward-bending (prochoanitic) in the ontogeny of primitive ammonites. Only the dorsal halves of the septal necks are illustrated. The arrow indicates the adoral (forward) direction. After Doguzhayeva and Mutvei (1986).

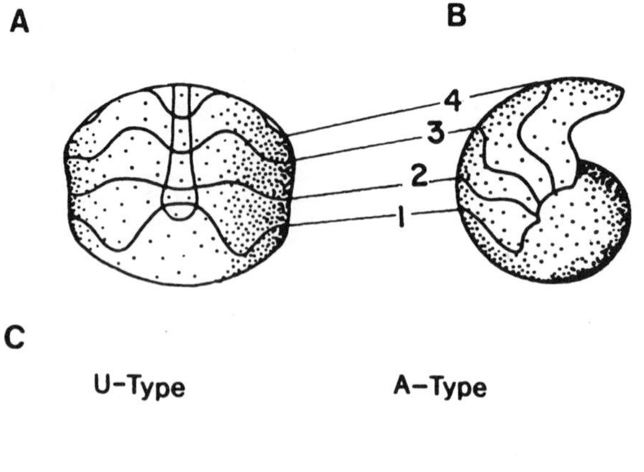

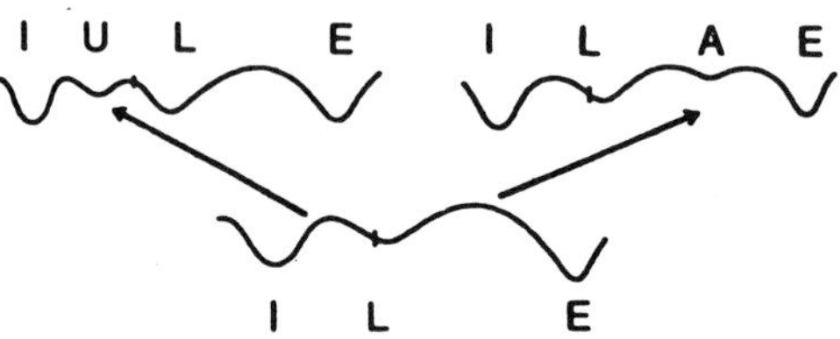

Figure 5. (A) Ventral and (B) side views of the early whorls of an ammonite showing the prosuture (1), primary suture (2), and third (3), and fourth (4) sutures. After Miller *et al.* (1957). (C) The primary suture consists of the external (E), lateral (L), and internal (I) lobes. In the ontogeny of ammonites, new lobes appear either between I and L (called umbilical lobes) or between L and E (called adventitious lobes). After Schindewolf (1954).

man and Bandel, 1985). The spacing of septa may vary through ontogeny, although similar patterns appear in taxonomically unrelated species (Bayer, 1972; Kulicki, 1974; Landman, 1987b).

The suture is the line of junction of the septum with the walls of the shell (Fig. 5A and 5B). It is only visible when the shell wall is removed. The first suture is the prosuture, corresponding to the proseptum, and forms during embryogenesis. It shares no homologous points with the second suture, called the primary suture. All subsequent sutures are based on elaborations of the primary suture and are numbered sequentially (3, 4, 5, . . .) in this chapter. Sutures consist of two kinds of elements, backwardly directed lobes and forwardly directed saddles. The three basic lobes in the primary suture consist of the external, lateral, and internal lobes (E, L, and I, respectively; Fig. 5C). New lobes form by the subdivision of these original lobes and appear either between lobes E and L (A type) or between lobes L and I (U type) (Wiedmann and Kullmann, 1980). The formation of A-type lobes occurs mainly in Paleozoic goniatites, whereas the formation of U-type lobes occurs in prolecanitids and all Mesozoic ammonites.

The suture generally becomes more complex through ontogeny due to the addition and frilling of new lobes and saddles. Sutural complexity has been defined quantitatively as the length of the suture or the ratio of sutural length to whorl circumference (Westermann, 1971; Hewitt, 1985). Sutural complexity does not increase uniformly throughout growth. In the early ontogeny of many Mesozoic ammonites, sutural complexity is constant (Hewitt, 1985). Thereafter, it increases exponentially with shell diameter. Among individuals within species, the degree of sutural complexity may covary with the shape of the whorl section and the degree of whorl overlap (Westermann, 1958, 1971).

The repetitive formation of septa in ontogeny has led to the impression that the time interval between septal formation is constant. However, during ontogeny, the ratio of the surface area of the siphuncle to the volume of the chambers decreases (Westermann, 1971; Chamberlain, 1978). Therefore, in analogy with *Nautilus*, increasingly longer periods of time were required for the removal of cameral liquid from successively larger chambers (Ward, 1985; Landman and Cochran, 1987). In other words, septal formation does not occur at equal time intervals. The time interval represented by a number of septa in late ontogeny is greater than that of a comparable number of septa in early ontogeny.

In general, except for the initial postembryonic and mature sutures, sutures are specified by reference to the shell diameter or whorl height at which they occur. In studies of heterochrony, we commonly compare the sutural ontogeny of the most advanced descendant with the adult sutures of its presumed ancestors. Comparisons of individuals at sutures intermediate between early and late ontogeny are more difficult because of the uncertainty in determining the respective ages or developmental stages that these sutures represent.

In summary, most morphological changes during postembryonic growth develop gradually. Commonly, a number of these changes coincide in their time of appearance. These changes may involve the expansion rate of the spiral, degree of whorl overlap, ornamentation, position of the siphuncle, complexity of the suture, thickness of the siphuncular wall, and thickness and spacing of the septa (Currie, 1942, 1943, 1944; Westermann, 1954; Obata, 1965; Tanabe, 1977; Landman, 1987b). The cooccurrence of such morphological changes may indicate the existence of a transitional stage prior to maturity. For example, in Upper Cretaceous scaphites, a number of these morphological changes cooccur at approximately 4–5 mm in shell diameter, suggesting a transition to another mode of life (Landman, 1987b).

3.3. Maturity

Dramatic changes in the morphology of the shell occur at the onset of maturity (Fig. 2C). One of the most conspicuous modifications involves the thickening and closer spacing (approximation) of the last few septa. According to Makowski (1962), this change is more strongly expressed in males than females. For example, in females, septal approximation may gradually develop over 20 septa, whereas in males, it may appear as an abrupt reduction in the spacing of the last few septa. This closer spacing may reflect both a decline in the rate of apertural growth as well as modifications in the thickness of the shell wall and overall density coincident with maturation.

The shape of the mature shell may depart substantially from that of earlier whorls. This change appears in most ammonites, although it is best expressed in uncoiled genera such as *Scaphites* (Fig. 2C). Ornamentation at maturity may become weaker or disappear. The mature aperture may become modified and develop elongate projections (lappets), commonly associated with males.

The rate of growth and age at maturity of ammonites are unknown. By analogy, modern *Nautilus* matures in 10–15 years and probably lives for several years afterward (Saunders, 1983; Landman and Cochran, 1987). Other cephalopods, such as squids and cuttlefish, rapidly reach maturity in 1–3 years, reproduce, and

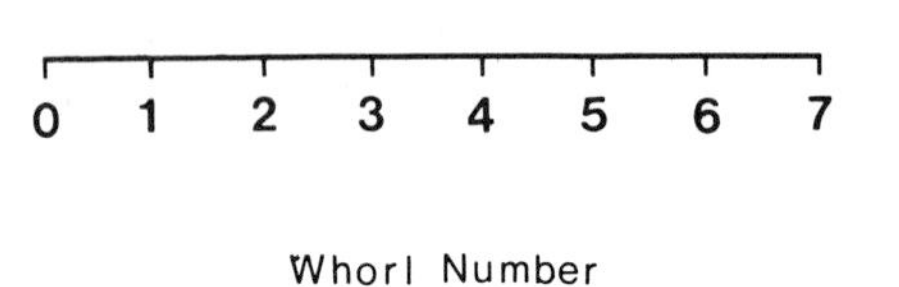

Figure 6. The approximate whorl number at which the septal neck changes from retrochoanitic to prochoanitic in the ontogenetic development of various ammonites.

die (Wells and Wells, 1983). Attempts to estimate the rate of growth and age at maturity of ammonites have been based on interpretations of morphological patterns of growth and identification of year classes (Trueman, 1941; Westermann, 1971; Kennedy and Cobban, 1976; Hirano, 1981; Doguzhayeva, 1982). They have also been based on the known growth rates of epizoans that grew on living ammonites (Schindewolf, 1934; Seilacher, 1960). These methods have yielded diverse estimates, but suggest a rate of growth and age at maturity more similar to that of *Nautilus* than that of most modern dibranchiates. This similarity may reflect the common costs of growing a large external shell (Ward, 1985).

The lack of information on the actual growth rates of ammonites precludes the comparison of ancestral and descendant ontogenies at the same absolute age [but see Dommergues (1988) for an innovative approach]. We are, therefore, generally limited to comparisons involving developmental stage or size. As a result, many patterns of heterochronic change are not fully resolvable (Jones, this volume; Gould, 1977; Dommergues *et al.*, 1986).

4. Examples of Heterochrony

We have now examined the ontogeny of ammonites from the start of postembryonic growth to maturity. Several morphological features exhibit allometric growth, particularly sutures, and provide the raw material for heterochrony. How is this potential realized? What is the case for recapitulation?

My first example, borrowed from Doguzhayeva and Mutvei (1986), traces the evolution of septal necks, suggesting a pattern of recapitulation among ammonites as a whole (Fig. 6). Recall that in ontogeny, septal necks may change from retro- to prochoanitic. Now, in Paleozoic goniatites, this transition occurs late in ontogeny. In *Agathiceras uralicum*, for example, the transition occurs in the seventh whorl and involves many septa. The transition is also incomplete in that only the dorsal part of the septal neck is transformed; the ventral part remains retrochoanitic. In Triassic ceratites, the transition may appear earlier in ontogeny,

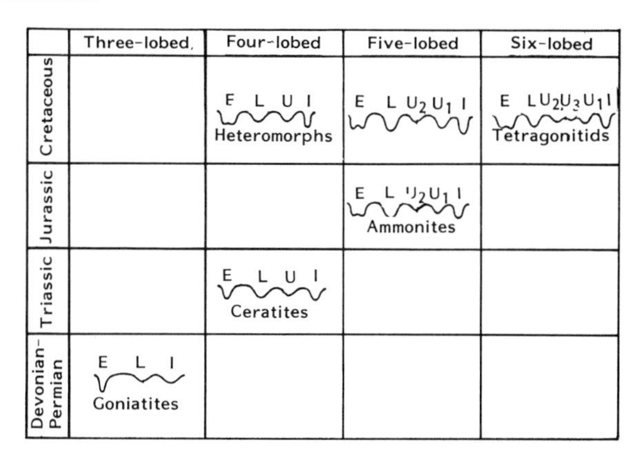

Figure 7. Predisplacement of sutural complexity occurs by the development of new lobes in the primary suture. Note a reduction in sutural complexity in the Cretaceous heteromorphs. Modified from Lehmann (1981), redrawn after Wiedmann (1970). Reprinted with permission of Cambridge University Press.

although it still may involve many septa and may only affect the dorsal side. For example, in mid-Triassic *Megaphyllites prometheus*, the transition occurs over two whorls (ten septa) between the second and fourth whorls. In all lytoceratids and most phylloceratids, the transition appears earlier and occurs over fewer septa, affecting both the dorsal and ventral portions of the septal neck simultaneously. For example, in the Late Cretaceous lytoceratid *Gaudryceras tenuiliratum*, the fifth septum is already prochoanitic. In ammonitids, prochoanitic necks appear at the beginning of postembryonic growth. Thus, in general, the development of prochoanitic necks is accelerated through ontogeny, appearing earlier, occurring over fewer septa, and affecting the entire neck. This acceleration results in recapitulation because the morphology of the septal neck in the early ontogeny of advanced descendants recapitulates the condition in the later ontogeny of more primitive ancestors. Nevertheless, this is a general pattern that requires further study.

Sutures have furnished the most examples of heterochrony, especially recapitulation, on all taxonomic levels because, according to Ruzhentsev (1962, p. 322) they "constitute peculiar ontograms which are hidden inside the shell and originate from the remote past." On the ordinal level, Kullmann and Wiedmann (1982) documented an increase in the sutural complexity of the primary suture, the initial suture for all subsequent differentiation (Fig. 7). In goniatites and prolecanitids the primary suture consists of three lobes, E, L, and I. However, in Triassic ceratites the primary suture becomes quadrilobate due to the development of an umbilical lobe. In most groups of Jurassic and Cretaceous ammonites, the primary suture is quinquelobate with a lobe formula ELU_2U_1I. Finally, a six-lobed primary suture develops in the upper Cretaceous tetragonitids. Thus, in general, the primary suture becomes progressively more complex, an example of predisplacement of sutural complexity. However, a reduction to a four-lobed primary suture reoccurs in one group of Cretaceous ammonites. This four-lobed primary suture is clearly derived from a five-lobed primary suture because both share a distinctive ventral saddle within the ventral lobe.

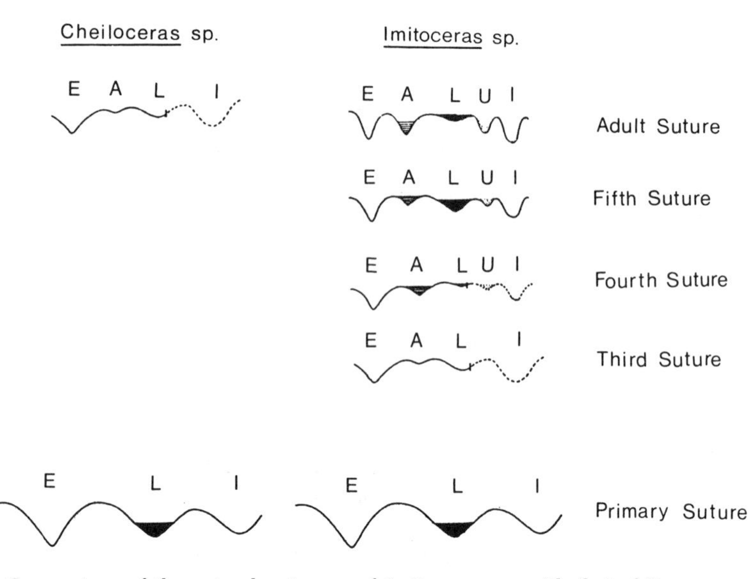

Figure 8. Comparison of the sutural ontogeny of *Imitoceras* sp. with that of its presumed ancestor *Cheilocaras* sp., indicating recapitulation. After Kullmann and Wiedmann (1982).

Recapitulation of sutural development has been documented in many groups of ammonites (Newell, 1949; Miller and Furnish, 1958; Ruzhentsev, 1962; Wiedmann, 1970; Tanabe, 1977; Kullmann and Wiedmann, 1982; Dommergues, 1986). Kullmann and Wiedmann (1982) have recently described a case among Paleozoic goniatites (Fig. 8). In *Cheiloceras* from the middle Upper Devonian, the primary suture is three-lobed. A four-lobed suture appears by the tenth septum and is retained to maturity (lobe formula EALI). Descendant *Imitoceras* from the highest Upper Devonian and lowest Lower Carboniferous develops a four-lobed suture by the third septum, that is, at a smaller size and earlier developmental stage, and eventually attains a five-lobed suture at maturity (lobe formula ELAUI). The sutural ontogeny of *Imitoceras*, therefore, recapitulates that of *Cheiloceras* due to the acceleration of sutural development relative to overall growth. However, septal modifications associated with maturity, for example, thickening and approximation, do not appear any earlier in the ontogeny of descendants.

Recapitulation is also observed in the sutural ontogeny of the ceratite genus *Megaphyllites* (Kullmann and Wiedmann, 1982, Fig. 6). In *M. prometheus* from the Middle Triassic, the primary suture consists of four lobes and subsequently five-lobed and six-lobed sutures appear at whorl heights of 0.5 and 0.6 mm. In descendant *M. robustus* from the Upper Triassic, the primary suture also consists of four lobes. However, five- and six-lobed sutures develop sequentially in the next two septa, at a much earlier stage than in *M. prometheus*. Thus, recapitulation occurs via the acceleration of sutural development relative to overall growth.

Similarities in the sutural patterns of ancestors and descendants may also reflect Von Baer's laws. For example, Fig. 9 depicts, on the right, the adult sutures of an evolutionary sequence of species and, on the left, the sutural ontogeny of the most advanced descendant, *Gargasiceras gargasense*. The primary suture of this species consists of four lobes, and a six-lobed suture is attained by a whorl

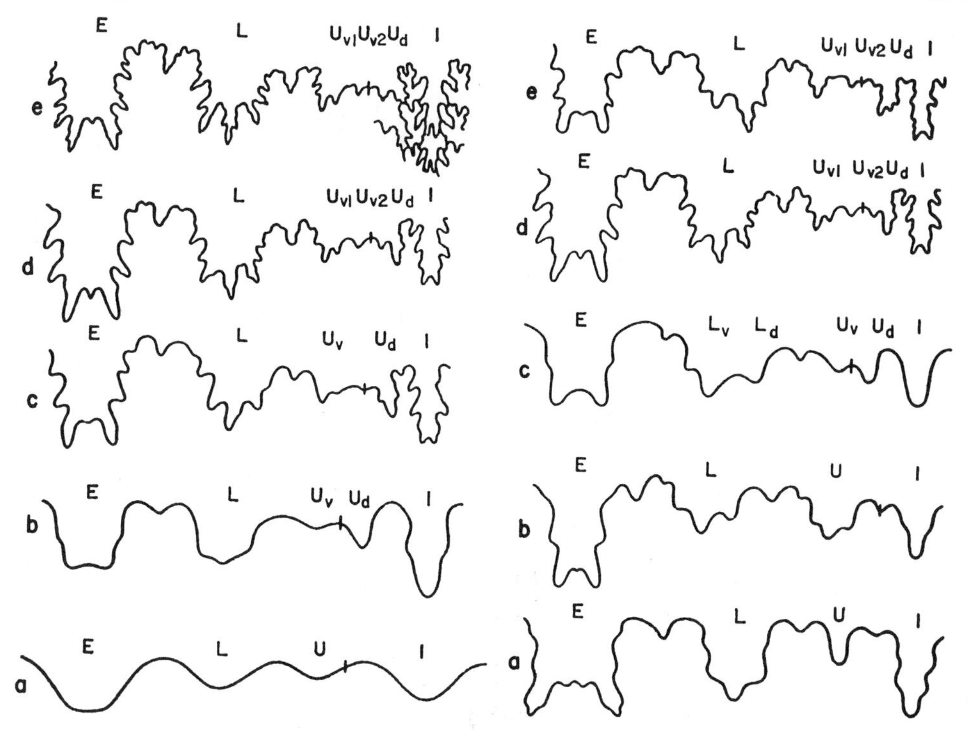

Figure 9. Comparison of (left) the sutural ontogeny of *Gargasiceras gargasense* with (right) the adult sutures of its presumed ancestors, from the most primitive to the most advanced (bottom to top). The primary suture of *G. gargasense* is represented on the lower left. The suture on the upper left is attained by a whorl height of 4.5 mm. From Wiedmann (1970). Reprinted with permission of the editor, *Eclogae Geologicae Helvetiae*.

height of 4.5 mm. The ontogeny of this species, therefore, recapitulates the adult sutures of all its ancestors. However, the similarity in the number of lobes between the primary suture of *G. gargasense* and the adult suture of its most distant ancestor merely reflects Von Baer's law. The primary suture of the ancestor is also four-lobed and reflects the common condition in the early ontogeny of both species, although in the ancestor, a four-lobed suture is retained to maturity. The ontogeny of the ancestor deviates less from this common condition than that of the descendant. The other similarities between the sutures of *G. gargasense* and those of its adult ancestors are due to the acceleration of sutural development relative to overall growth.

In all these examples of recapitulation, descendants develop more complex sutures than those of their ancestors. However, more complex sutures do not represent independent acquisitions added to the end of ontogeny, but, rather, predictable extensions of the sutural patterns already present in ancestors. As Ruzhentsev (1962, p. 299) pointed out "the suture line firmly maintains its particular type of ontogeny in phylogenetically related groups even in the presence of an accelerated development. Once established, the particular pattern of complication of the suture line may develop at different rates but always regularly."

Recapitulation of sutural patterns may be more closely examined by quan-

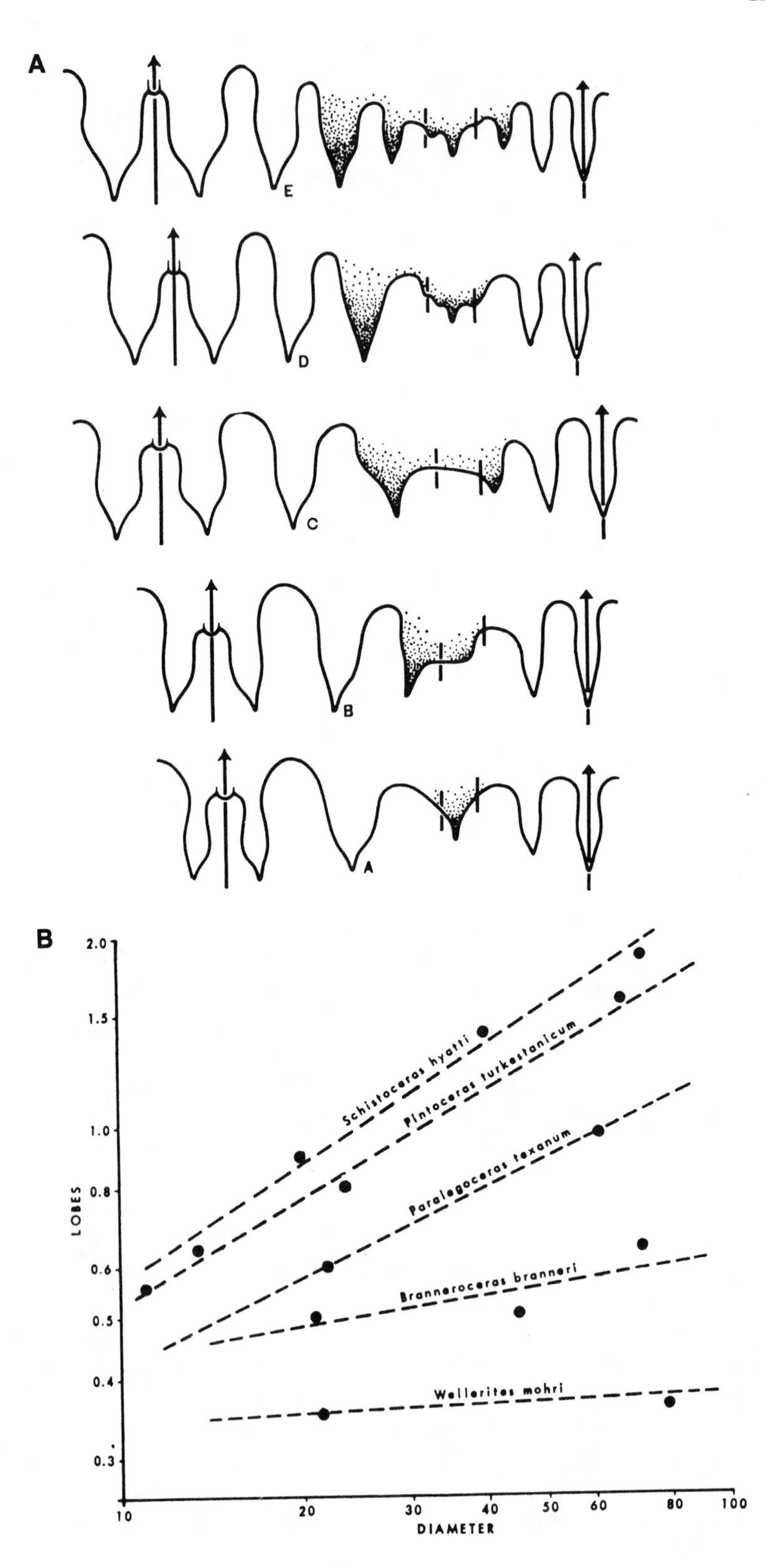
A
E
D
C
B
A
B
LOBES
2.0
1.5
1.0
0.8
0.6
0.5
0.4
0.3
Schistoceras hyatti
Pintoceras turkestanicum
Paralegoceras texanum
Branneroceras branneri
Wellerites mohri
10
20
30
40
50
60
80
100
DIAMETER

tifying sutural complexity. For example, within the Pennsylvania goniatite family Schistoceratidae, the number of sutural lobes increases in ontogeny and phylogeny (Fig. 10A) (Miller and Furnish, 1958). Descendants exhibit more sutural lobes than do ancestors at comparable shell diameters. In addition, all these species exhibit a median saddle within the external lobe (E). The relative size of this saddle may be expressed as a ratio of the height of the saddle to the depth of the first lateral lobe (L). Figure 10B represents a plot of this ratio vs. shell diameter for five species comprising an evolutionary sequence. Within each species, the ratio increases through ontogeny. Descendants exhibit progressively higher slopes, indicating acceleration in the rate of development of the median saddle. Therefore, this feature is relatively larger in descendants than ancestors at comparable shell diameters.

Newell (1949) also employed a measure of sutural complexity in describing recapitulation in an evolutionary sequence of Paleozoic prolecanitids. He measured the half length of the suture vs. shell diameter for five successive genera. The slope of the regression line for individual ontogeny exceeds 1.0, indicating positive allometry; in other words, sutures become increasingly more complex through ontogeny. Alberch *et al.* (1979) inferred that recapitulation within this sequence is due to a combination of three processes: (1) acceleration of sutural development relative to overall size, as indicated by higher slopes for the regression lines of some descendants, (2) hypermorphosis because descendants reach maturity at larger sizes and, as a result, exhibit a degree of sutural complexity beyond that in ancestors, and (3) predisplacement of sutural complexity, as indicated by progressively higher y intercepts. However, the values of the y intercepts are extrapolations. In fact, all species begin with the same primary suture, although differentiation from this basic suture may occur more rapidly in the ontogeny of descendants.

Recapitulation of sutural development due to these same processes also occurs in the evolution of *Otoscaphites puerculus*, an Upper Cretaceous ammonite (Tanabe, 1977). Tanabe compared the sutural ontogeny of ancestors and descendants in two ways. First, he expressed sutural complexity as the logarithm of the half length of the suture (Fig. 11A). He plotted sutural complexity vs. whorl number in specimens from successively higher horizons and calculated their individual regression lines. Descendants exhibit progressively higher slopes and y intercepts, indicating both acceleration and predisplacement. Second, Tanabe plotted the number of sutural lobes vs. the number of whorls in these same specimens (Fig. 11B). He observed that the primary suture in these specimens always consists of four lobes (ELUI). Therefore, predisplacement, as indicated by progressively higher y intercepts, is an artifact of extrapolation. In fact, after the formation of the primary suture, the number of lobes remains the same for the next few whorls. The development of new lobes occurs earlier in the ontogeny

Figure 10. (A) Comparison of sutures at comparable shell diameters in an evolutionary sequence of Pennyslvanian goniatites, indicating an increase in the number of lobes. Note the growth of the median saddle within the external lobe. (B) Ratio of the height of the median saddle to the depth of the first lateral lobe vs. shell diameter for many of these same goniatites. From Miller and Furnish (1958). Reprinted with permission of the Paleontological Society.

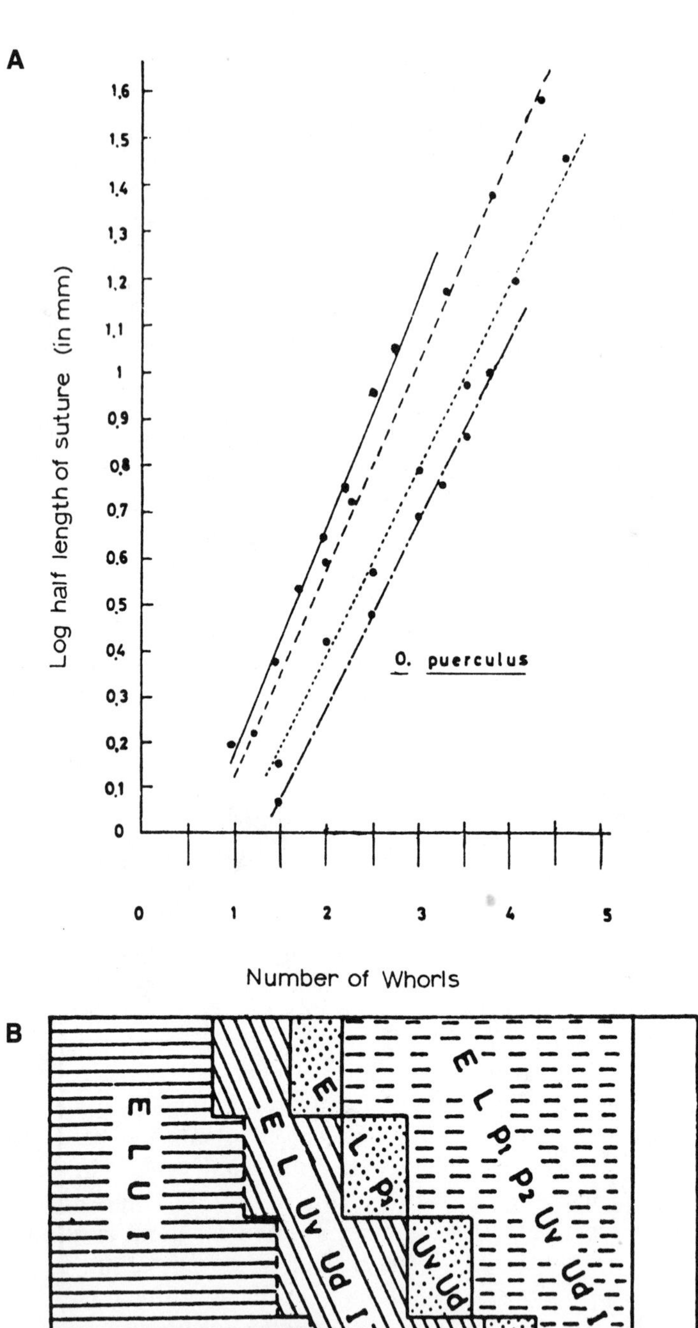
A
Log half length of suture (in mm)
1,6
1,5
1,4
1,3
1,2
1,1
1
0,9
0,8
0,7
0,6
0,5
0,4
0,3
0,2
0,1
0
O. puerculus
0
1
2
3
4
5
Number of Whorls
B
E L U I
E L Uv Ud I
E
E L P1
Uv Ud
I
E L P1 P2 Uv Ud I
0
1
2
3
4
5
Number of Whorls

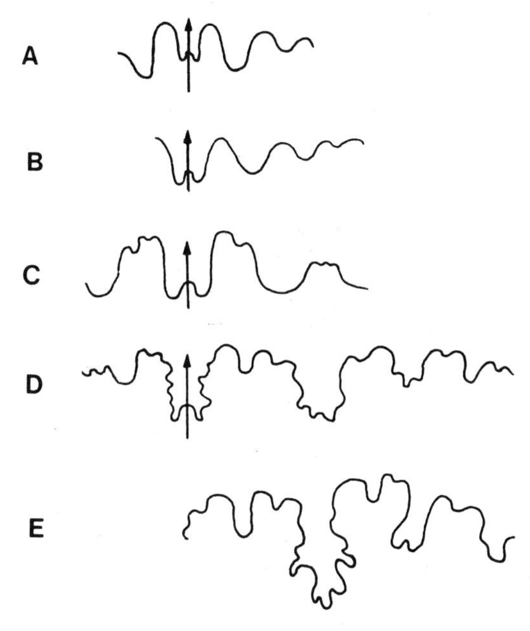

Figure 12. Sutural simplification in an evolutionary sequence of Cretaceous ammonites as revealed by a comparison of their adult sutures. (A) *Flickia simplex*, (B) *Ficheuria pusilla*, (C) *Ficheuria pernoni*, (D) *Ficheuria kiliani*, (E) *Salaziceras salazacense*. After Wright and Kennedy (1979). Reprinted with permission of the author and the Palaeontological Association.

of descendants and recapitulation results via acceleration and condensation. Descendants exhibit more sutural lobes than do ancestors at comparable whorl numbers. The size at maturity also increases, although no new lobes appear in the adult sutures of larger descendants.

Sutural recapitulation usually entails the development of more complex sutures in the ontogeny of descendants. However, an evolutionary reduction in sutural complexity may also occur. Within the family Flickiidae from the Upper Cretaceous, Wright and Kennedy (1979) have documented paedomorphosis expressed as a reduction in sutural complexity and ornamentation. The adult sutures of descendants resemble those of juvenile ancestors, although in the ontogeny of any one individual, sutural complexity generally increases (Fig. 12). Mature specimens of *Flicka simplex*, the most advanced descendant, exhibit sutures consisting of entire simple lobes and saddles. Similarly, Glenister (1985) has described an evolutionary reduction in sutural complexity in the prolecanitid subfamily Sicanitinae. At maturity, the most advanced descendant is less than one-quarter the size of its presumed ancestor and exhibits a suture characteristic of its ancestor at a juvenile stage of development.

Heterochrony may also involve the shape of the shell and its ornament (see, e.g., Ruzhentsev, 1962; Kennedy, 1977; Dommergues, 1982, 1986, 1987; Westermann and Riccardi, 1985). Kennedy and Wright (1985) have described numerous examples of paedomorphosis within the Cretaceous family Acanthoceratidae. Ac-

←

Figure 11. (A) Logarithm of the half length of the suture vs. whorl number for four individuals of *Otoscaphites puerculus* from successively higher horizons (from bottom to top). (B) Sutural complexity expressed by the number of lobes vs. the number of whorls for the same four specimens. Letters refer to different lobes. Modified from Tanabe (1977).

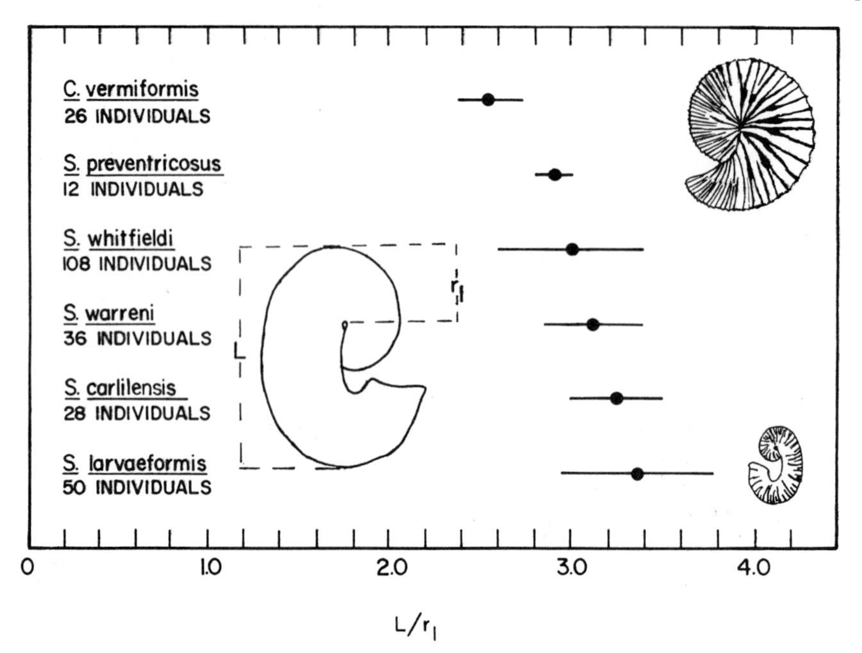

Figure 13. (A) Evolutionary trend toward adult recoiling in an evolutionary sequence of scaphitid ammonites from the Upper Cretaceous of the Western Interior of North America. The shape ratio (L/r_1) indicates the extent to which a hooklike body chamber is developed at maturity. This trend also coincides with an increase in adult size.

cording to them, this family evolved from Stoliczkaiinae by the retention of juvenile ornament into maturity, although new features developed in conjunction with an overall increase in size. *Acanthoceras*, in turn, gave rise via paedomorphosis to *Protacanthoceras*. The most primitive species, *P. tuberculatum*, is progenetic and attains maturity at less than 10% the size of its ancestor. Another species, *P. tegulicium*, evolved via neoteny from *P. arkelli*. Maturity occurs at the same size in both species, but constrictions in the juvenile shells of ancestors are retained into the mature shells of descendants.

In an evolutionary sequence of Paleozoic clymeniids, Schindewolf (1950, Fig. 226) described an example of neoteny in which triangular coiling in the juvenile shells of ancestors appeared in the adult shells of decendants. Similarly, in an evolutionary sequence of Cretaceous ammonites ranging from *Brancoceras* to *Mortoniceras*, neoteny occurs in which a keel on the juvenile shells of ancestors is retained on the adult shells of descendants (Kullmann and Wiedmann, 1982, Fig. 10).

Heterochrony involving the shape of the shell also occurs among endemic scaphites in the Upper Cretaceous of the Western Interior of North America. For example, species of *Scaphites* and the closely related genus *Clioscaphites* exhibit an evolutionary trend toward recoiling (Cobban, 1951). In the ontogeny of *Scaphites*, the shell is initially tightly coiled, but uncoils at maturity, forming a hooklike body chamber. The extent to which the adult hook develops may be expressed as a simple shape ratio, as illustrated in Fig. 13 (Landman, 1987b). This ratio progressively decreases, indicating that descendants are more tightly coiled at maturity than ancestors. This trend also coincides with an evolutionary increase in adult size and whorl number. Thus, adult descendants are larger, but exhibit a shape characteristic of their ancestors at a juvenile stage of development. This pattern suggests neoteny in which shape is retarded relative to overall growth, with an associated delay in maturity. However, a new ornamentation may develop on the adult body chamber.

On the other hand, the micromorph genus *Pteroscaphites* Wright may have evolved via progenesis from *Scaphites* (Landman, 1987b). The ammonitellas of the two genera and their ontogenetic trajectories in early postembryonic growth, until a diameter of 4–5 mm are identical. At this size, species of both genera exhibit changes in whorl expansion, degree of whorl overlap, ornamentation, septal spacing, and siphuncular position. Thereafter, *Scaphites* species secrete two to three additional whorls before attaining maturity. However, species of *Pteroscaphites* accelerate into maturity at this diameter and develop a typical uncoiled scaphitid body chamber. This body chamber also bears lateral projections (lappets), which are a new feature absent in any of the *Scaphites* species.

Finally, the Upper Cretaceous scaphite species *Hoploscaphites nicolleti* exhibits heterochrony in the development of mature ornamentation (Waage, 1968). Figure 14 represents individuals of the same species over a time interval of several million years. The first appearance of fine ridges on the adult body chamber is progressively delayed or postponed.

5. Conclusion

In ammonite ontogeny, the growth of many characters is allometric and provides the potential for heterochronic change. For example, the development of sutures is positively allometric. Sutures, in contrast to shell shape and ornament, furnish numerous examples of heterochrony, especially recapitulation, on all taxonomic levels. Recapitulation of sutural development occurs via a combination of predisplacement and acceleration. This may result from a selective advantage for the precocious development of more complex sutures (Gould, 1977). Recapitulation may also occur via hypermorphosis or the extension of ancestral allometries into the prolonged ontogenies of larger descendants. In all of these examples, however, more complex sutures are not independent acquisitions added to the end of ontogeny, but extrapolations of patterns already present in ancestors. Nevertheless, many of the similarities attributed to recapitulation merely reflect Von Baer's law.

Kennedy and Wright (1985) cite many examples of heterochrony in the evolution of Upper Cretaceous ammonites. It is a common pattern and may imply an underlying flexibility in the ontogenetic development of these and many other cephalopods. For example, *Nautilus* in aquaria have been reported to reach maturity at an accelerated rate of development (Ward, 1985; Arnold, J. M., personal communication, 1986). The adults are abnormally small, but exhibit full mature modifications, including septal approximation and thickening. This acceleration may result from a combination of favorable conditions, for example, surface water pressure and abundant food, and suggests a dissociability between growth and

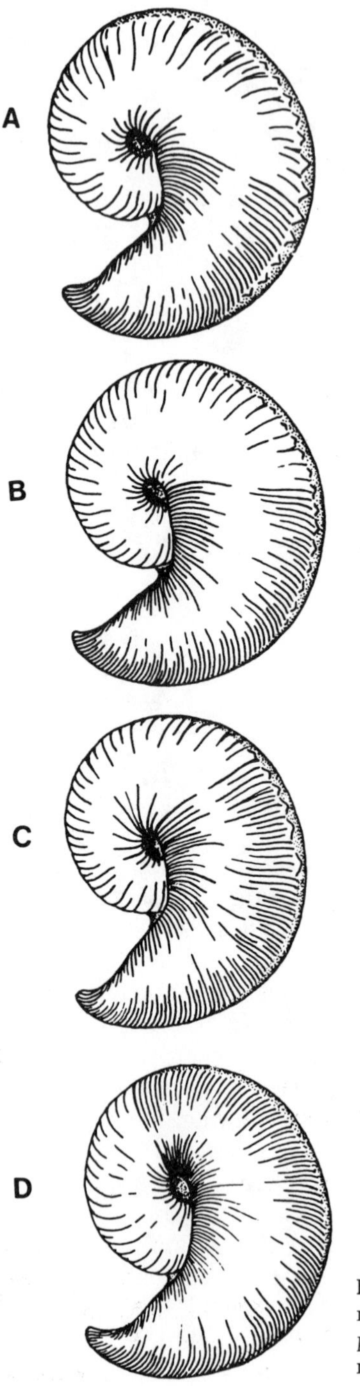

Figure 14. Progressive delay in the first appearance of fine ornamentation on the mature body chamber of specimens of *Hoploscaphites nicolleti* from successively higher stratigraphic horizons. Modified from Waage (1968).

development. A similarly facile dissociability between growth and development may have also accounted for the high incidence of heterochrony in the evolution of ammonites.

References

Alberch, P., Gould, S. J., Oster, G. F., and Wake, D. B., 1979, Size and shape in ontogeny and phylogeny, *Paleobiology* **5**(9):296–317.

Arkell, W. J., 1957, Introduction to Mesozoic Ammonoidea, in: *Treatise on Invertebrate Paleontology*, Part L (R. C. Moore, ed.), pp. 81–129, Geological Society of America and University of Kansas Press, Lawrence, Kansas.

Arnold, J. M., and Carlson, B. A., 1986, Living *Nautilus* embryos: Preliminary observations, *Science* **232**:73–76.

Bandel, K., 1982, Morphologie und Bildung der frueontogenetischen Gehaeuse bei conchiferen Mollusken, *Facies* **7**:1–198.

Bandel, K., 1986, The ammonitella: A model of formation with the aid of the embryonic shell of archaeogastropods, *Lethaia* **19**:171–180.

Bandel, K., and Boletzky, S. V., 1979, A comparative study of the structure, development, and morphological relationships of chambered cephalopod shells, *Veliger* **21**:313–354.

Bandel, K., Landman, N. H., and Waage, K. M., 1982, Micro-ornament on early whorls of Mesozoic ammonites: Implications for early ontogeny, *J. Paleontol.* **56**(2):386–391.

Bayer, U., 1972, Zur ontogenie und variabilitaet der Jurassischen Ammoniten *Leioceras opalinum*, *Neues Jahrb. Geol. Paläontol. Abh.* **140**(3):306–327.

Birkelund, T., 1965, Ammonites from the Upper Cretaceous of West Greenland, *Medd. Gronl.* **171**:1–192.

Birkelund, T., 1980, Ammonoid shell structure, *in: The Ammonoidea* (M. R. House and J. R. Senior, eds.), pp. 177–214, Academic Press, New York.

Bogoslovsky, B. T., 1976, Early ontogeny and origin of clymenid ammonoids, *Paleontol J.* **10**(2):150–158.

Boletzky, S. V., 1974, The "larvae" of Cephalopoda: A review, *Thalassia Jugoslavica* **10**:45–76.

Buckman, S. S., 1887–1907, A monograph of the ammonites of the Inferior Oolite Series, *Palaeontographr. Soc.* **40–61**:1–456.

Buckman, S. S., 1909, *Yorkshire Type Ammonites*, Vol. 1, No. 1, pp. 1–12, Wesley, London.

Buckman, S. S., 1918, Jurassic chronology: I—Lias, *Geol. Soc. Lond. Q. J.* **73**:257–327.

Callomon, J. H., 1963, Sexual dimorphism in Jurassic ammonites, *Trans. Leicester Litt. Philos. Soc.* **57**:21–66.

Callomon, J. H., 1980, Dimorphism in Ammonoids, in: *The Ammonoidea* (M. R. House and J. R. Senior, eds.), pp. 257–273, Academic Press, New York.

Chamberlain, J. A., 1978, Permeability of the siphuncular tube of *Nautilus*: Its ecologic and paleoecologic implications, *Neues Jahrb. Geol. Paläontol. Monatsch.* **3**:129–142.

Cobban, W. A., 1951, Scaphitoid Cephalopods of the Colorado Group, U. S. Geological Survey Professional Paper 239.

Currie, E. D., 1942, Growth changes in the ammonite *Promicroceras marstanense*, Spath., *Proc. Soc. Ed. B* **61**:344–367.

Currie, E. D., 1943, Growth stages in some species of *Promicroceras*, *Geol. Mag.* **80**:15–22.

Currie, E. D., 1944, Growth stages in some Jurrassic ammonites, *Trans. R. Soc. Ed.* **61**(6):171–198.

Doguzhayeva, L., 1982, Rhythms of ammonoid shell secretion, *Lethaia* **15**:385–394.

Doguzhayeva, L., and Mutvei, H., 1986, Retro- and prochoanitic necks in ammonoids, and transition between them, *Palaeontogr. Abt. A* **195**:(1–3):1–18.

Dommergues, J. L., 1982, L'évolution des Liparoceratidae "capricornes" (Ammonitina, Jurassique, Lias moyen): Diversité des rythmes évolutifs, in: *Modalités, Rythmes et Mécanismes de l' Evolution Biologiques*, pp. 233–236, CNRS, Paris.

Dommergues, J. L., 1986, Les Dactylioceratidae du Carixien et du Domérian basal, un groupe monophylétique. Les Reynesocoeloceratinae nov. subfam., *Bull. Sci. Bourg.* **39**:1–26.

Dommergues, J. L., 1987, *L'évolution chez les Ammonitina du Lias moyen (Carixien, Domérian basal) en Europe Occidentale, Documents des Laboratories de Géologie, Lyon* No. 98.
Dommergues, J. L., 1988, Can ribs and septa provide an alternative standard for age in ammonite ontogenetic studies. *Lethaia* (in press).
Dommergues, J. L., David, B., and Marchand D., 1986, Les relations ontogenèse–phylogenèse: Applications paléontologiques, *Géobios* **19**(3):335–356.
Donovan, D. T., 1973, The influence of theoretical ideas on ammonite classification from Hyatt to Trueman, pp. 1–16, University of Kansas Paleontological Contributions, Paper 62.
Druschits, V. V., Doguzhayeva, L. A., and Mikhaylova, I. A., 1977, The structure of the ammonitella and the direct development of ammonites, *Paleontol. J.* **2**:188–199.
Glenister, B. F., 1985, Terminal progenesis in Late Paleozoic ammonoid families, in: *Abstracts, 2nd International Cephalopod Symposium, Tübingen*, p. 9.
Gould, S. J., 1977, *Ontogeny and Phylogeny*, Harvard University Press, Cambridge.
Hewitt, R. A., 1985, Numerical aspects of sutural ontogeny in the Ammonitina and Lytoceratina, *Neues Jahrb. Geol. Paläontol. Abh.* **170**(3):273–290.
Hewitt, R. A., and Westermann, G. E. G., 1986, Function of complexly fluted septa in ammonoid shells I. Mechanical principals and functional models, *Neues Jahrb. Geol. Paläontol. Abh.* **172**(1):47–69.
Hirano, H., 1981, Growth rates in *Nautilus macromphalus* and ammonoids: Its implications, in: *International Symposium on Concepts and Methods in Paleontology Barcelona 1981* (J. Martinell, ed.), pp. 141–146, University of Barcelona, Barcelona.
House, H. R., 1965, A study in the Tornoceratidae: The succession of *Tornoceras* in the North American Devonian, *Philos. Trans. R. Soc. Lond. B* **250**:79–130.
Hyatt, A., 1866, On the agreement between the different periods in the life of the individual shell and the collective life of the tetrabranchiate cephalopods, *Boston Soc. Nat. Hist. Proc.* **10**:302–303.
Hyatt, A., 1883, Fossil cephalopods in the museum of comparative zoology, *Am. Assoc. Adv. Sci. Proc.* **32**:323–361.
Hyatt, A., 1889, *Genesis of the Arietidae*, Smithsonian Contributions to Knowledge, Vol. 26, No. 673.
Hyatt, A., 1894, Phylogeny of an acquired characteristic, *Am. Phil. Soc. Proc.* **32**(143):349–647.
Karpinski, A. P., 1889, Über die Ammoneen der Artinsk-Stufe und einige mit denselben verwandte carbonische Formen, *Acad. Imp. Sci. St. Petersbourg Mém. 7e Ser.* **37**(2):1–104.
Karpinski, A. P., 1890, Zur Ammoneen-Fauna der Artinsk-Stufe, *Acad. Imp. Sci. St. Petersbourg Bull. (Mélanges Géol. Paléontol.)* **1**:65–80.
Kennedy, W. J., 1977, Ammonite evolution, in: *Patterns of Evolution* (A. Hallam, ed.), pp. 251–304, Elsevier, Amsterdam.
Kennedy, W. J., and Cobban, W. A., 1976, Aspects of ammonite biology, biogeography, and biostratigraphy, *Spec. Pap. Palaeontol.* **17**:1–33.
Kennedy, W. J. and Wright, C. W., 1985, Evolutionary patterns in late Cretaceous ammonites, *Spec. Pap. Paloeontol.* **33**:131–143.
Kulicki, C., 1974, Remarks on the embryogeny and postembryonal development of ammonites, *Acta Palaeontol. Pol.* **20**:201–224.
Kulicki, C., 1979, The ammonite shell: Its structure, development and biological significance, *Acta Palaeontol. Pol.* **39**:97–142.
Kullmann, J., and Wiedmann, J., 1982, Bedeutung der Rekapitulationsentwicklung in der Paläontologie, *Verh. Naturwiss. Ver. Hamburg (N.F.)* **25**:71–92.
Landman, N. H., 1985, Preserved ammonitellas of *Scaphites* (Ammonoidea, Ancyloceratina), *Am. Mus. Novit.* **2815**:1–10.
Landman, N. H., 1987a, Early ontogeny of Mesozoic ammonites and nautilids, in: *Cephalopods—Present and Past* (J. Wiedmann and J. Kollmann, eds.), pp. 215–228, Schweizerbart'sche Verlagsbuchhandlung, Stuttgart.
Landman, N. H., 1987b, Ontogeny of Upper Cretaceous (Turonian–Santonian) scaphitid ammonites from the Western Interior of North America: Systematics, developmental patterns, and life history, *Am. Mus. Bull.* **185**(2):117–241.
Landman, N. H., and Bandel, K., 1985, Internal structures in the early whorls of Mesozoic ammonites, *Am. Mus. Novit.* **2823**:1–21.
Landman, N. H., and Cochran, J. K., 1987, Growth and longevity of *Nautilus*, in: *Nautilus: Biology and Paleobiology of a Living Fossil* (W. B. Saunders and N. H. Landman, eds.), pp. 401–420, Plenum Press, New York.

Lehmann, U., 1981, *Ammonites: Their Life & Their World*, Cambridge University Press, New York.
Makowski, H., 1962, Problems of sexual dimorphism in ammonites, *Acta Palaeontol. Pol.* **12**:1–92.
Miller, A. K., and Furnish, W. M., 1958, Middle Pennsylvanian Schistoceratidae (Ammonoidea), *J. Paleontol.* **32**(2):253–268.
Miller, A. K., Furnish, W. M., and Schindewolf, O. H., 1957, Paleozoic Ammonoidea, in: *Treatise on Invertebrate Paleontology* L (R. C. Moore, ed.), pp. 11–79, Geological Society of America and University of Kansas Press, Lawrence.
Newell, N. D., 1949, Phyletic size increase, an important trend illustrated by fossil invertebrates, *Evolution* **3**:103–124.
Obata, I., 1965, Allometry of *Reesidites minimus*, a Cretaceous ammonite species, *Trans. Proc. Paleontol. Soc. Japan N. S.* **58**:39–63.
Pavlov, A. P., 1901, Le Crétacé inférieur de la Russie et sa faune, *Nouv. Mém. Soc. Imp. Nat. Moscou Livr. 21* (*sér. nouv.*, 16).
Raup, D. M., 1967, Geometric analysis of shell coiling: Coiling in ammonoids, *J. Paleontol.* **41**:43–65.
Ruzhentsev, V. E., 1962, Superorder Ammonoidea. General section, in: *Fundamentals of Paleontology*, Vol. 5, *Mollusca—Cephalopoda I.* (V. E. Ruzhentsev, ed.), pp. 243–328, Izdatel'stvo Akademi Nauk USSR, Moscow.
Saunders, W. B., 1983, Natural rates of growth and longevity of *Nautilus belauensis*, *Paleobiology* **9**:280–288.
Schindewolf, O. H., 1929, Ontogenie und Phylogenie, *Palaeontol. Z.* **11**:54–67.
Schindewolf, O. H., 1934, Über Epöken auf Cephalopoden-Gehäuse, *Palaeontol. Z.* **16**:15–31.
Schindewolf, O. H., 1942, Evolution im Lichte der Paläontologie, *Jen. Z. Med. Naturwiss.* **75**:324–386.
Schindewolf, O. H., 1950, *Grundfragen der Paläontologie*, Schweizebart'sche Verlagsbuchhandlung, Stuttgart.
Schindewolf, O. H., 1954, On development, evolution, and terminology of ammonoid suture line, *Harvard Univ. Mus. Comp. Zool. Bull.* **112**(3):217–237.
Seilacher, A., 1960, Epizoans as a key to ammonoid ecology, *J. Paleontol.* **34**:189–193.
Smith, James P., 1898, The development of *Lytoceras* and *Phylloceras*, *Calif. Acad. Sci. Proc. 3rd Ser. Geol.* **1**(4):129–161.
Smith, James P., 1914, Acceleration of Development in Fossil Cephalopoda, Stanford University Publications University Series, Stanford, California.
Spath, L. F., 1924, The ammonites of the Blue Lias, *Geol. Assoc. Proc.* **35**:186–211.
Spath, L. F., 1938, *A Catalogue of the Ammonites of the Liassic Family Liparoceratidae*, British Museum (Natural History), London.
Tanabe, K., 1977, Functional evolution of *Otoscaphites puerculus* (Jimbo) and *Scaphites planus* (Yabe), Upper Cretaceous ammonites, *Mem. Fac. Sci. Kyushu Univ. Ser. D Geol.* **23**(3):367–407.
Thompson, D'A. W., 1917, *On Growth and Form*, Cambridge University Press, London.
Trueman, A. E., 1919, The evolution of the Liparoceratidae, *Geol. Soc. Lond. Q. J.* **74**:247–298.
Trueman, A. E., 1922, Aspects of ontogeny in the study of ammonite evolution, *J. Geol.* **30**:140–143.
Trueman, A. E., 1941, The ammonite body chamber with special reference to the buoyancy and mode of life of the living ammonite, *Q. J. Geol. Soc. Lond.* **96**:339–383.
Waage, Karl M., 1968, The Type Fox Hills Formation, Cretaceous (Maestrichtian), South Dakota. Part 1. Stratigraphy and Paleoenvironments, Peabody Museum of Natural History Bulletin 27.
Ward, P. D., 1985, Periodicity of chamber formation in chambered cephalopods: Evidence from *Nautilus macromphalus* and *Nautilus pompilius*, *Paleobiology* **11**(4):438–450.
Wells, M. J., and Wells, J., 1983, Cephalopods do it differently, *New Sci.* **100**:332–338.
Westermann, G. E. G., 1954, Monographie der Otoitidae (Ammonoidea), *Beih. Geol. Jahrb. Heft* **15**:1–364.
Westermann, G. E. G., 1958, The significance of septa and sutures in Jurassic ammonite systematics, *Geol. Mag.* **95**:441–455.
Westermann, G. E. G., 1966. Covariation and taxonomy of the Jurassic ammonite *Sonninia adicra* (Waagen), *Neues Jahrb. Geol. Paläontol. Abh.* **24**:389–412.
Westermann, G. E. G., 1971, Form, structure and function of shell and siphuncle in coiled Mesozic ammonoids, *Life Sci. Contrib. R. Ont. Mus.* **78**:1–39.
Westermann, G. E. G., and Riccardi, A. C., 1985, Middle Jurassic ammonite evolution in the Andean province and emigration to Tethys, in: *Lecture Notes in Earth Sciences* (U. Bayer, ed.), pp. 6–34, Springer–Verlag, Berlin.

Wiedmann, J., 1970, Problems der Lobenterminologie, *Eclog. Geol. Helv.* **63**:909–922.
Wiedmann, J., and Kullmann, J., 1980, Ammonoid sutures in ontogeny and phylogeny, in: *The Ammonoidea* (M. R. House and J. R. Senior, eds.), pp. 215–255, Academic Press, New York.
Wright, C. W., and Kennedy, W. J., 1979, Origin and evolution of the Cretaceous micromorph ammonite family Flickiidae, *Paleontology* **22**(3):685–704.

Chapter 10

Heterochrony in Gastropods

A Paleontological View

DANA GEARY

1. Introduction

Fossil gastropods seem to be ideally suited for studies of heterochrony. The description and analysis of heterochrony in evolutionary sequences require a detailed understanding of the ontogeny of the organisms involved. Not only must the ontogeny be well-preserved in its entirety, but the basic parameters underlying growth must be amenable to quantitative description. In a gastropod shell, most, or frequently all, of an individual's ontogeny is recorded. The gastropod shell grows by continued accretion of new material onto the previously existing shell, so that the form of the juvenile is generally intact and part of the adult shell. Furthermore, there is no question about which juvenile form grew into which adult form, as there might be with organisms that molt, such as ostracodes and trilobites.

The gastropod shell is also well-suited to quantitative analysis (Thompson, 1942; Raup, 1961, 1966; Raup and Michelson, 1965). Raup has established a basic set of four parameters that, taken together, describe much of the variation possible in gastropod shells. In practice, Raup's parameters cannot always be measured

DANA GEARY • Department of Geology and Geophysics, University of Wisconsin, Madison, Wisconsin 53706.

[resorption of the columella in *Melanopsis*, for example, makes the generating curve of early whorls impossible to trace (Geary, 1986)], but they provide a basic dynamic framework for studying shell ontogeny and can be modified to fit particular situations. In addition to basic shell form, shell ornamentation often exhibits conspicuous changes during ontogeny, which, if modified by heterochrony, would be relatively easy to detect.

A typical problem in the study of heterochrony concerns the age of individuals. Most fossil gastropods, like many other fossil organisms, lack obvious markers that indicate the onset of sexual maturity. This means that adult shells are usually identified on the basis of size (see McKinney, this volume). Size is an imperfect proxy for age, however, since growth rates may vary. This uncertainty means that specifying the type of heterochronic process may be difficult in some cases. For example, a descendant may be clearly paedomorphic, but without knowing the actual ages of both ancestor and descendant as adults, it is difficult to assess the relative contributions of neoteny and progenesis. This is not an intractable problem for fossil gastropods, however. The age of individuals, or at least relative growth rates, may be determined in many cases (see discussion of Gould's work on *Poecilozonites*, below; Jones, this volume). Overall, because of the accessibility of their ontogeny to description and quantification, fossil gastropods have great potential in studies of heterochrony.

2. Previous Work

Despite all of the advantages to the study and detection of heterochrony in gastropod shells, I have found only three reported cases in the fossil gastropod literature. This is surprising in light of the abundance of cases described in some other fossil invertebrate groups (see McNamara, this volume). What follows is a discussion of some of the strengths and shortcomings of these published cases, and a brief look at a potentially fruitful set of cases that upon closer examination may yield additional examples of heterochrony in fossil gastropods.

The terminology employed here is the same as that used by Alberch *et al.* (1979) and McNamara (1986). Briefly, the adults of *paedomorphic* descendants resemble the *juvenile* stages of their ancestors. Paedomorphosis may arise via one or more processes: *neoteny* (overall retardation of developmental rate); *progenesis* (precocious attainment of maturity); and *postdisplacement* (delay in onset age of growth). Conversely, the morphology of *peramorphic* descendants "goes beyond" that of their adult ancestors. Peramorphic processes include *acceleration* (acceleration of developmental rate), *hypermorphosis* (delayed attainment of maturity or delayed cessation of development), and *predisplacement* (earlier onset age of growth).

2.1. *Zygopleura neotenica*

Elias (1958) describes the species *Zygopleura (Cyclozyga) neotenica* from the Late Mississippian of Oklahoma. This species is characterized by revolving keels on the third to sixth whorls, which "nearly fade out" in the last two to three whorls. Elias (1958) concludes (p. 16): "the present example indicates the development of such keels in the adolescent stage, from which apparently this evolutionary innovation spread over the adult stages in a later geological time; hence the specific name *neotenica*." *Zygopleura neotenica* is not itself neotenic, but is thought to be ancestral to other members of the subfamily that are. Without accurate information on the age of individuals or their growth rates, however, the relative contributions of neoteny and progenesis cannot be assessed. Elias' use of the term neoteny rather than the more general term paedomorphosis is not surprising, given the confusion of terminology that has generally surrounded studies of heterochrony. A more fundamental problem with this example concerns the phylogenetic relationship between *Z. neotenica* and its purported descendant species; no descendants are specifically identified, nor is any independent justification for the relationship given. Finally, the description of *Z. neotenica* is based upon two rather poorly preserved external molds, and the species has elsewhere been referred to as "a questionable form" (Hoare and Sturgeon, 1978). *Zygopleura neotenica* must be considered a possible, but unconfirmed, example of paedomorphosis.

2.2. *Glossaulax*

Majima (1985) describes a complex situation involving three closely related species of *Glossaulax* (Naticidae) in which he believes heterochrony has been important. Majima utilizes four measures of the shape of the umbilical and parietal calluses, supplemented with qualitative descriptions of these same features, to describe intra- and interspecific differences. Though his data are intriguing, there appear to be several reasons to view his claim of heterochrony with caution.

The first difficulty is Majima's use of an "imaginary ancestral species," which he places at the base of a paedomorphocline. The ontogeny of this imaginary ancestor is constructed from various morphological stages of the three real species supposed to be its descendants. These stages do not necessarily correspond to adjacent ontogenetic stages of the real species. For example, the "intermediate" and "adult" stages of the "imaginary" ancestor correspond to two different morphs of one of the real species. It is conceivable that the differences between these two morphs are due to intraspecific variation in developmental timing. However, it hardly seems plausible to define an imaginary ancestral ontogeny by piecing together the various stages of real species and then to call the changes between these real species heterochronic based on comparison with this imaginary ancestor.

Aside from the discussion of this imaginary ancestor, which is both unnecessary and unconvincing, Majima presents some interesting data on three "real" species. Majima believes that *G. hyugensis*, a late Miocene species, gave rise to two late Pliocene species, *G. nodai* and *G. hagenoshitensis*. Majima emphasizes that these three species share certain features of the umbilical wall, which distinguish them from all other species in the genus.

Qualitative descriptions of the callus shapes seem to support a paedomorphic and a peramorphic interpretation for *G. nodai* and *G. hagenoshitensis*, respectively. Juvenile specimens of the ancestral species *G. hyugensis* are characterized by a "wedge-shaped anterior lobe of parietal callus" and a "subtrigonal" umbilical callus. As an adult, *G. hyugensis* exhibits a "subquadrate" umbilical callus. *Glossaulax nodai* is thought to be paedomorphic because adults of this species exhibit the callus morphology characteristic of juvenile *G. hyugensis*.

Majima argues that neoteny has been the mechanism of paedomorphosis in this case, because the descendant species attains a larger adult size than does the ancestor (50 mm and 36 mm, respectively). In general, the relationship between heterochronic process and the relative adult size of ancestor and juvenile is based on the typical pattern of changing growth rates during ontogeny (McNamara, 1986). Growth rates tend to be relatively high in juveniles, and to slow considerably or stop altogether with the attainment of sexual maturity. This means that an increase in the time spent in the juvenile phase (as in neoteny) will often result in a larger individual. Conversely, progenetic individuals, due to their early maturation, will spend a shorter time in the juvenile phase of rapid growth, and will attain correspondingly smaller adult sizes (McNamara, 1986). For these reasons, neoteny and progenesis are often associated with relatively larger and smaller descendants, respectively. These relationships, however, cannot be considered absolute rules. Various paedomorphic processes may act together in some cases (Alberch *et al.*, 1979) thereby eliminating any dependable correlation between size and process. For the same reasons, the various processes resulting in peramorphosis (acceleration, hypermorphosis, and predisplacement) will not exhibit a perfect correlation with size change. It seems certain only that "pure" progenesis must result in a relatively smaller descendant, while hypermorphosis, acting alone, can only result in a descendant larger than its ancestor. Potentially informative instances of combined heterochronic factors will be overlooked if we use size as a simple indicator of process. In the case of *G. nodai* then, postdisplacement, alone or in addition to neoteny, may have been involved. Particularly since juvenile forms of *G. nodai* are lacking, it is not possible to specify which paedomorphic process(es) are represented in this transition.

Glossaulax hagenoshitensis is thought to be peramorphic because juveniles and the "unusual morph" (the less common of two adult morphs) of this species exhibit a subquadrate umbilical callus like that of adult specimens of *G. hyugensis*. Majima claims that acceleration was involved in this transition because of the increased frequency of the "developmentally more advanced" of two morphs early in ontogeny. Yet the developmental relationship between the two morphs is based exclusively on the ontogeny of the "imaginary ancestor," and is therefore only a possibility, rather than fact. Majima also claims that predisplacement occurred, because the subquadrate umbilical callus appears earlier in ontogeny in the descendant. Without better knowledge of the early development of this character, it is impossible to know if this represents predisplacement or acceleration. In general, then, Majima's discussion of specific heterochronic processes is unconvincing, though both paedomorphosis and peramorphosis seem to have occurred.

2.3. *Poecilozonites*

The clearest case of heterochrony documented for fossil gastropods is described by Gould (1968, 1969, 1970, 1977). *Poecilozonites bermudensis* is a Pleistocene–Recent pulmonate gastropod from Bermuda. At least four times during the Pleistocene, the basic stock of *P. bermudensis* gave rise to paedomorphic populations. Individuals in these samples show the color patterns, thickness, and external shape characteristic of juveniles, but are scaled up in size. Furthermore, each incidence of paedomorphosis appears to have been an independent event in a peripheral isolate, based on geographic and morphological evidence. (In three of the four instances of paedomorphosis, the paedomorphs share distinctive morphological characteristics with contemporaneous nonpaedomorphs.)

Gould utilizes information on growth rates and the intensity of color banding to distinguish among the possible mechanisms of paedomorphosis. Young nonpaedomorphs are characterized by high growth rates, and exhibit less intense shell coloration than that found in adults. Given that the growth rate in pulmonates tends to decline at sexual maturity (Boettger, 1952) and that the intensity of shell pigmentation is thought to vary inversely with growth rate (Comfort, 1951), Gould argues that the less intense coloration characteristic of paedomorphs indicates that they retain these rapid juvenile growth rates to a larger size. In other words, the retardation of growth rates that typically occurs during ontogeny is itself retarded; paedomorphs spend a longer time in this juvenile phase and typical adult features never appear. This, then, is an example of neoteny. By utilizing coloration as an indicator of growth rate, Gould can specify the type of paedomorphic process.

Aside from the intensity of color banding, significant allometry occurs in many other variables in *P. bermudensis*, including thickness of the parietal callus, overall shell thickness, spire height, relative apertural height, and umbilical width. In the paedomorphs, the rate of ontogenetic retardation is similar for all of these characters, indicating that "paedomorphosis involves the entire developmental patterns of the shell, not just a few of its features" (Gould, 1968). Gould concludes that a relatively simple genetic change in some basic developmental mechanism was probably responsible for these neotenic events.

In *Poecilozonites*, then, we have an example of iterative evolution at the intraspecific level. Interestingly, these events seem to be correlated with environment; the most neotenic subspecies developed in red soils, while most nonpaedomorphs were found in carbonate dunes. Presumably, the thin paedomorphic shells were adaptive in the low-calcium environment of the red soils (Gould, 1968).

2.4. Dwarf Gastropods

Dwarf faunas, often including gastropods, have been described and discussed by many authors (Tasch, 1953; Hallam, 1965; Snyder and Bretsky, 1971; Mancini, 1978a,b; and references therein). Several types of explanation have been advanced to account for the diminutive size of these faunas. In many cases, atypically small size may represent stunting, produced by environmental factors such as low food supply, abnormal salinity, temperature, or any of numerous other factors important in growth (Crabb, 1929; Tasch, 1953; Hallam, 1965). Purely sedimentological factors such as winnowing may also result in abnormally small faunas.

Aside from these nongenetic influences, dwarfed faunas may evolve through

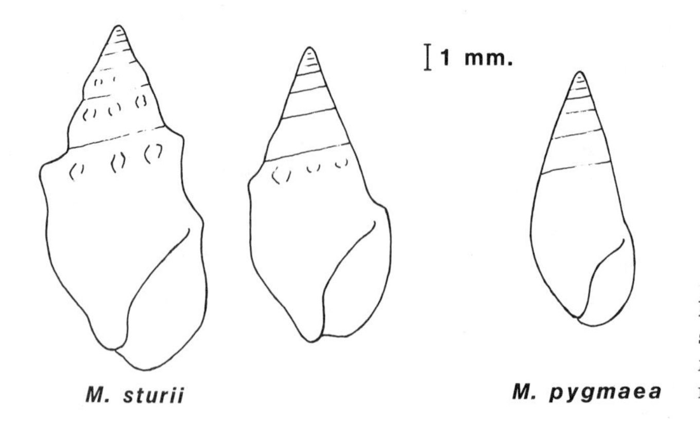

Figure 1. Sketch of *Melanopsis sturii*, its descendant *M. pygmaea*, and a specimen of intermediate morphology.

heterochrony, particularly via progenesis. Progenesis may result from selection pressures for small size or for early maturation. Early maturation is thought to be advantageous when resources are in abundance, or in a high-stress environment where high fecundity is important (Snyder and Bretsky, 1971; Gould, 1977). The challenge with the interpretation of dwarf faunas is deciding which environmental, sedimentological, or evolutionary (genetic) factors apply in any particular case (Mancini, 1978a,b).

There are several reports of dwarf gastropods (Ball, 1935; Stauffer, 1937; Britton and Stanton, 1973; Ginda, 1976; Pampe, 1979), though no case has been attributed to heterochrony. These cases remain possible examples, which upon detailed reinvestigation may turn out to involve heterochrony. The occurrence of dwarf gastropods is a phenomenon bearing a closer look if the general importance of heterochrony in fossil gastropods is to be assessed.

3. Recognizing Heterochrony in Sequences of Fossil Gastropods: Examples from *Melanopsis*

The mesogastropod genus *Melanopsis* radiated in the brackish-fresh waters of the Pannonian Basin of Eastern and Central Europe in the late Miocene. Detailed study of this radiation (Geary, 1986) provides us with two new examples of heterochrony in fossil gastropods.

3.1. *Melanopsis sturii–M. pygmaea*

The first example involves the ancestral species *M. sturii* and its paedomorphic descendant *M. pygmaea* (Geary, 1986). These species are illustrated in Fig. 1. *Melanopsis sturii* begins its ontogeny with a smooth shell, growing with a high rate of translation. At some point between the fifth and seventh whorls, the shell begins to develop its noded ornament, and the component of growth away from the axis increases slightly, resulting in a more inflated shell. *Melanopsis pygmaea* begins its ontogeny in the same way; the shell is smooth and

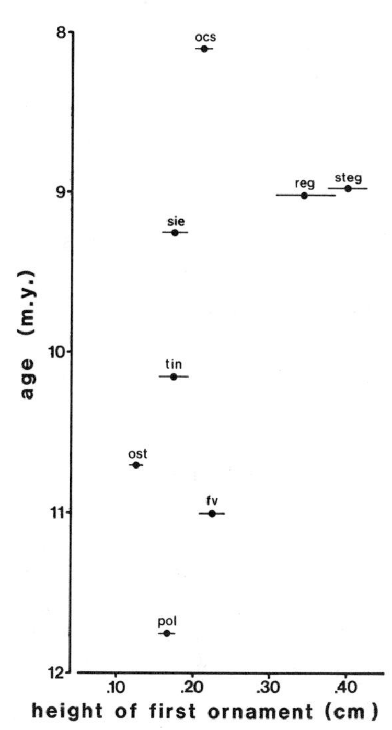

Figure 2. Height of first ornament (measured from apex of shell) vs. geologic age for samples of *Melanopsis sturii*. Dots represent sample means; error bars are plus or minus two standard errors about the mean. Letters correspond to locality abbreviations (see Geary, 1986).

growth is mostly translational. *Melanopsis pygmaea* simply retains this juvenile growth program into adulthood, yielding its characteristically smooth and relatively narrow shell. Unfortunately, without more information on the absolute age of the individual gastropods, it is impossible to know which paedomorphic process(es) were involved.

One of the most interesting aspects of this transition lies in the patterns of variation within the ancestral species *M. sturii*. "Intermediate" forms occur (see middle sketch of Fig. 1), which appear to be just that: the onset of ornamentation is delayed so that only the final whorl or so is ornamented. There are scattered occurrences of this intermediate form in at least five separate samples, taken from a variety of localities across the basin and from various times in the 3 million year history of *M. sturii* after the branching event that led to *M. pygmaea*. These scattered occurrences differ from the repeated paedomorphic events discussed by Gould (1968) for *Poecilozonites*. His evidence for the independence of each paedomorphic event was both geographic and morphological. In the case of *M. sturii*, no such evidence exists. The intermediates occur in various localities across the basin, but are always associated with "normal" specimens of *M. sturii*. I believe these intermediates simply represent rather extreme intraspecific variation. This variation is of particular interest because it is oriented along the same development pathway as the one that led to *M. pygmaea*.

Aside from the scattered occurrence of these intermediate specimens, the height of onset of ornamentation is a highly variable character among samples of *M. sturii*. Shown in Fig. 2 are the sample means (plus or minus two standard errors about the mean) for the height of onset of ornamentation, plotted against the absolute age of each sample. There is no real temporal trend for this character; what is of interest is the relatively high degree of variation among samples. In a one-way ANOVA for this character, I found that 80% of all possible pairwise comparisons among samples were significantly different in *a posteriori* contrasts ($\alpha = 0.05$). There were far more significant differences among samples of *M. sturii* for the height of onset of ornamentation than there were for any other character measured. Thus, in the speciation event that gave rise to *M. pygmaea*, in the occasional appearance of the intermediate morphology with only the final whorl ornamented, and in the high degree of variability among samples of *M. sturii*, variation occurs along the same developmental pathway: always involving a delay in the onset of ornamentation.

3.2. *Melanopsis impressa–M. fossilis–M. vindobonensis*

The second example from *Melanopsis* involves three closely related species: *M. impressa*, the most ancient melanopsid species in the basin, its daughter species *M. fossilis*, and *M. vindobonensis*, the daughter species of *M. fossilis* (Geary, 1986). The ontogeny of shell form in each of these species is best described by a sequence of comparable points on the right side of the shell (see Fig. 3). These points are the upper-rightmost "corner" of each whorl, marked by an inflection in the whorl outline, and can be seen easily in X-ray pictures of the shell. The angle between successive points (measured parallel to the axis of coiling as shown in Fig. 3) describes shell growth during that interval of ontogeny. A low angle indicates mostly translational growth down the axis; higher angles correspond to an increase in the whorl expansion component of growth, growth perpendicular to the axis of coiling. The series of angles for each shell describes the basic ontogenetic sequence of shape changes.

Each angle was plotted against the absolute height along the axis from the bottom of the angle to the apex of the shell. Figures 4–6 are examples of these plots, for *M. impressa*, *M. fossilis*, and *M. vindobonensis*, respectively. Figure 5, for example, includes all of the angles measured for all of the individuals (adults and subadults) in a single sample of *M. fossilis* from Stegersbach, Austria. Moving from left to right across the diagram corresponds to moving down the axis of coiling. Earliest growth, represented by points at the left of the plot, yields angles between 0 and 25°, corresponding to a high rate of translation down the axis. This phase of growth produces the high-spired tip of the shell. Following this, the angles quickly increase to between 25 and 50°, occasionally even higher. These high angles mark an increase in the component of growth away from the axis. This phase of growth results in the broad, sloping "shelf" on the upper part of the shell. Finally, the angles taper off, representing the final phase of shell elongation. Other populations of *M. fossilis* were plotted in the same way, and all yielded the same pattern.

It is of interest here to compare the patterns typical of each species. Figure 4 is a typical pattern for *M. impressa*, the ancestral species. Earliest angles are low, and in this case remain so through ontogeny, with only a slight increase and then a gradual decrease. This pattern yields the fairly straight-sided *M. impressa* form. The pattern for *M. impressa*'s descendant species *M. fossilis* (Fig. 5) was

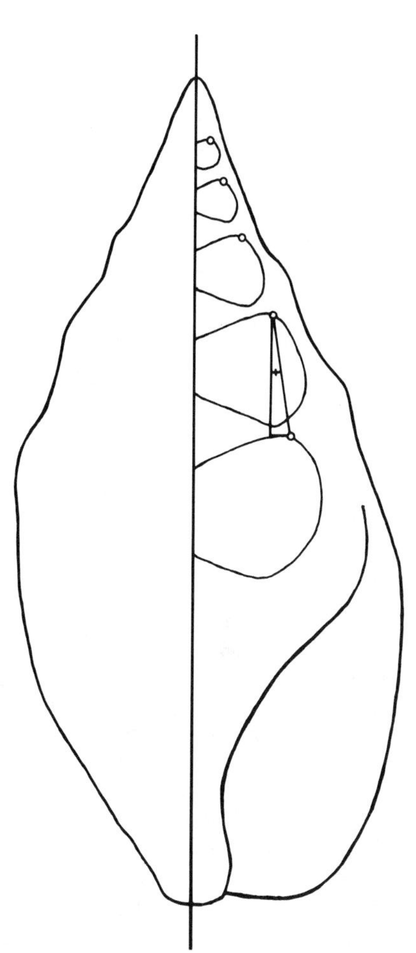

Figure 3. Internal angles measured on specimens of *Melanopsis*. Comparable points marked on each whorl are indicated by open circles. Angles were measured between successive whorls and the line drawn parallel to the coiling axis as shown: low angles indicate high translation; high angles (>45 deg) indicate mostly expansional growth. Occasionally, the lowermost point fell to the left of the line parallel to the coiling axis. These angles were recorded as negative (fall below 0 in Figs. 5 and 6).

described above. *Melanopsis fossilis'* descendant, *M. vindobonensis*, is shown in Fig. 6. In the case of *M. vindobonensis*, angles are low in earliest ontogeny, corresponding to the tip of the spire. (X-rays of relatively high voltage are required to penetrate the extremely thick shells of *M. vindobonensis*, so proper exposure of the bulk of the shell results in overexposure of the relatively thin apex. For this reason, the angles in the earliest whorls were only rarely visible or recorded, yet I am confident that earliest growth in all cases was highly translational.) The angles then jump quickly to over 50°, and then fall almost as quickly to very low values, corresponding, respectively, to the early period of widening, and then the rounding off of the shell.

Comparison of the ontogenetic patterns of ancestors and descendants reveals an interesting sequence of changes. The simple pattern characteristic of *M. impressa* is exaggerated and extended in its descendant species *M. fossilis*. The change from *M. impressa* to *M. fossilis* is important because it represents the development of a complex pattern of allometric growth from a rather simple growth sequence.

The change from *M. fossilis* to its descendant species *M. vindobonensis* is

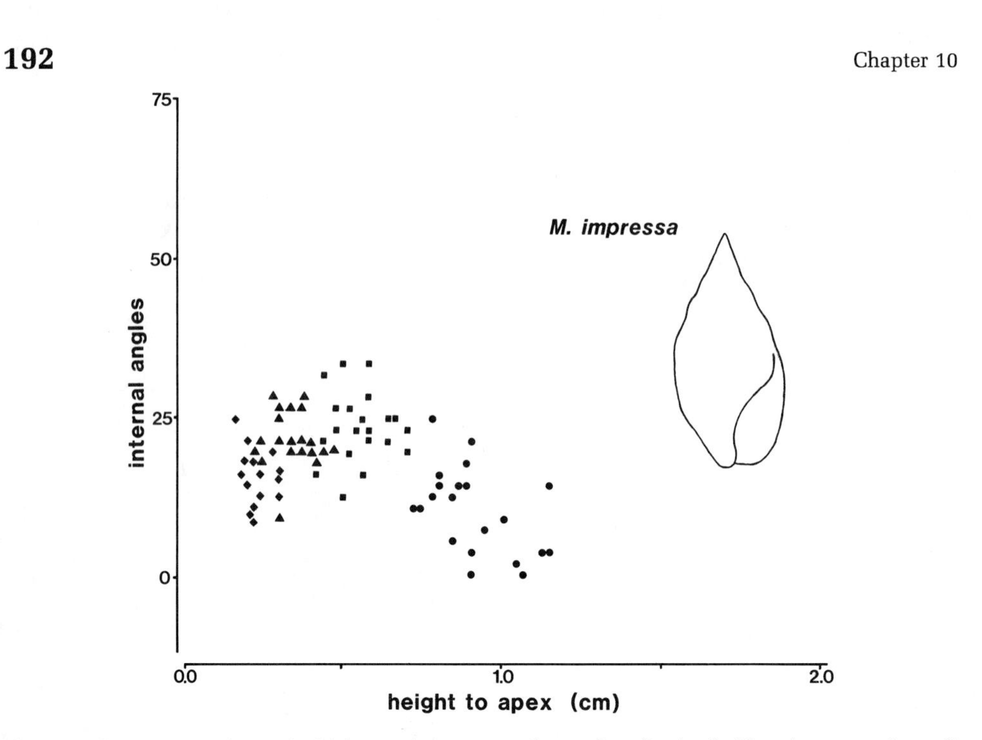

Figure 4. Ontogenetic change in *Melanopsis impressa.* Internal angles (as in Fig. 3) measured on all individuals from one sample are plotted against the height from the bottom of the angle (point on lower whorl) to apex of shell. A maximum of four angles was recorded from each specimen. The various symbols represent angles between particular whorls: diamonds, triangles, squares, and circles represent, respectively, the uppermost (younger) to lowermost (older) angles (see Fig. 3).

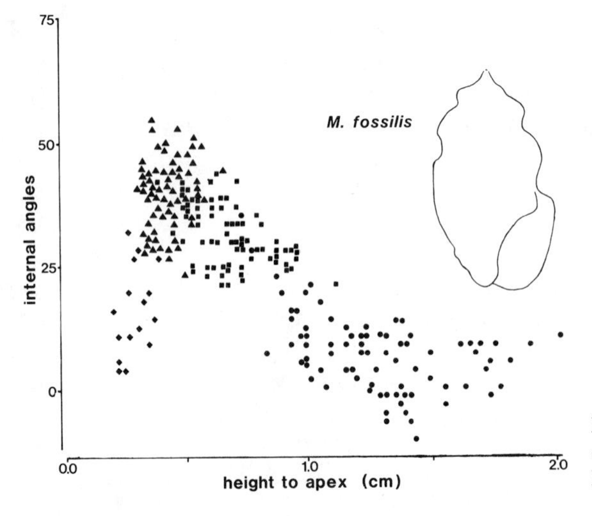

Figure 5. Ontogenetic change in *Melanopsis fossilis.* Axes and symbols as in Fig. 4.

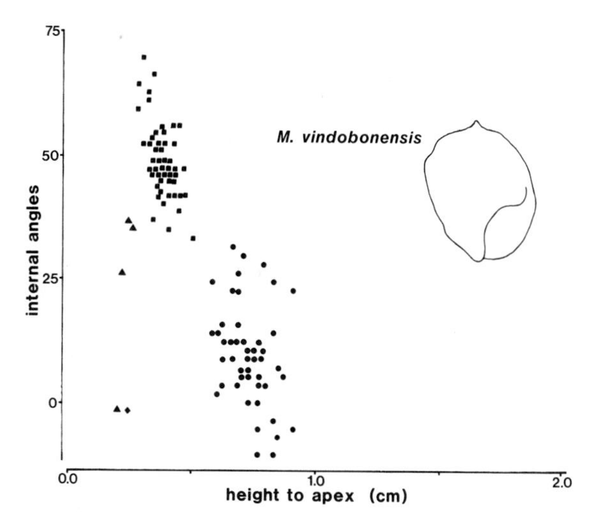

Figure 6. Ontogenetic change in *Melanopsis vindobonensis*. Axes and symbols as in Fig. 4.

achieved by simple variations on a similar ontogenetic theme. Both species have a three-phase allometry, but the descendant sequence is compressed and exaggerated with respect to that of the ancestor. It appears that a peramorphic change in the timing of the developmental process occurred between *M. fossilis* and *M. vindobonensis*, so that the same process is accelerated with respect to shell size in the descendant. Accompanying this acceleration, however, we do not see a recapitulation of ancestral stages in the ontogenetic sequence of the descendant. *Melanopsis vindobonensis* does not pass through a "*fossilis*" stage in its ontogeny, because the *fossilis* morphology is not a stage in the ontogenetic sequence. Instead, the *fossilis* morphology represents the static outcome of the combination of various growth parameters that have been implemented in morphogenesis. Alberch (1985) argues that the proper dynamic view of development will involve no expectation of recapitulation. In his view, morphology emerges "as the combination of a set of specific pattern-generating algorithms with a set of initial boundary conditions." The relationship between two morphologies should be sought in terms of changes in the parameters of these algorithms, rather than in the sequence of ontogenetic stages (Alberch, 1985). The transition between *M. fossilis* and *M. vindobonensis* provides an example of this type of developmental change; it is the underlying process involving the basic parameters of shell growth that is accelerated, rather than a sequence of morphological stages.

4. Discussion

Several important conclusions can be drawn from the case studies discussed above. First, it is clear that a detailed study of heterochrony, as with most types of evolutionary inquiries, depends on an accurate phylogeny. Elias' work, for instance, suffers from a lack of phylogenetic detail; without a more clearly drawn set of interspecific relationships, his argument for heterochrony remains unconfirmed. In Majima's study, the relationship of *G. hyugensis*, *G. nodai*, and *G.*

hagenoshitensis appears to be more carefully examined. Some of the claims for heterochrony are based not on these real species, however, but on their relationship to an imaginary ancestor. This imaginary ancestor cannot be used to construct a phylogenetic hypothesis *and* as the basis for statements about heterochrony.

A wide range of gastropod shell characters exhibit heterochronic change, including coiling parameters, shell thickness, ornamentation, coloration, and callus morphology. In *Poecilozonites*, the entire suite of allometric shell characters was affected by paedomorphosis. In the *Melanopsis sturii–M. pygmaea* sequence, both allometric characters (ornament and shape) were affected by paedomorphosis. In these cases, then, the correlations among ontogenetic characters remained intact through phylogenetic changes. Heterochrony may not always work in this way, however; correlations among ontogenetic characters can be broken, leaving some characters affected by heterochrony and others not (see Lindberg, this volume). It remains to be seen with what frequency the correlations among various gastropod shell characters can be broken. It is important to understand to what extent the gastropod shell is an integrated developmental whole, rather than a set of independently varying characters [see Gould (1984) for a discussion of covariance sets in gastropod shells].

These studies also point out the difficulties in distinguishing among the various heterochronic processes. In the majority of cases discussed, the overall morphological result can be described as paedomorphic or peramorphic, but, for lack of information on the age of individuals, the specific process or processes that led to this result cannot be identified. These difficulties may be only part of a deeper issue concerning mechanisms of developmental change. The nomenclature of heterochrony, which now seems to have been clarified to the point where discussion can proceed on a common ground (Gould, 1977; Alberch *et al.*, 1979; McNamara, 1986), may nonetheless be inadequate, or at least overly rigid. At least two general problems exist. First, when two or more heterochronic processes have influenced the same transition, the results may be difficult to interpret in the current framework. Interpretations based on an overly simplistic view of heterochronic processes and their results will lead to erroneous conclusions. Second, when ontogeny and heterochrony are viewed in a dynamic context, our current terminology may not be adequate to describe the mechanisms of change (Alberch, 1985). In the sequence involving *Melanopsis impressa–fossilis–vindobonensis*, for instance, changes in developmental timing are evident in ontogenetic plots (Figs. 4–6), yet comparison of actual morphologies at any "stage" (adults with adults, or adults with juveniles) yields no obvious indication of heterochrony. Changes in the developmental program will not always yield predictable morphological results.

Given the relative abundance of cases of heterochrony described in some other invertebrate groups, how can we account for the scarcity of cases in fossil gastropods? It seems likely that gastropod workers have, in their attention to other issues, sometimes missed instances of heterochrony in fossil lineages. One reason for this may be that gastropod shell allometries are often quite subtle. Unless ontogeny is studied carefully, with the identification of allometry as an objective, ontogenetic variation in coiling parameters is likely to be missed. To date, most applications of Raup parameters have involved examination of theoretical morphospace based on all possible variations of these parameters, or comparison of parameter values among different coiling types. These studies have generally focused on coiling parameters that do not change during ontogeny [though see Raup (1966)]. With increased attention to ontogenetic change in coiling parameters, and with awareness of the fact that Raup parameters may often have to be modified to fit the situation at hand, the potential for the recognition of heterochrony in fossil gastropods will grow.

Heterochrony is a type of evolutionary change, but it is not an explanation in itself. It provides a means to an end, often a selective end. It is therefore appropriate to ask about the nature of the selective advantage in any particular case. In so doing, the study of heterochrony can move from a purely descriptive phase to a more explanatory one. (The current paleontological literature on heterochrony involves both types of approaches.) As Gould (1977) has emphasized, the adaptive significance of heterochrony may involve life history characteristics, as well as size itself or other aspects of morphology. Ecological correlations (Gould, 1968, 1977; Snyder and Bretsky, 1971; McKinney, 1986) seem to provide a promising avenue toward uncovering the impetus for heterochronic change in particular cases, and toward establishing more general principles regarding the frequency and direction of heterochronic change.

Our explanations, however, must involve factors other than selective ones. In our descriptions of heterochronic change lie a wealth of information on correlations among characters, within species and within evolutionary lineages. These sets of correlated characters reflect the influence of both phylogenetic and structural constraints, and define, to a large extent, the potential pathways of evolutionary change. The *Melanopsis sturii–M. pygmaea* sequence illustrates this type of pathway; variation at several levels appears to be channeled along the same developmental lines.

The study of heterochrony, then, offers insights into both extrinsic and intrinsic mechanisms of evolutionary change. By evaluating the morphological and life history consequences of heterochrony in the light of environmental or ecological setting, we may discover patterns that suggest what sorts of selective mechanisms are at work in various situations. Selective mechanisms are only part of the story, however. The other, equally important part involves the actual path of change taken, which depends on intrinsically defined pathways. Through studies of heterochrony, we can hope to enhance our understanding of both the nature of the selective forces operating in particular cases and the nature of an organism's responses to these forces.

ACKNOWLEDGMENTS. I would like to thank Robert Bleiweiss, David Lindberg, and Michael McKinney for reviews of this chapter. Stephen Jay Gould, Peter Williamson, and Ruth Turner provided helpful comments on my study of melanopsid gastropods. My work in eastern Europe was funded and facilitated by International Research and Exchanges Board (IREX) in Princeton, NJ.

References

Alberch, P., 1985, Problems with the interpretation of developmental sequences, *Syst. Zool.* **34**:46–58.

Alberch, P., Gould, S. J., Oster, G. F., and Wake, D. B., 1979, Size and shape in ontogeny and phylogeny, *Paleobiology* **5**:296–317.

Ball, J. R., 1935, Dwarfed gastropods in the basal Guttenberg, Southwestern Wisconsin, (abs.), *Geol. Soc. Am. Proc.* **1935**:384.

Boettger, C. R., 1952, Grossenwachstum und Geschlechtsreife bei Schnecken und pathologischer Riesenwuchs als Folge einer gestorten Wechselwirkung beider Faktoren, *Zool. Anz.* **17**(suppl.):468–487.

Britton, E. R., and Stanton, R. J., Jr., 1973, Origin of "dwarfed" fauna in the Del Rio Formation, Lower Cretaceous, east central Texas, in: *Geological Society of America, Abstracts with Program*, Vol. 5, pp. 248–249.

Comfort, A., 1951, The pigmentation of molluscan shells, *Biol. Rev.* **26**:285–301.

Crabb, E. D., 1929, Growth of a pond snail *Lymnaea stagnalis appressa* as indicated by increase of shell-size, *Biol. Bull.* **56**:41–63.

Elias, M. K., 1958, Late Mississippian fauna from the Redoak Hollow Formation of Southern Oklahoma, *J. Paleontol.* **32**:1–57.

Geary, D. H., 1986, The evolutionary radiation of melanopsid gastropods in the Pannonian Basin (Late Miocene, Eastern Europe), Ph.D. dissertation, Harvard University, Cambridge, Massachusetts.

Ginda, V. A., 1976, The dwarf gastropods in the Ordovician Baltic Basin, *Paleontol. Sb.* (*L'vov*) **13**:51–55.

Gould, S. J., 1968, Ontogeny and the explanation of form: An allometric analysis, *Paleontol. Soc. Mem.* **2**:81–98.

Gould, S. J., 1969, An evolutionary microcosm: Pleistocene and Recent history of the land snail *P.* (*Poecilozonites*) in Bermuda, *Bull. Mus. Comp. Zool.* **138**:407–532.

Gould, S. J., 1970, Land snail communities and Pleistocene climates in Bermuda: A multivariate analysis of microgastropod diversity, in: Proceedings North American Paleoontology Convention, Part E, pp. 486–521.

Gould, S. J., 1977, *Ontogeny and Phylogeny*, Harvard University Press, Cambridge.

Gould, S. J., 1984, Morphological channeling by structural constraint: Convergence in styles of dwarfing and gigantism in *Cerion*, with a description of two new fossil species and a report on the discovery of the largest *Cerion*, *Paleobiology* **10**:172–194.

Hallam, A., 1965, Environmental causes of stunting in living and fossil marine benthonic invertebrates, *Palaeontology* **8**:132–155.

Hoare, R. D., and Sturgeon, M. T., 1978, The Pennsylvanian gastropod genera *Cyclozyga* and *Helminthozyga* and the classification of the Pseudozygopleuridae, *J. Paleontol.* **52**:850–858.

Majima, R., 1985, Intraspecific variation in three species of *Glossaulax* (Gastropoda, Naticidae) from the Late Cenozoic strata in central and southwest Japan, *Trans. Proc. Palaeontol. Soc. Japan* (N. S.) **0**(138):111–137.

Mancini, E. A., 1978a, Origin of micromorph faunas in the geologic record, *J. Paleontol.* **52**:311–322.

Mancini, E. A., 1978b, Origin of the Grayson micromorph fauna (Upper Cretaceous) of North Central Texas, *J. Paleontol.* **52**:1294–1314.

McKinney, M. L., 1986, Ecological causation of heterochrony: A test and implications for the study of chronoclines, *Paleobiology* **12**:282–289.

McNamara, K. J., 1986, A guide to the nomenclature of heterochrony, *J. Paleontol.* **60**:4–13.

Pampe, W. R., 1979, A dwarfed fauna from the Grayson Formation near Lake Waco, Texas, *Earth Sci. Bull.* **12**:18–32.

Raup, D. M., 1961, The geometry of coiling in gastropods, *Proc. Natl. Acad. Sci. USA* **47**:602–609.

Raup, D. M., 1966, Geometric analysis of shell coiling: General problems, *J. Paleontol.* **40**:1178–1190.

Raup, D. M., and Michelson, A., 1965, Theoretical morphology of the coiled shell, *Science* **147**:1294–1295.

Snyder, J., and Bretsky, P. W., 1971, Life habits of diminutive bivalve molluscs in the Maquoketa Formation (Upper Ordovician), *Am. J. Sci.* **271**:227–251.

Stauffer, C. R., 1937, A diminutive fauna from the Shakopee Formation at Cannon Falls, Minn., *J. Paleontol.* **11**:55–60.

Tasch, P., 1953, Causes and paleoecological significance of dwarfed fossil marine invertebrates, *J. Paleontol.* **27**:356–444.

Thompson, D'A. W., 1942, *On Growth and Form*, Cambridge University Press, Cambridge.

Chapter 11

Heterochrony in Gastropods

A Neontological View

DAVID R. LINDBERG

1. Introduction: Taxa, Methods, and Materials

All members of monophyletic taxa share a common ancestor whose development pathway has been modified to produce descendant morphologies. One important class of development modifications consists in changes in the timing of the development of existing characters; this process is called heterochrony (Gould, 1977, p. 4). The morphological expression of heterochronic changes at the developmental level is limited, and the forms of expression have been described qualitatively and quantitatively by de Beer (1951), Gould (1977, p. 209), Alberch *et al.* (1979), Alberch (1980), Bonner (1982), Fink (1982), McNamara (1982, 1986), McKinney (1986), and references in these works. In summary, heterochrony pro-

DAVID R. LINDBERG • Museum of Paleontology, University of California, Berkeley, California 94720.

duces two forms of morphological expression: *paedomorphosis*, the retention of ancestral juvenile characters by later ontogenetic stages of descendants (Gould, 1977, p. 484), and *peramorphosis*, new descendant characters produced by additions to the ancestral ontogeny (Alberch *et al.*, 1979). McNamara (1986) has defined three processes for each form of expression as follows. The paedomorphic processes are (1) *progenesis*, precocious sexual maturation, (2) *neoteny*, reduced rate of morphological development, and (3) *postdisplacement*, delayed onset of growth. The peramorphic processes are (1) *hypermorphosis*, delayed sexual maturation, (2) *acceleration*, increased rate of morphological development, and (3) *predisplacement*, earlier onset of growth (see also McKinney, this volume).

The role of heterochrony in biotic evolution has received renewed interest and study since the appearance of Gould's (1977) seminal treatment of ontogeny and phylogeny. Today the search for heterochrony has been taken from its foundations in developmental biology and ontogenetic time scales into the fossil record, where data may consist of only adult morphology and time is measured in millions of years rather than in hours, days, or weeks.

In this chapter, developmental pathways in extant molluscan species and heterochronic processes inferred from them are examined in the context of evolutionary trends in morphology and life history within the class Gastropoda, particularly marine members of the subclass Prosobranchia and the order Patellogastropoda (= former superfamily Patellacea).

1.1. Phylogenetics

The phylogenetic history among taxa must be known before the role of heterochrony can be determined (Fink, 1982; de Queiroz, 1985). The class Gastropoda is a monophyletic group diagnosed by torsion: the counterclockwise rotation of the visceral mass 180° relative to the anteroposterior axis of the head–foot complex (Spengel, 1881). The phylogenetic relationships of the gastropod subclades used here is primarily based on the work of Fretter and Graham (1962), Graham (1985), and Haszprunar (1985b). These workers considered the Archaeogastropoda (includes Patellogastropoda) to be the most primitive of the prosobranch gastropods, and Caenogastropoda the most derived. In the order Patellogastropoda, Lindberg (1988) considers the Patellidae to be the most primitive and the Lepetidae the most derived taxon.

Polarity of character states has been determined by: (1) outgroup comparisons using the Monoplacophora and Polyplacophora (e.g., Table I); (2) comparisons of outgroup and ingroup ontogenies (e.g., Fig. 1); and (3) the fossil record (e.g., Fig. 2); these data sets are not entirely independent [see de Queiroz (1985) for a critique of this methodology]. The distribution of character states discussed herein is presented in Table I.

1.2. Fossil Record

Recent taxa have a decided advantage over fossil species in studies of heterochrony because of the availability of whole organisms and complete life cycles.

TABLE I. General Character State Polarity

Character	Outgroup	Prosobranch gastropods		Patellogastropoda	
		Primitive	Derived	Primitive	Derived
Radula	Docoglossate	Docoglossate	Taxoglossate	Docoglossate	Docoglossate
Radular teeth	Numerous	Numerous	Few	Numerous	Few
Tooth fields	Three	Three	One	Three	One
Gut looping	Complex	Complex	Simple	Complex	Simple
Style	Present	Present	Absent	Absent	Absent
Gills	Many	Two	One or none	Many	One or none
Auricles	Many	Two	One	One	One
Fertilization	External	External	Internal	External	Internal
Spawn	Broadcast	Broadcast	Brood	Broadcast	Brood
Larval nourishment	Yolk	Yolk	Plankton	Yolk	Yolk
Protoconch	Noncoiled	Coiled	Coiled	Noncoiled	Noncoiled
Operculum	Absent	Present	Present	Present	Present

Moreover, when fossils are incorporated into a cladistically and stratigraphically derived phylogeny (Paul, 1982, 1985), clues of the role of heterochrony in their evolution may be uncovered. Examining patterns of morphological change in living taxa can also link the patterns of morphological change in the fossil record and the inferred processes of heterochrony.

The Gastropoda fossil record is one of the best available and provides an added historical perspective for many patterns seen in living snails. Although its time of origin is unknown, most workers consider Gastropoda to have appeared in the Late Cambrian (Signor, 1985; Runnegar and Pojeta, 1985). In the fossil record extant "primitive" prosobranch taxa, which are diagnosed primarily on anatomical criteria, appear before extant "derived" groups (Sepkoski, 1982). Other orders, such as Opisthobranchia and Pulmonata, are believed to date from the Carboniferous (Kollmann and Yochelson, 1976; Solem and Yochelson, 1979); however, because their preservable hard parts are fragile and few, the history of these groups is not as well known as that of the prosobranchs.

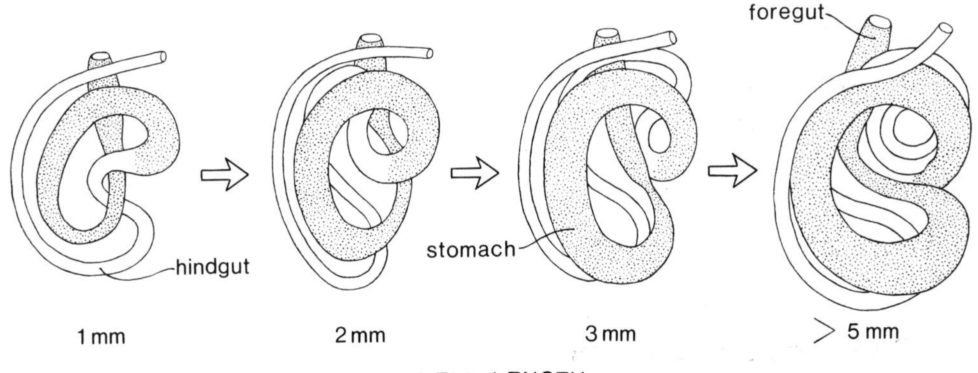

Figure 1. Ontogeny of the alimentary system of *Lottia*. Dorsal views. Note that the relative size and shape of the foregut (ectoderm) and stomach (endoderm) (both stippled) remain unchanged, while the hindgut (endoderm and mesoderm) lengthens. After Walker (1968).

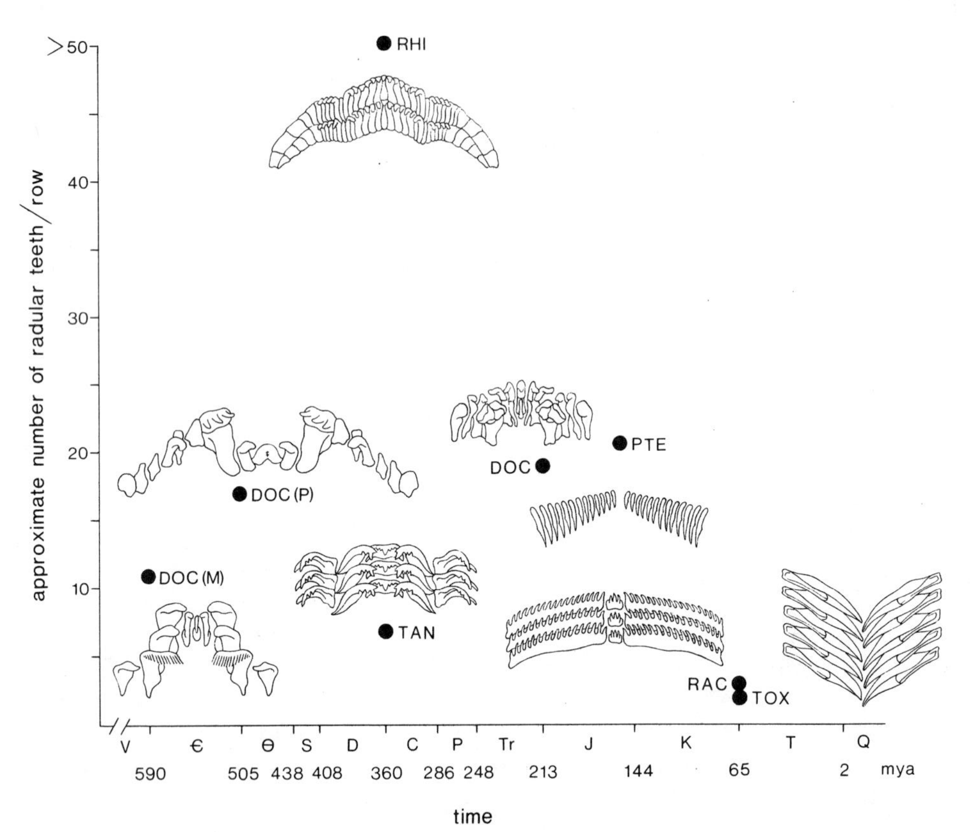

Figure 2. Times of appearance of major extant radula types. Family origin times from Sepkoski (1982). Pre-Silurian data represents the outgroups Monoplacophora (M) and Polyplacophora (P). DOC, Docoglossate; PTE, ptenoglossate; RAC, rachiglossate; RHI, rhipidoglossate; TAN, taenioglossate; and TOX, toxoglossate. Gastropod radula (RHI, TAN, DOC, PEN, RAC) redrawn from Fretter and Graham (1962), (TOX) from McLean (1971); monoplacophoran radula [DOC (M)] redrawn from McLean (1979); polyplacophoran [DOC (P)] radula drawn by C. Niemi.

1.3. Development

Gastropod development has been documented and discussed by Fretter and Graham (1962), Raven (1966), Hyman (1967), Bandel (1982), Verdonk *et al.* (1983), Jablonski (1985), and in references cited in these works. The overview presented here is summarized from these treatments and focuses primarily on the development of organ systems and associated structures, thus providing a frame of reference for the timing of heterchronic changes discussed below.

2. Gastropod Development

The germ layers differentiate very early in gastropod development. The ectoderm is derived from the first three quartets of micromeres, the endoderm from the fourth quartet and macromeres, and the mesoderm from two sources: (1) the ectomesoderm from the second or third quartet of micromeres, and (2) the endomesoderm from the 4d cell (Verdonk and van den Biggelaar, 1983).

After gastrulation, either by invagination or epiboly (the latter being found in species with yolk-rich eggs), further cell divisions and reorganization produce a trochophore larval stage (Fig. 3, bottom left). At this stage the ectoderm is divided into two distinct regions: the pretrochal region above the prototroch and the posttrochal region below. From the posttrochal ectoderm will be derived the shell gland (and ultimately the mantle), mantle cavity, gill, body trunk, foot, and stomodaeum (which differentiates into the buccal cavity, pharynx, radular sac, and esophagus). The head region is derived from the pretrochal ectoderm. The nervous system (ganglia and commissures) is derived from both the pre- and posttrochal regions.

Endoderm makes up the archenteron, which is formed during gastrulation. Endoderm structures include the stomach and digestive gland. The endoderm fuses anteriorly with the stomodaeum (ectoderm) to complete the anterior portion of the alimentary system. The hindgut is formed from the posterior portion of the archenteron (Moor, 1983) and some of the enteroblasts that originate from 4d cell descendants (mesoderm) (Verdonk and van den Biggelaar, 1983).

After these cell divisions produce the precursors of the hindgut, the two mesodermal teloblasts (also derived from the 4d cell) begin dividing and produce two mesodermal bands on either side of the archenteron. From these two bands the musculature, kidneys, pericardium, and heart will be derived (Fig. 3, schematics). The gonad typically develops last and may be derived from the pericardium, independent mesodermal cells, or by differentiation of the primordial germ cells (Moor, 1983). In most prosobranchs, the gonoduct, which connects the gonad with the mantle cavity, has a proximal mesoderm section and a distal ectoderm section.

Developmental rates (egg to settlement) within the Gastropoda vary among and within taxa. In general, the Caenogastropoda tend to have longer developmental periods than the Patellogastropoda, Archaeogastropoda, and Opisthobranchia (Amio, 1963, Fig. 11).

3. Gastropod Character Patterns and Heterochrony

Several distinct evolutionary trends are represented in Recent Gastropoda (Fretter and Graham, 1962; Hyman, 1967; Golikov and Starobagatov, 1975; Salvini-Plawen, 1980; Haszprunar, 1985a,b; Graham, 1985), some of which may have been produced by heterochrony.

3.1. Previous Work

Heterochrony in gastropod shell evolution has been proposed by numerous workers, including Garstang (1929, p. 95) for the Zygobranchia (= superfamilies Pleurotomariacea and Fissurellacea), Gould (1969) for pulmonate snails, Batten (1975) for the evolution of the Scissurellidae, McLean (1984, p. 17) for the origin

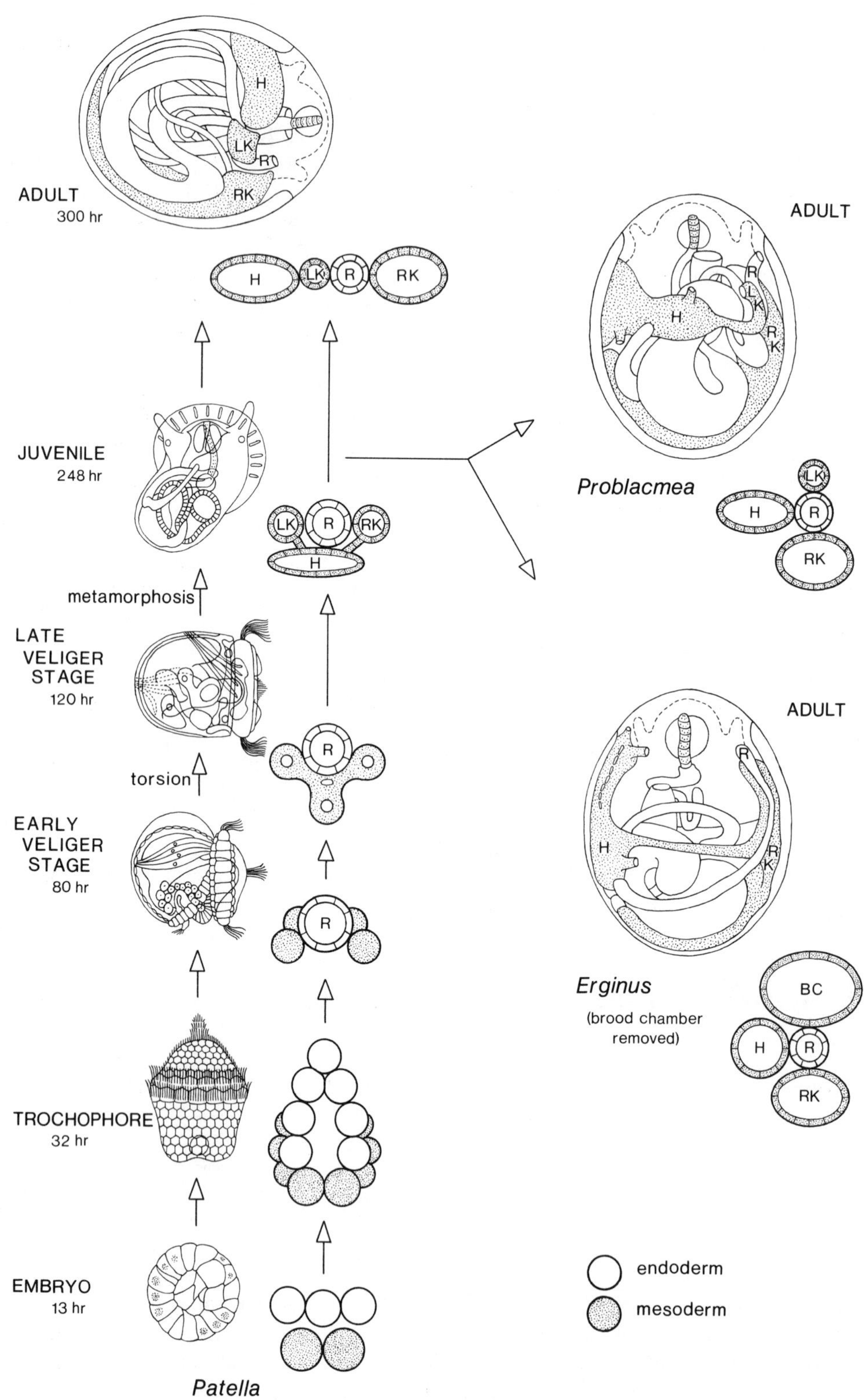
ADULT
300 hr
H
LK
R
RK
JUVENILE
248 hr
metamorphosis
LATE
VELIGER
STAGE
120 hr
torsion
EARLY
VELIGER
STAGE
80 hr
TROCHOPHORE
32 hr
EMBRYO
13 hr
Patella
ADULT
Problacmea
ADULT
Erginus
(brood chamber removed)
BC
endoderm
mesoderm

of the Fissurellacea, Robertson (1985) for the origin of the mesogastropods, and Tissot (1988) for species of *Cypraea*. Heterochrony in shell characters is further treated by Geary and Tissot in this volume.

The molluscan veliger larva (Fig. 3, center left), and even torsion, have been argued as resulting from heterochronic change. Garstang (1929) suggested that the veliger larva was the product of adult molluscan characters (e.g., shell, operculum, adult organs) appearing earlier and earlier in the primitive larva (trochophore stage). Although Garstang also argued that torsion originated in the veliger larva and was not an adult character pushed back into the larval stage, other workers before (Naef, 1911) and since (Runnegar and Pojecta, 1985) have argued that torsion arose as an adaptation to solve adult problems, but was subsequently pushed back into the larval stage (Ghiselin, 1966). Also associated with torsion is the reduction or loss of many of the organs and structures on the right side (posttorsional left) of the body, including gills, kidneys, tentacles, and auricles (Crofts, 1937, 1955; Moor, 1983). The differential development of these structures (which arise from equal, paired precursors) in gastropod subclades may result from heterochronic processes in some of these groups.

3.2. Gastropoda

In prosobranch gastropods most organs and structures show a progressive reduction or loss of elements as one proceeds from primitive to derived groups (Fig. 2; Fig. 5, left to right; Tables I and II). For example, in the alimentary system there is reduction in the number of loops that the hindgut makes, the number of radular teeth, the complexity of the stomach (loss of crystalline style, gastric cecum, etc.). In the circulatory system, a reduction in the number of gills is accompanied by increased fusion of the remaining gill with other components of the mantle cavity and a reduction in the size of the kidneys and secondary modification for a variety of other functions, including respiration and reproduction. On the other hand, sensory structures, such as eyes and osphradia (mantle cavity sense organs), and reproductive structures have become more complex. There are also increases in egg size, larval complexity, length of larval development, and even chromosome numbers.

←

Figure 3. Morphological benchmarks in Patellogastropoda ontogeny, and the derivation of brooding morphology. (Left) Ancestral ontogeny with cross-sectional schematic representation of the kidney–heart complex. (Right) Descendant adult morphology of the genera *Erginus* and *Problacmaea* with cross-sectional schematics of the adult kidney–heart complex in brooders. H, heart; LK, left kidney; R, rectum; RK, right kidney; mesodermal cells and organs are stippled, endodermal cells and organs are solid. Hindgut development in the brooding genera is arrested at the juvenile stage of the ancestor, approximately 248 hr. A 90° rotation of the pericardial rudiment at this stage in the ancestral pathway produces the configuration of the kidney–heart complex seen in the genus *Problacmaea*. Rotation followed by predisplacement of the left kidney rudiment, to produce the brood chamber, and elongation and modification of the heart to redirect blood flow around the chamber give rise to the adult morphology of the genus *Erginus*. Sources: Patten (1885), Boutan (1899), Wilson (1904), Smith (1935), Crofts (1955), Raven (1966), Proctor (1968), and Lindberg (1983).

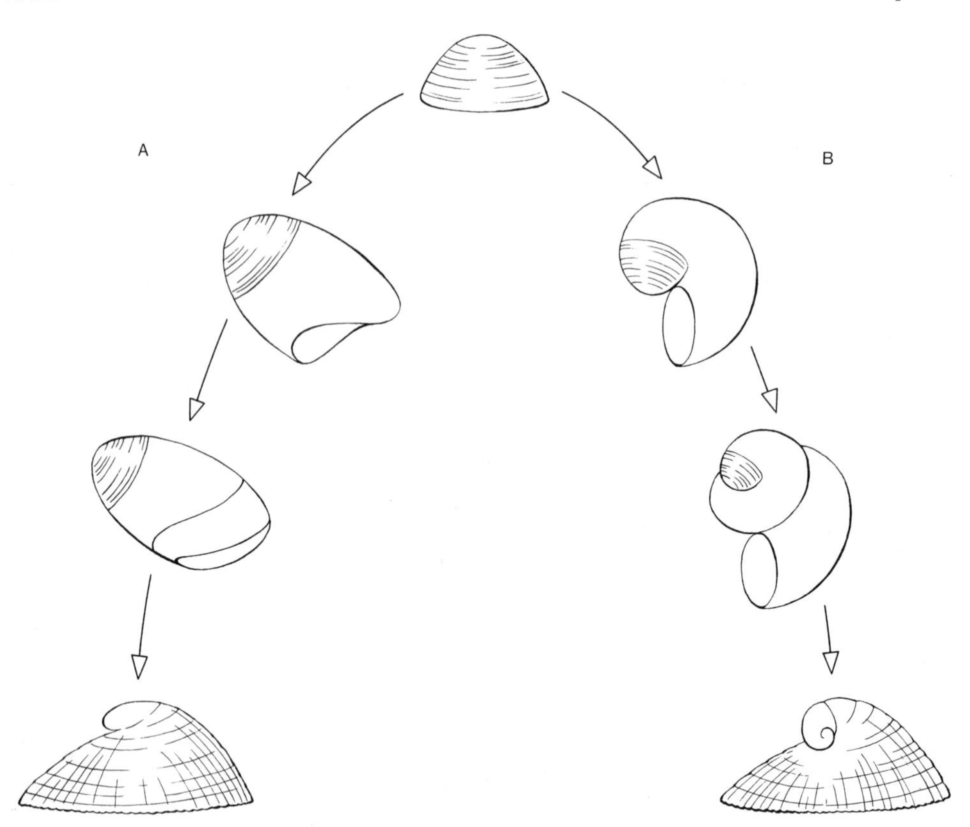

Figure 4. Ontogeny of limpet shell morphology in the prosobranch gastropods. (A) Patellogastropoda, (B) all other prosobranch limpets (cocculiniform limpets excepted).

3.3. Patellogastropoda

Many of the same patterns and trends within Gastropoda also occur in subclades such as the patellogastropods. The patellogastropods are limpets (noncoiled, conical, or cap-shaped shells), but the presence of an operculum in their ontogeny and the asymmetry of the larval shell (protoconch) on the early adult shell suggests descent from coiled ancestors (Lindberg, 1981b). They show a phyletic size decrease (Golikov and Starobagatov, 1975), but within subclades there are often size increases within faunas through time (Lindberg, 1983).

Unlike limpets in other gastropod subclades, Patellogastropoda lack a spirally coiled protoconch and have no extant coiled sister taxa. In patellogastropod ontogeny the protoconch does not coil and differential growth at the protoconch margins produces the adult conical form (Fig. 4A). In the other prosobranch limpet groups (e.g., Fissurellacea, Hipponacea, Siphonariacea, cocculiniform limpets excepted), the protoconch and early juvenile shell are coiled and the limpet form results from substantial increases in the whorl expansion rate later in ontogeny (Raup, 1966) (Fig. 4B). In Patellogastropoda the limpet shell form is achieved by neoteny; the earliest form of the gastropod protoconch is retained into the adult stage of the life cycle. In most other prosobranch limpets, changes in coiling parameters produce the limpet form *after* a typical coiled larval shell has been passed through. Noncoiled larval shells are also known in Opisthobranchia (Amio, 1963).

Patellogastropod anatomical evolution shows several patterns similar to those of Gastropoda as a group; patterns in other prosobranch subclades may differ. Alimentary systems show reduction in the number of intestinal loops and in the number of radular teeth in the lineage leading to Lottiidae (= former Acmaeidae) from the ancestral Patellacea (Fig. 5[1–2]*). In Gastropoda this trend has been associated with transition from herbivory (archeogastropods) to carnivory (caenogastropods) (Fretter and Graham, 1962). However, this cannot be the selective impetus in the Patellogastropoda, because all members of this clade are herbivores. After studying the radular ontogeny of several pulmonate gastropod species, Kerth (1979) suggested that the order of tooth appearance in derived species recapitulated ancestral radular features. In contrast to this pulmonate example, tooth fields and the number of teeth are reduced in the evolution of Patellogastropoda (Table II; Fig. 5[2,1–4]), and strongly suggests a paedomorphic process operating. In patellogastropod clades, gills are also reduced and even lost (Figs. 5[3,1–4] and 5[4,1–3]; however, the number of bundles that form the shell attachment muscle increases in more derived taxa (Fig. 5[3,1–4]) (Thiem, 1917). Hearts and kidneys are fairly similar among patellogastropod lineages; however, there are substantial modifications to mesodermally derived organ systems and structures in brooding species and these are discussed in the next section.

3.4. Summary

Many gastropod character distributions, at taxonomic levels ranging from class to genus, can be interpreted as morphological expressions of heterochronic processes, and many of these patterns suggest that identical heterochronic processes are occurring in the clade Gastropoda and in subclades such as Prosobranchia, Opisthobranchia, and Patellogastropoda.

4. Life History Traits and Heterochrony

4.1. Gastropoda

Jägersten (1972), using gastropod examples, argued that feeding larvae were primitive in the mollusks and that acceleration was associated with the evolution of "direct development" or brooding. Lindberg (1985a) and Chaffee and Lindberg (1987) examined larval developmental mode polarities using outgroup comparisons and data from the fossil record, and concluded that life history trends in the Mollusca (and Gastropoda) are from external to internal fertilization, non-

* References to specific character states or transformations in Fig. 5 are referred to in the form Fig. 5[row, column], where the row equals the characters (from top) and the column equals taxa (from left).

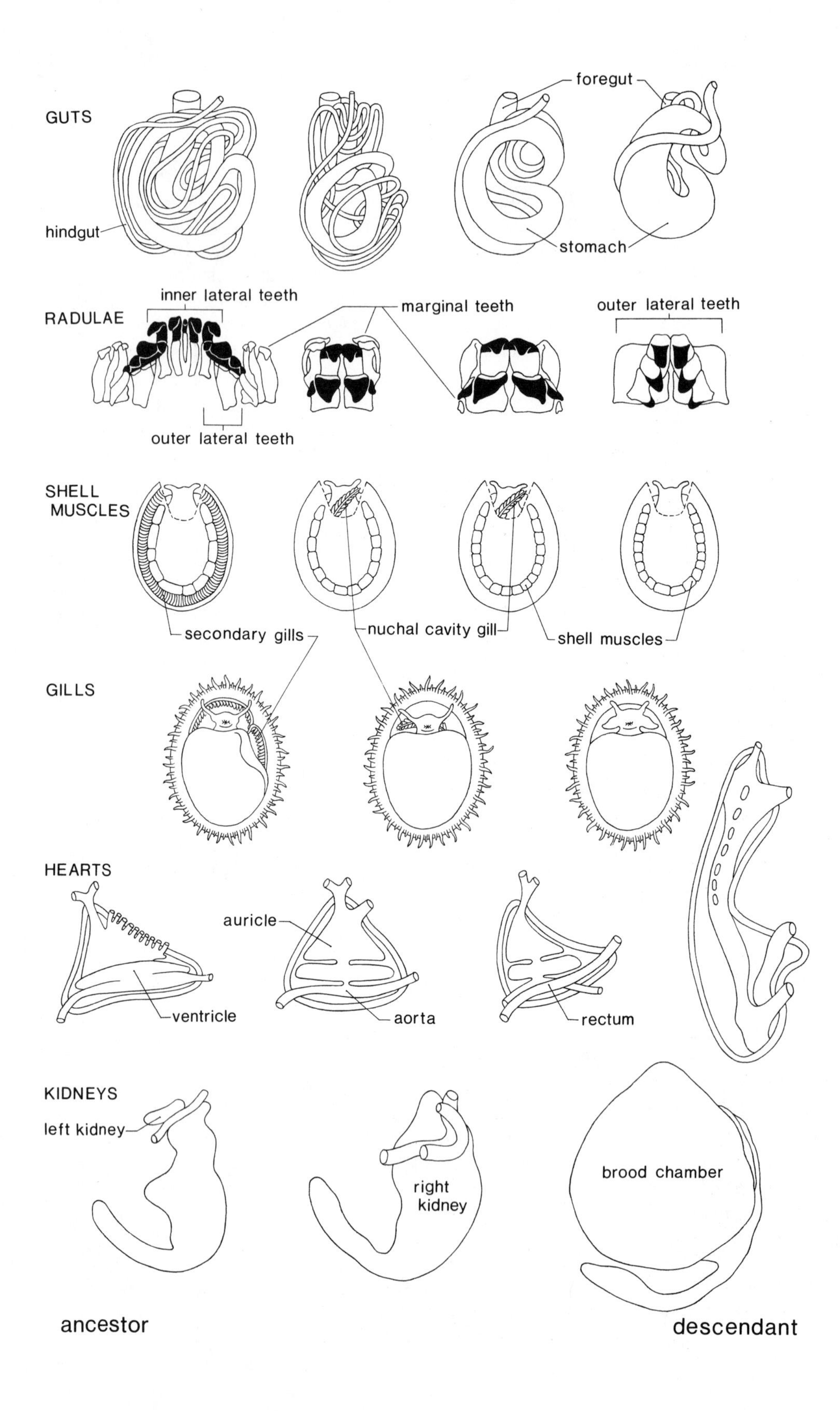
GUTS
foregut
hindgut
stomach
RADULAE
inner lateral teeth
marginal teeth
outer lateral teeth
outer lateral teeth
SHELL
MUSCLES
secondary gills
nuchal cavity gill
shell muscles
GILLS
HEARTS
auricle
ventricle
aorta
rectum
KIDNEYS
left kidney
right
kidney
brood chamber
ancestor
descendant

TABLE II. Radular Tooth Field and Numbers in the Patellogastropoda

Taxon	Inner lateral field (number of pairs of teeth)[a]	Outer lateral field (number of pairs of teeth)	Marginal field (number of pairs of teeth)
Outgroups			
Polyplacophora	2.5	1 (2–3 teeth fused)	6
Monoplacophora	1.5	2	2
Archaeogastropoda	4.5	1	~30
Patellogastropoda			
Patellidae	2.5	1 (3–4 teeth fused)	3
Nacellidae	0	3	3
Lottiidae			
Patelloidinae	0	3	2
Lottiinae			
Lottia	0	3	1
Tectura	0	3	0
Acmaeidae	0	3	0
Lepetidae	0	2 (partial fusion)	2

[a] Half a tooth pair (0.5) indicates the presence of a central tooth.

feeding to feeding larvae, and broadcast spawning to passive (e.g., egg capsules) and active brood protection (e.g., brood chambers) (Table I).

The life history traits discussed above typically represent extremes on models of the "r–K continuum" (Table III) (see also Stearns, 1976, and references therein), and the possibility that heterochronic processes are in part responsible for these patterns is likely. Gould (1977, pp. 293, 327) has suggested a relationship between the evolution of life history traits (in the context of *r*- and *K*-selection) and heterochronic change in brooding gastropods and bivalves, and Stearns (1982), Lindberg (1983), and McKinney (1986) have further explored and documented the relationship of heterochrony with life history evolution. There is also the possibility that both paedomorphic and peramorphic processes may be associated with the evolution of the complex pulmonate and opisthobranch reproductive tracts (Ghiselin, 1965; Moor, 1983; Schmekel, 1985).

4.2. Patellogastropoda

Life history evolution in the Patellogastropoda follows the general pattern for gastropods discussed above. Moreover, because brooding has evolved in only one

←

Figure 5. Morphological character states of patellogastropod organs and systems. The organs and structures on the left are the primitive conditions as found in *Patella* species; those in the center, *Patelloida* and *Lottia* species; and those on the far right represent the derived states as found in members of the genus *Erginus*. References to specific character states or transformations in this figure are given as Fig. 5[row, column], where the row = the character row (from top), and column = taxa (from left). Most near-vertical combinations of character states produce real taxa. For example, [1,3] + [2,3] + [3,3] + [4,2] + [5,2] +[6,1] = the genus *Lottia*. Redrawn from the following sources: Fleure (1904), Thiem (1917), Graham (1932), Walker (1968), Golikov and Kussakin (1972), Powell (1973), and Lindberg (1983).

TABLE III. Advantages and Disadvantages of Broadcasting and Brooding Reproductive Strategies for Marine Invertebrates[a]

Trait	Advantages	Disadvantages
Broadcasting	High dispersal	High mortality, disperse to poor areas
	Maximum outbreeding	—
	"Cheap" eggs	Need large numbers of eggs
	Larvae do not require large amounts of yolk	Larvae dependent on plankton
	Fecundity not limited by brood space	—
	Can successfully compete with brooders in recruitment	Bad years cause large losses and competitive disadvantages
Brooding	Low larval mortality	Low larval dispersal
	Able to use proven resources	Sibling, parent competition
	Maintain proven genotype	Increases inbreeding
	Low fecundity	"Expensive" eggs
	Larvae independent of plankton	More yolk per egg
	—	Brood space limits clutch size

[a] Compiled by J. S. Pearse *et al.* (1981).

genus, outgroup comparisons at lower taxonomic levels are possible, and they provide additional examples of life history characters with heterochronic patterns.

In Patellogastropoda, the age of first reproduction, a trait that is important in both heterochrony and life history evolution (Cole, 1954; Lewontin, 1965; Gould, 1977, pp. 290, 292), decreases substantially. In the primitive Patellidae, species are typically older and larger at maturation than in the Nacellidae and Lottiidae. For example, in the Patellidae most species attain sexual maturity at 40–50% of full-grown size, while in derived lottiid species, sexual maturity is often achieved before the limpet has reached 20% of full size (Lindberg, D.R., unpublished data). There also appears to be a corresponding decrease in developmental times in more derived clades. The time from fertilization to the completion of torsion in the Patellidae is about 120 hr (Smith, 1935; Crofts, 1955; Dodd, 1955). Although this is relatively fast compared to other primitive gastropod families (Crofts, 1955), developmental time continues to decrease, independent of temperature, in the more derived patellogastropod families [e.g., Nacellidae, 57 hr (Anderson, 1962; Rao, 1975); Lottiidae, 50 hr (Anderson, 1966; Kessel, 1964; Proctor, 1968)] (Lindberg, 1983).

Evolution of active brood protection occurs in both the Nacellidae (Lindberg, 1981a) and Lottidae (Golikov and Kussakin, 1972; Lindberg, 1983); evolution of internal fertilization accompanies evolution of brooding in the Lottiidae (Golikov and Kussakin, 1972; Lindberg, 1983). In the Lottiidae, brooding is restricted to five species in a single relict genus in the North Pacific (Lindberg, 1983). The anatomy of the brooding species is a mosaic of heterochronic processes. The development of the hindgut is truncated, and sexually mature brooders have guts that resemble the guts of broadcasting patellid species at metamorphosis (Fig. 3, top right). Radular development is also retarded and the radula lacks a marginal tooth field (Fig. 5[2,4]); gills are also lacking in some brooders (Fig. 5[4,3]). However, the structures derived from the mesoderm are bizarre, morphological innovations (Lindberg, 1983). In some species, kidneys are rotated 90° and the left (dorsal) one has been coopted for a brood chamber (Fig. 5[6,3]). The hearts can appear juvenilized in some taxa (Fig. 5[5,3]), whereas entirely new pericardial arrangements have appeared in others (Fig. 5[5,4]). None of these characters is necessarily correlated with simple size reduction, because sister taxa contain numerous smaller species with typical lottiid anatomies.

Although brooding species appear to have drastically reorganized the patellogastropod *Bauplan*, most of these new anatomical character states can be derived by simple topographic changes in patellogastropod development. The 90° rotation of kidney rudiments around the rectum during the veliger stage places them and the pericardium rudiment in a position that would explain most of the anatomical variation seen in brooding species (Fig. 3, right) (Lindberg, 1983). Subsequent development of the heart either departs totally from the typical patellogastropod developmental pathway and produces a heart unlike that known from any molluscan group (Fig. 5[5,4]), or, early development is arrested and the adult heart is so paedomorphic that it has characters resembling the more general characters of groups outside Patellogastropoda (Fig. 5[5,3]). In the latter case, the gross morphology of this heart was illustrated by Fleure (1904) as a theoretical, intermediate condition in the derivation of the Patellidae from an archaeogastropod ancestor.

The amount of dissociation among characters affiliated with the evolution of brooding in Patellogastropoda is readily apparent in some organs and structures. There is little variation in patellogastropod external foregut and stomach morphology during late ontogeny (Fig. 1) or in phylogeny (Fig. 5[1,1–4]). Juvenile characters in the alimentary system arise because the hindgut and radula are paedomorphic. The foregut, stomach, and hindgut are dissociated by germ layers. The foregut (which produces the radula) is derived from ectoderm, the stomach is derived from endoderm, and the hindgut is derived from endoderm (gut lining) and mesoderm (muscles and connective tissue). However, they form a single functional system. In contrast, organs derived from mesoderm (shell muscle, kidneys, hearts) are clearly associated from single-cell stages to the adult organ state by a shared developmental pathway, and changes in this pathway to the kidney–pericardial rudiments appear to produce the morphological innovation in patellogastropod body plans (Fig. 3, right) and changes in developmental mode. The heterochronic process appears to be progenesis, and, as Gould (1977, p. 337) predicted, it has provided a substantial release from the constraints of adult morphology, particularly when restricted to a single germ line that could accumulate changes at early developmental stages.

4.3. Evolution of Brooding

4.3.1. Ecological Setting

The ecological setting in which brooding evolved in the Patellogastropoda can be reconstructed from the fossil record, and gradual regional cooling seems to be the climatic condition under which the brooding patellogastropods appear (Lindberg, 1983). The brooding patellogastropod taxa are derived from two tropical–subtropical clades (Nacelliidae and Patelloidinae) that dominated the northeastern Pacific for about 90 million years. Regional cooling in the late Neogene is correlated with regional extinctions of broadcast-spawning members of these clades (Lindberg and Hickman, 1986). As the habitat cooled, developmental times for planktonic members of these clades may have lengthened substantially. For example, a decrease of 10°C almost triples time to hatching in the Patelloidinae (Amio, 1963). A longer developmental period in the plankton would contribute to greater larval mortality. To counter this increased mortality, selection on life history traits would be expected to raise the intrinsic rate of increase r of the taxon.

4.3.2. Scenario

One of the quickest and most efficient ways to increase r is to lower the age of first reproduction (Cole, 1954; Lewontin, 1965). Lowering the age of first reproduction may have unleashed progenesis in this clade, providing a heterochronic process that produced the innovative morphologies needed to transform a broadcasting body plan into that of a brooder. Gould (1977, p. 294) stated that he was "struck by how many of the smallest progenetic organisms, living in obvious r environments and possessing all other classical attributes of an r strategist, brood a few relatively large eggs." Correlation between small size and brooding in marine invertebrates has often been noted (see Strathmann and Strathmann, 1982, and references therein). If progenesis is often the mechanism by which broadcasting marine invertebrate species increase r during times of intense selection, then its association with small size and brooding is not unexpected. The inability of smaller animals to produce enough eggs to sustain the high attrition of planktonic development has been discussed by several workers, including Thorson (1950), Ghiselin (1963), Chia (1974), Hoagland (1975), and Chaffee and Lindberg (1987). The acquisition of brooding may represent the only evolutionary solution available for many taxa. Moreover, progenesis would also provide the necessary release from adult morphology (Gould, 1977, pp. 337, 339) that such a switch in reproductive mode would require.

4.4. Summary

Aspects of the evolution of gastropod life history traits, such as the origin of feeding larvae, and passive and active brood protection, have been interpreted as the results of heterochronic processes. The distributions of these traits suggest that identical heterochronic processes are occurring in parallel in different clades. In the Patellogastropoda, dissociation among characters affiliated with the evolution of brooding is readily apparent in some organs and structures, and is restricted to specific germ layers. The evolution of brooding in marine invertebrates may be associated with lowering of the age of first reproduction.

5. Phylogeny, Systematics, and Heterochrony

When autapomorphic characters that diagnose clades and terminal taxa are the products of heterochrony, it can be difficult to determine character polarity (Fink, 1982; Alberch, 1985; de Queiroz, 1985). Character polarity is typically determined by outgroup comparisons and developmental data (Wiley, 1981), but, as pointed out by Fink (1982), Alberch (1985), and de Queiroz (1985), there are both methodological and conceptual problems with inferring phylogenetic character polarity from the sequence of ontogenetic transformation.

Recognizing the presence of heterochronic processes plays an important role in determining the direction of evolutionary trends in taxa, and mistaken assumptions of character polarity can reverse entire phylogenies. W. H. Dall (1845–1927), proposed a classification for the patellogastropods that was strictly Haeckelian (simple to complex forms by terminal addition). Although Dall never formally stated his assumptions, in three different papers on the phylogeny of Docoglossa (= Patellogastropoda) he discussed "a certain geographical agreement in regard to generic characters which favors the hypothesis of a development of the various forms from a few more simple . . . ancestors" (Dall, 1871, p. 233), and noted that derived taxa had "changed or added to their original characters" Dall (1876, p. 244). In his final paper on the subject, he concluded that "we find therefore in Lepetidae the greatest number of archaic characters (somewhat masked by degeneration of other organs) which remain in any of the three groups" (Dall, 1893). The direction of Dall's evolutionary trends was to add characters and increase complexity. An alternative, cladistic hypothesis reverses these trends and argues that the order exhibits mostly paedomorphic characters (Lindberg, 1988).

Like Dall's classification of Patellogastropoda, most contemporary classification schemes for Gastropoda have been gradistic, but the number of cladistic studies at a variety of taxonomic levels has recently grown. In most cladistic studies, an outgroup is used to determine character polarity and only adult morphology is considered. False symplesiomorphies can result when workers use only characters from the adult mollusk rather than from the complete ontogeny. For example, the absence of the operculum and epipodial tentacles in the patellogastropods is scored in most character matrices as primitive (e.g., Graham, 1985); however, both structures are present earlier in ontogeny and are lost before metamorphosis.

6. Conclusions

The paleontological, fossil, developmental, anatomical, phylogenetic, and ecological data for Gastropoda suggest that heterochrony has had an important role in the evolution of some members of the class. Although both paedomorphic and peramorphic expressions have been identified, most of the processes appear to be paedomorphic (i.e., progenesis and neoteny). In some clades these forms of heterochrony appear to predominate, and thus constrain, and thereby direct, morphological evolution in the group. For example, progenetic paedomorphosis operating in Patellogastropoda produces trends and character states that range across entire superfamilies or are restricted to single genera.

As summarized above, molluscan development is mosaic and cell determination occurs early in ontogeny. Heterochronic changes that occur early in development in specific cell lineages may have large-scale effects on subsequent morphology. However, these changes will typically be restricted to single systems or organs that share this common precursor.

Examples of heterochrony must be examined carefully. Even large-scale trends, such as the paedomorphic appearance of alimentary systems belonging to derived gastropod taxa, must be studied from as many levels in the ontogenetic and phylogenetic hierarchies as possible. For example, in most gastropod groups all available evidence suggests that radular evolution has proceeded from numerous tooth fields (marginals, outer laterals, inner laterals) with numerous teeth to single fields with few teeth. In gut evolution, the stomach has lost the gastric shield and style, and the long, looped hindgut has been shortened. Although these three character sources interact to produce the alimentary system, the sources are originally dissociated because they belong to different germ layers. Thus, the products of three different germ layers, some through different processes of heterochrony, are induced to form a single anatomical system that shows its own heterochronic trend at the level of the system (see also Wake and Larson, 1987).

6.1. Dissociability

6.1.1. Introduction

Dissociability in gastropod character states is readily apparent in several molluscan organ systems, and sometimes produces unique combinations of character states in closely related taxa (e.g., the genera *Erginus* and *Problacmaea* in Fig. 3 or the various combinations of vertical arrangements of the character states in Fig. 5 that produce "real" taxa). Dissociation in brooding patellogastropods produces mosaic beasts with ectodermally derived structures that differ little from other taxa, endodermally derived structures that become more juvenilized in derived taxa, and mesodermally derived structures that produce new structures and bizarre morphologies. In the Patellogastropoda, and perhaps other gastropod groups, patterns restricted to germ layers provide important data for identifying heterochronic processes (see also Gould, 1977, p. 234, and Fink, 1982).

6.1.2. Implications for Systematics

In gastropod systematics the entire ontogeny of a character state needs to be considered, or, as de Queiroz (1985) has argued, the characters we use should really be character ontogenies. Taxa must be viewed as mosaics of character states and the direction of morphological evolution within a character suite should be based on outgroup comparisons of character ontogenies. Interpretations of character polarity using ontogenetic sequences are almost always complicated by heterochronic processes (acceleration and hypermorphosis excepted), and even determining polarity by outgroup comparison can be confounded by the presence of paedomorphic processes (Fink, 1982; Alberch, 1985; de Queiroz, 1985).

6.2. Life History Evolution

The evolution of some life history traits in the Gastropoda can be correlated with heterochronic processes, as Gould (1977, p. 303) first pointed out with examples from several molluscan taxa. Primitive gastropod taxa have external fertilization and planktonic, nonfeeding development; derived taxa have internal fertilization, passive or active brood protection, and feeding larvae. Brooding species that occur in predominantly broadcasting clades are often progenetic, and many of these brooders are members of clades that have undergone geologically recent ecological perturbations (Lindberg, 1982, 1983, 1985b).

ACKNOWLEDGMENTS. I thank C. Chaffee, D. Geary, D. Jablonski, G. Oster, and J. Pearse for thoughtful discussions; R. Houbrick, K. de Queiroz, J. McLean, K. Padian, M. Russell, J. Pearse, D. Wake, and K. Warheit for commenting on earlier drafts of this chapter; M. Taylor for preparing the illustrations; and M. McKinney and the Paleontological Society for the invitation to participate in the symposium for which this chapter was originally prepared.

References

Alberch, P., 1980, Ontogenesis and morphological diversification, *Am. Zool.* **20**:653–667.

Alberch, P., 1985, Problems with the interpretation of developmental sequences, *Syst. Zool.* **34**(1):46–58.

Alberch, P., Gould, S. J., Oster, G., and Wake, D., 1979, Size and shape in ontogeny and phylogeny, *Paleobiology* **5**:296–317.

Amio, M., 1963, A comparative embryology of marine gastropods, with ecological considerations, *J. Simonoseki Univ. Fish.* **12**:229–358 [in Japanese].

Anderson, D. T., 1962, The reproduction and early life-histories of the gastropods, *Bembicium auratum* (Quoy and Gaimard) (Fam. Littorinidae), *Cellana tramoserica* (Sowerby) (Fam. Patellidae) and *Melanerita melanotragus* (Smith) (Fam. Neritidae), *Proc. Linn. Soc. N. S. Wales* **87**:62–68.

Anderson, D. T., 1966, The reproductive and early life-histories of the gastropods *Notoacmea petterdi* (Ten. Woods), *Chiazacmaea flammea* (Quoy and Gaimard) and *Patelloida alticostata* (Angus) (Fam. Acmaeidae), *Proc. Linn. Soc. N. S. Wales* **90**:106–114.

Bandel, K., 1982, Morphologie und Bildung der Frühontogenetischen Gehäuse bei conchiferan Mollusken, *Facies* **7**:1–198.

Batten, R. L., 1975, The Scissurellidae—Are they neotenously derived fissurellids? (Archaeogastropoda), *Am. Mus. Novit.* **2567**:1–37.

Bonner, J. T. (ed.), 1982, *Evolution and Development*, Springer-Verlag, Berlin.

Boutan, L., 1899, La cause principale de l'asymétrie des mollusques gastéropodes, *Arch. Zool. Exp. Gen.* **7**:203–342.

Chaffee, C., and Lindberg, D. R., 1987, Larval biology of early Cambrian mollusks: The implications of small size, *Bull. Mar. Sci.* **39**:536–549.

Chia, F.-S., 1974, Classification and adaptive significance of developmental patterns in marine bottom invertebrates, *Thalassia Jugoslavica* **10**:121–130.

Cole, L. C., 1954, The population consequences of life history phenomena, *Q. Rev. Biol.* **29**:103–137.

Crofts, D. R., 1937, The development of *Haliotis tuberculata*, with special reference to the organogenesis during torsion, *Philos. Trans. R. Soc. Lond. B* **228**:219–268.

Crofts, D. R., 1955, Muscle morphogenesis in primitive gastropods and its relation to torsion, *Proc. Zool. Soc. Lond.* **125**:711–750.
Dall, W. H., 1871, On the limpets with special reference to the species of the west coast of America, and to a more natural classification of the group, Am. J. Conch. **6**:227–282.
Dall, W. H., 1876, On the extrusion of the seminal products in limpets, with some remarks on the phylogeny of the Docoglossa, *Proc. Acad. Nat. Sci. Phila.* **1876**:239–247.
Dall, W. H., 1893, The phylogeny of the Docoglossa, *Proc. Acad. Nat. Sci. Phila.* **1893**:285–287.
De Beer, G. R., 1951, *Embryos and Ancestors*, Oxford University Press, London.
De Queiroz, K., 1985, The ontogenetic method for determining character polarity and its relevance to phylogenetic systematics, *Syst. Zool.* **34**(3):280–299.
Dodd, J. M., 1955, Artificial fertilisation, larval development and metamorphosis in *Patella vulgata* L. and *Patella coerulea* L., *Pub. Staz. Zool. Napoli* **29**:172–186.
Fink, W. L., 1982, The conceptual relationship between ontogeny and phylogeny, *Paleobiology* **8**(3):254–264.
Fleure, H. J., 1904, On the evolution of topographical relationships among the Docoglossa, *Trans. Linn. Soc. Lond.* **9**:269–290.
Fretter, V., and Graham, A., 1962, *British Prosobranch Molluscs*, Ray Society, London.
Garstang, W., 1929, The origin and evolution of larval forms, *Br. Assoc. Adv. Sci. Rep.* **1928**:77–98.
Ghiselin, M. T., 1963, On the functional and comparative anatomy of *Runcina setoensis* Baba, an opisthobranch gastropod, *Publ. Seto Mar. Biol. Lab.* **11**:219–228.
Ghiselin, M. T., 1965, Reproductive function and the phylogeny of opisthobranch gastropods, *Malacologia* **3**(3):327–378.
Ghiselin, M. T., 1966, The adaptive significance of gastropod torsion, *Evolution* **20**(3):337–348.
Golikov, A. N., and Kussakin, O. G., 1972, Sur la biologie de la reproduction des patelles de la famille Tecturidae (Gastropoda: Docoglossa) et sur la position systématique des ses subdivisions, *Malacologia* **11**(2):287–294.
Golikov, A. N., and Starobagatov, Y. I., 1975, Systematics of prosobranch gastropods, *Malacologia* **15**(1):185–232.
Gould, S. J., 1969, An evolutionary microcosm: Pleistocene and Recent history of the land snail *P. (Poecilozonites)* in Bermuda, *Bull. Mus. Comp. Zool.* **138**:407–532.
Gould, S. J., 1977, *Ontogeny and Phylogeny*, Harvard University Press, Cambridge.
Graham, A., 1932, On the structure and function of the alimentary system of the limpet, *Trans. R. Soc. Ed.* **57**:287–308.
Graham, A., 1985, Evolution within the Gastropoda: Prosobranchia, in: *The Mollusca*, Vol. 10, *Evolution* (E. R. Trueman and M. R. Clark, eds.), pp. 151–186, Academic Press, New York.
Haszprunar, G., 1985a, The fine morphology of the osphradial sense organs of Mollusca. Part 1: Gastropoda–Prosobranchia, *Philos. Trans. R. Soc. Lond. B* **307**:457–496.
Haszprunar, G., 1985b, The Heterobranchia—A new concept of the phylogeny and evolution of the higher Gastropoda, *Z. Zool. Syst. Evolutionsforsch.* **23**:15–37.
Hoagland, K. E., 1975, Reproductive strategies and evolution in the genus *Crepidula* (Gastropoda: Prosobranchia), Ph. D. dissertation, Harvard University, Cambridge.
Hyman, L. H., 1967, *The Invertebrates, Vol. VI, Mollusca I*, McGraw-Hill, New York.
Jägersten, G., 1972, *Evolution of the Metazoan Life Cycle*, Academic Press, London.
Jablonski, D., 1985, Molluscan development, in: *Mollusks. Notes for a Short Course* (T. W. Broadhead, ed.), pp. 33–49, University of Tennessee Department of Geological Sciences Studies in Geology 13.
Kerth, K., 1979, Phylogenetische Aspekte der Radulamorphogenese von Gastropoden, *Malacologia* **19**:103–108.
Kessel, M. M., 1964, Reproduction and larval development of *Acmaea testudinalis* (Müller), *Biol. Bull.* **127**:294–303.
Kollmann, H. A., and Yochelson, E. L., 1976, Survey of Paleozoic gastropods possibily belonging to the subclass Opisthobranchia, *Ann. Naturhist. Mus. Wien* **80**:207–220.
Lewontin, R. C., 1965, Selection for colonizing ability, in: *The Genetics of Colonizing Species* (H. G. Baker and G. L. Stebbins, eds.), pp. 77–91, Academic Press, New York.
Lindberg, D. R., 1981a, Rhodopetalinae, a new subfamily of Acmaeidae from the boreal Pacific: Anatomy and systematics, *Malacologia* **20**:291–305.
Lindberg, D. R., 1981b, Is there a coiled ancestor in the docoglossan phylogeny?, *Ann. Rep. W. Soc. Malacol.* **12**:15.

Lindberg, D. R., 1982, Tertiary biogeography and evolution of the genus *Problacmaea* in the North Pacific (Acmaeidae: Patelloidinae), *J. Paleontol.* **56**(2 Suppl.):16.

Lindberg, D. R., 1983, Anatomy, systematics, and evolution of brooding Acmaeid limpets, Ph. D. dissertation, University of California, Santa Cruz.

Lindberg, D. R., 1985a, Aplacophorans, monoplacophorans, polyplacophorans and scaphopods: The lesser classes, in: *Mollusks. Notes for a Short Course* (T. W. Broadhead, ed.), pp. 230–247, University of Tennessee Department of Geological Sciences Studies in Geology 13.

Lindberg, D. R., 1985b, The evolution of brooding in North Pacific marine invertebrates: Clade-specific patterns in space and time, in *Program and Abstracts, Western Society of Naturalists*, p. 50.

Lindberg, D. R., 1988, The Patellogastropoda, in: *Prosobranch Phylogeny, Proceedings of the 9th International Malacological Congress, Edinburgh, 1986* (W. F. Ponder, ed.), pp. 35–63, Malacological Review, Supplement, Ann Arbor.

Lindberg, D. R., and C. S. Hickman, 1986, A new anomalous giant limpet from the Oregon Eocene (Mollusca: Patellida), *J. Paleontol.* **60**(3):661–668.

McKinney, M. L., 1986, Ecological causation of heterochrony: A test and implications for evolutionary theory, *Paleobiology* **12**:282–289.

McLean, J. H., 1971, A revised classification of the family Turridae, with the proposal of new subfamilies, genera, and subgenera from the eastern Pacific, *Veliger* **14**(1):114–130.

McLean, J. H., 1979, A new monoplacophoran limpet from the continental shelf off southern California, *Contrib. Sci. Nat. Hist. Mus. Los Ang. Co.* **307**:1–19.

McLean, J. H., 1984, A case of derivation of the Fissurellidae from the Bellerophontacea, *Malacologia* **25**(1):3–20.

McNamara, K. J., 1982, Heterochrony and phylogenetic trends, *Paleobiology* **8**(2):130–142.

McNamara, K. J., 1986, A guide to the nomenclature of heterochrony, *J. Paleontol.* **60**:4–13.

Moor, B., 1983, Organogenesis, in: *The Mollusca*, Vol. 3, Development (N. H. Verdonk, J. A. M. van den Biggelaar, and A. S. Tompa, eds.), pp. 123–177, Academic Press, New York.

Naef, A., 1911, Studien zur generellen Morphologie der Mollusken. 1. Teil: Uber Torsion und Asymmetrie der Gastropoden, *Ergeb. Fortsch. Zool.* **3**:73–164.

Paul, C. R. C., 1982, The adequacy of the fossil record, in: *Problems of Phylogenetic Reconstruction* (K. A. Joysey and A. E. Friday, eds.), pp. 75–117, Academic Press, New York.

Paul, C. R. C., 1985, The adequacy of the fossil record reconsidered, *Spec. Paper Palaeonotol.* **33**:7–15.

Patten, W., 1885, Artificial fecundation in the Mollusca, *Zool. Einz.* **8**:236–237.

Powell, A. W. B., 1973. The patellid limpets of the world (Patellidae); *Indo- Pac. Mollusca* **3**(15):75–206.

Proctor, S. J., 1968, Studies on the stenotopic marine limpet *Acmaea insessa* (Mollusca: Gastropoda: prosobranchia) and its *algal host Egregia menziesii* (Phaeophyta), Ph. D. dissertation, Stanford University, Stanford, California.

Rao, B. M., 1975, Some observations on spawning behavior and larval development in the limpet *Cellana radiata* (Born) (Gastropoda: Prosobranchia), *Hydrobiologia* **47**:265–272.

Raup, D. M., 1966, Geometric analysis of shell coiling: general problems, *J. Paleontol.* **40**:1178–1190.

Raven, C. P., 1966, *Morphogenesis: The Analysis of Molluscan Development*, Pergamon Press, Oxford.

Robertson, R., 1985, Archaeogastropod biology and the systematics of the genus *Tricolia* (Trachacea: Tricoliidae) in the Indo-West-Pacific, *Monog. Mar. Mollusca* **3**:1–103.

Runnegar, B., and Pojeta, J., Jr., 1985, Origin and diversification of the Mollusca, in: *The Mollusca*, Vol. 10, Evolution (E. R. Trueman and M. R. Clarke, eds.), pp. 1–57, Academic Press, New York.

Salvini-Plawen, L. V., 1980, A reconsideration of systematics in the Mollusca (phylogeny and higher classification), *Malacologia* **19**:249–278.

Schmekel, L., 1985, Aspects of evolution within the Opisthobranchs, in: *The Mollusca*, Vol. 10, *Evolution* (E. R. Trueman and M. R. Clarke, eds.), pp. 221–267, Academic Press, New York.

Sepkoski, J. J., Jr., 1982, A compendium of fossil marine families, *Milwaukee Pub. Mus. Contrib. Biol. Geol.* **51**:1–125.

Signor, P. A., 1985, Gastropod evolutionary history, in: *Mollusks. Notes for a Short Course* (T. W. Broadhead, ed.), pp. 157–173, University of Tennessee Department of Geological Sciences Studies in Geology 13.

Smith, F. G. W., 1935, The development of *Patella vulgata*, *Philos. Trans. R. Soc. Ed. B* **225**:95–125.

Solem, A., and Yochelson, E. L., 1979, North American Paleozoic Land Snails, with a Summary of other Paleozoic Nonmarine Snails, U. S. Geological Survey Professional Paper 1072.

Spengel, J. W., 1881, Die Geruchsorgane und das Nervensystem der Mollusken, *Z. Wiss. Zool.* **35**:333–383.

Stearns, S. C., 1976, Life-history tactics: A review of the ideas, *Q. Rev. Biol.* **51**:3–47.

Stearns, S. C., 1982, The role of development in the evolution of life histories, in: *Evolution and Development* (J. T. Bonner, ed.), pp. 237–258, Springer-Verlag, Berlin.

Strathmann, R. R., and Strathmann, M. F., 1982, The relationship between adult size and brooding in marine invertebrates, *Am. Nat.* **119**(1):91–101.

Thiem, H., 1917, Beiträge zur Anatomie und Phylogenie der Docoglossen. II. Die Anatomie und Phylogenie der Monobranchen (Akmäiden und Scurriiden nach der Sammlung Plates), *Jen. Z.* **54**:405–630.

Thorson, G., 1950, Reproductive and larval ecology of marine bottom invertebrates, *Biol. Rev.* **25**:1–45.

Tissot, B. N., 1988, Geographic variation and heterochrony in two species of cowries (Genus *Cypraea*), *Evolution* **42**(1):103–117.

Verdonk, N. H., and van den Biggelaar, J. A. M., 1983, Early development and the formation of the germ layers, *The Mollusca*, Vol. 3, *Development* (N. H. Verdonk, J. A. M. van den Biggelaar, and A. S. Tompa, eds.), pp. 91–122, Academic Press, New York.

Verdonk, N. H., van den Biggelaar, J. A. M., and Tompa, A. S., 1983, *The Mollusca*, Vol. 3, *Development*, Academic Press, New York.

Wake, D. B., and Larson, A., 1987, Multidimensional analysis of an evolving lineage, *Science* **238**:42–48.

Walker, C. G., 1968, Studies on the jaw, digestive system, and coelomic derivatives in representatives of the genus *Acmaea*, *Veliger* **11**(Suppl.):88–97.

Wiley, E. O., 1981, *Phylogenetics. The Theory and Practice of Phylogenetic Systematics*, Wiley, New York.

Wilson, E. B., 1904, On germinal localization in the egg. II. Experiments on the cleavage mosaic in *Patella*, *J. Exp. Zool.* **1**:197–268.

Chapter 12

Heterochrony in Rodents

JOHN C. HAFNER and MARK S. HAFNER

1. Introduction

The mammalian order Rodentia is by far the largest order of mammals (approximately 1700 species), and rodents show ranges in body size, body plan, and ecological diversity that far exceed those seen in any other group of mammals, including bats and cetaceans. Living rodents inhabit all continents except Antarctica, and they are found in nearly every terrestrial habitat throughout their geographic range. Rodents usually play integral roles in the terrestrial ecosystems they inhabit, and they are often the most abundant and diverse of all vertebrates in a terrestrial community.

For these reasons and others, the order Rodentia seems the obvious mammalian group in which to examine patterns of morphological and ecological differentiation (usually termed evolutionary trends) in an effort to reveal causal evolutionary processes that generate organismal diversity. To be recognized as such, a higher order evolutionary trend must cut across phyletic boundaries (i.e., be lineage-independent), and it must generate predictable results in the groups it

JOHN C. HAFNER • Moore Laboratory of Zoology and Department of Biology, Occidental College, Los Angeles, California 90041. MARK S. HAFNER • Museum of Natural Science and Department of Zoology and Physiology, Louisiana State University, Baton Rouge, Louisiana 70803.

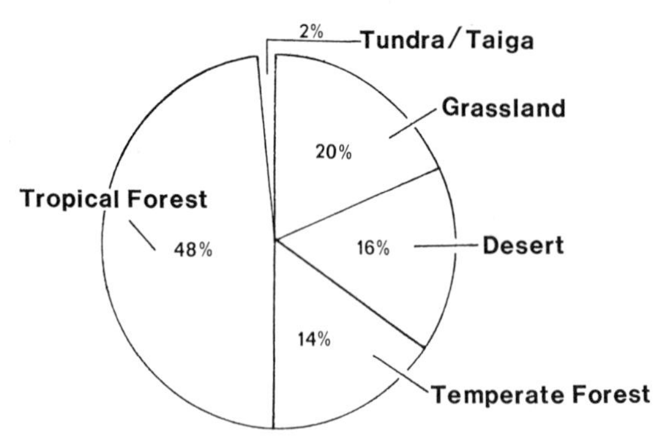

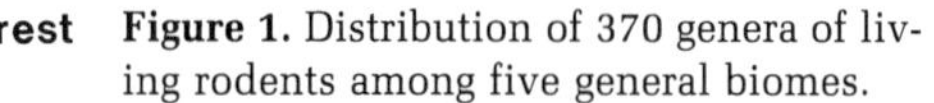
Figure 1. Distribution of 370 genera of living rodents among five general biomes.

affects. For example, the evolutionary trend leading to the many cases of gigantism in normally small mammals inhabiting oceanic islands and the trend toward reduction (or loss) of lateral digits in cursorial mammals are both lineage-independent phenomena that lead to predictable morphological consequences. Once the evolutionary trend is identified, one begins the search for an underlying causal factor, the elucidation of which is generally the major focus of interest in any evolutionary study.

The initial sections of this chapter focus on the identification of higher order evolutionary trends within the Rodentia. The remaining sections focus on the much more difficult (but infinitely more interesting) challenge of explaining the observed trends. In the latter effort, we find that we are only partially successful; but we do offer hypotheses to be tested and potential avenues for fruitful research that may bring future workers closer to understanding the higher order evolutionary principles that orchestrate broad patterns of mammalian differentiation.

2. A General Overview of Rodent Diversity

2.1. Ecological Diversity

Figure 1 provides a broad overview of ecological diversity among the 370 genera of living rodents recognized by Nowak and Paradiso (1983). Although subdivision of the earth's habitats into tropical forest, temperate forest, desert, grassland, and tundra/taiga is necessarily crude, it is clear that almost half of all living rodent genera inhabit tropical forest environments. This same pattern is evident when rodents are considered at the species level; for example, approximately 41% of 1673 living species of rodents inhabit tropical forest environments. Approximately equal numbers of rodent taxa, whether considered by genus or species, inhabit grassland, desert, and temperate forest environments. Finally, only 2% of all rodent genera (also 2% by species) inhabit the combined tundra/taiga habitat. Importantly, each ecological group contains rodents from distantly related lineages (i.e., the pattern is not lineage dependent), and the pattern could not have been predicted solely from the relative abundance of habitat types on earth. Thus, a higher order evolutionary trend is evident, and this trend has been

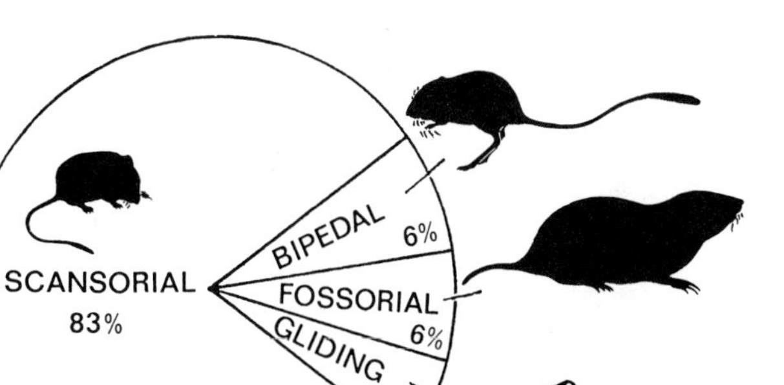

Figure 2. Diversity in body plan among extant genera of rodents.

explained in adaptive terms by zoogeographers such as Darlington (1957) and Keast (1972).

2.2. Diversity in Body Size

Living rodents range in body size from approximately 6 g (e.g., the mouse genera *Micromys* and *Delanymys*) to more than 60 kg (the capybara, *Hydrochaeris*). This range spans four orders of magnitude and exceeds the size range of any other mammalian order, including even the cetaceans. Not surprisingly, the distribution of individual rodent taxa within this vast size range is heavily skewed to the right; the median body size (by mass) among 370 genera of living rodents is approximately 80g, and 75% of all rodents (whether considered by genus or species) weigh less than 240 g. Given that the earliest rodents were also generally small (Romer, 1966), it is reasonable to conclude that there has been a major evolutionary trend toward retention of small body size in most rodent lineages. The adaptive explanation for this trend has involved a mixture of metabolic, trophic, and reproductive considerations (see Eisenberg, 1981, and references therein).

2.3. Diversity in Body Plan

Living rodents are easily subdivided into four discrete body plan categories (*Baupläne*): scansorial (= scampering); bipedal (= hopping); fossorial (= obligate burrowing); and gliding (= modified for nonpowered flight). Based on comparison with nonrodent mammals and vertebrates in general, the scansorial body plan is both generalized and primitive within the Rodentia. The three remaining categories (bipedal, fossorial, and gliding) include highly specialized and derived body plans that appear to represent evolutionary commitments to novel ways of life.

Figure 2 illustrates the distribution of rodent genera among the four body plan categories. Not surprisingly, most taxa (83%) are scansorial (or typically

TABLE I. Relationship between Habitat and Three Specialized Body Plans in Rodents

Specialized body plan	Habitat[a] Desert Exp	Desert Obs	Grassland Exp	Grassland Obs	Tropical forest Exp	Tropical forest Obs	Temperate forest Exp	Temperate forest Obs	Tundra/taiga Exp	Tundra/taiga Obs
Bipedal	4*	18*	5	3	12*	1*	3	2	0	0
Fossorial	4	4	5*	12*	12*	2*	3	6	0	0
Gliding	2	0	3	0	7	11	2	4	0	0

[a] Exp, Expected frequency based on a survey of 370 genera of rodents. Obs, observed. An asterisk denotes a significant departure from expectation.

mouselike in body plan), whereas only 6% are bipedal, 6% are fossorial, and 4% are gliders. These same patterns hold when taxa are analyzed by species.

None of the three specialized body plan categories is restricted to a single rodent lineage. For example, the bipedal rodents include representatives from the distantly related superfamilies Muroidea, Dipodoidea, and Geomyoidea; the fossorial rodents include representatives from distinct rodent groups (e.g., caviomorphs, geomyoids, bathyurgids); and gliding species are found in the distantly related families Sciuridae and Anomaluridae. This survey indicates, therefore, that each of these specialized body plan categories is lineage independent, which suggests that each may represent a higher order evolutionary trend. Although trends, such as these, which lead to morphological or ecological similarity among distantly related taxa are generally termed evolutionary convergence, it should be recognized that "convergence" merely describes the trend; it does not explain it in causal terms.

The most striking and best-studied example of locomotory specialization among rodents is bipedality (e.g., Bartholomew and Caswell, 1951; Hatt, 1932; Howell, 1932). Bipedal rodents include 24 genera representing eight families, and all are remarkably similar morphologically and ecologically. Table I shows that bipedal rodents are not distributed randomly among the five habitat types, but occur more often than expected in desert habitats and less often than expected in tropical forest habitats.

In addition to elongated hindlimbs and reduced forelimbs, many bipedal rodents possess unusually large auditory bullae (the boney capsule housing the middle ear). Our examination of rodent crania [plus additional information provided in Nowak and Paradiso (1983)] reveals that roughly 40 of 370 rodent genera have grossly enlarged auditory bullae. If this character (bullar enlargement) were distributed randomly among rodents irrespective of body type, only two bipedal genera (6% of 40; see Fig. 2) would be expected to possess the feature. Our survey reveals, however, that 12 of the 40 rodent genera with enlarged auditory bullae are also bipedal forms ($\chi^2 = 54.4$; d.f. $= 3$; $p < 0.001$). Thus, there appears to be a nonrandom association among bipedality, enlarged auditory bullae, and desert habitation in rodents. Whether or not this association is biologically meaningful (i.e., whether or not the factors are causally linked) remains to be seen.

3. Functional Descriptions and Causal Explanations

Conventional explanations for the existence of novel body plans, such as those seen in certain rodents, focus on the adaptive aspects (advantages) of the functional design [e.g., see review by Eisenberg (1975)]. For example, gliding rodents [including sciurid "flying" squirrels (e.g., *Glaucomys*), and anomalurid scaly-tailed squirrels (e.g., *Anomalurus*)] have "wing" membranes because these small airfoils are selectively advantageous in that they allow for protracted leaps and thus may aid in predator avoidance. Fossorial rodents [e.g., tucu-tucos (*Ctenomys*), pocket gophers (*Thomomys*), and mole rats (*Spalax*)] have massive bodies equipped with powerful forelimbs and rugose, dense skulls because this morphology is adaptive in burrowing through the soil. Bipedal rodents [e.g., jerboas (*Dipus*) and kangeroo rats (*Dipodomys*)] have delicate, gracile bodies with long hindlimbs and long tails because natural selection seems to favor this "antipredator morphology" (Kotler, 1985) in rodents inhabiting open desert environments. Suffice it to say, the adaptive significance of the novel morphologies seen in certain rodents has received much attention from several generations of biologists. The emphasis on the current use (immediate adaptive significance) of the novelties has been so overwhelming that few workers have sought to understand the evolution of the novelties. Actually, the extreme adaptationist tradition has been so alluring in its simplicity that most workers have assumed (usually implicitly) that they are explaining the evolution of a feature when they explain its present use (see also Gould and Vrba, 1982, p. 13). In this tradition, morphological evolution is often explained as the result of long-term, directional selection (orthoselection) wherein natural selection favors certain adaptations present in ancestral species in response to long-term selective pressures on individual characters of atomized individuals. Caution should be exercised when viewing morphology from this kind of adaptationist perspective, inasmuch as it may lead to pseudoexplanatory inferences about morphological evolution [for review, see Gould and Lewontin (1979), Mayr (1983)].

A distinction should be made here between two kinds of perspectives which are important in evolutionary morphology: functional descriptions and causal explanations. Functional descriptions, such as those discussed above, explain morphology in terms of its purpose to the animal; prior morphological states of the feature are irrelevant in this context. Causal explanations, however, focus on prior morphological states and attempt to predict future states from earlier ones. Hence, functional descriptions address the current use (adaptive significance) of a feature, whereas causal explanations seek to address only the evolution of a feature, regardless of its present use. It is important to note that functional descriptions and causal explanations are not directly competitive. Conversely, the two must not be automatically treated as one and the same; functional descriptions that entail unwarranted extrapolations as to a character's origin (i.e., scenarios of adaptation) may belie the actual historical genesis of the feature.

It seems that virtually all of the explanations for the extreme morphological variation among rodents stem from functional descriptions. As noted by Brookfield (1982), an emphasis on present function may obfuscate our attempts to understand the evolution of evolutionary novelties. Our intent in the remainder of this chapter, then, is to focus on the evolution of the novel body plans seen in

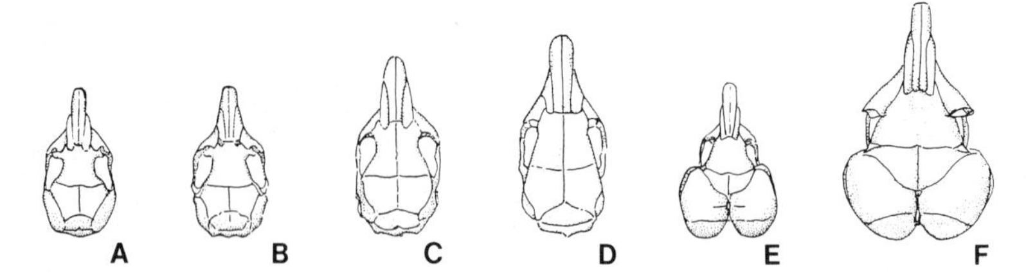

Figure 3. Representative crania of the Heteromyidae, showing diversity of morphological types. (A) *Perognathus*; (B) *Chaetodipus*; (C) *Liomys*; (D) *Heteromys*; (E) *Microdipodops*; (F) *Dipodomys*.

certain rodents. As such, we seek causal explanations for the evolution of morphological features, regardless of their present adaptive significance.

4. Geomyoid Rodents As a Case Study

4.1. Morphological and Ecological Breadth

We have shown that rodents, as a whole, show extreme morphological diversity (Fig. 2). Even so, one rodent group in particular, the superfamily Geomyoidea, provides an exceptional opportunity for studies in evolutionary morphology; three of the four major rodent body plans (Fig. 2) are found in this one superfamily (no gliding forms). No other mammalian superfamily, extant or extinct, is known to have the high degree of morphological differentiation of the geomyoids.

The history of geomyoid rodents is restricted to the New World, and extant forms include two families: the Heteromyidae (includes scansorial and bipedal forms) and the Geomyidae (all are fossorial forms). Although this superfamily is geographically restricted when compared with most other major rodent groups, geomyoids occupy virtually all kinds of terrestrial habitats, from sea level to well over 3000 m. For example, certain geomyoids, such as spiny pocket mice (*Heteromys*), inhabit moist, tropical forests. Other geomyoids, including kangaroo mice (*Microdipodops*) and certain species of kangaroo rats (*Dipodomys*), are restricted to extremely xeric, sand dune habitats in the desert.

The Heteromyidae (an appropriate epithet meaning "different mice") shows a surfeit of morphological types (Fig. 3) when compared to the typical range of morphological diversification seen in most other rodent families. Four of six heteromyid genera represent varied types of quadrupedal (scansorial) rodents: the pocket mice (*Perognathus* and *Chaetodipus*) and the spiny pocket mice (*Liomys* and *Heteromys*). This scansorial *Bauplan* is generalized for the Geomyoidea, as well as for the Rodentia (and also the Mammalia). Extreme morphological diversity is seen in the bipedal forms: kangaroo mice (*Microdipodops*) and kangaroo rats (*Dipodomys*). These bipedal ricochetors (saltators) are highly derived in morphology; notable features include large head (in proportion to body size) with greatly expanded auditory bullae (see Fig. 3) and extremely elongated hindlimbs. Kangaroo mice are tiny, bipedal rodents (approximately 10 g) and kangaroo rats are large, bipedal animals (approximately 40–170 g).

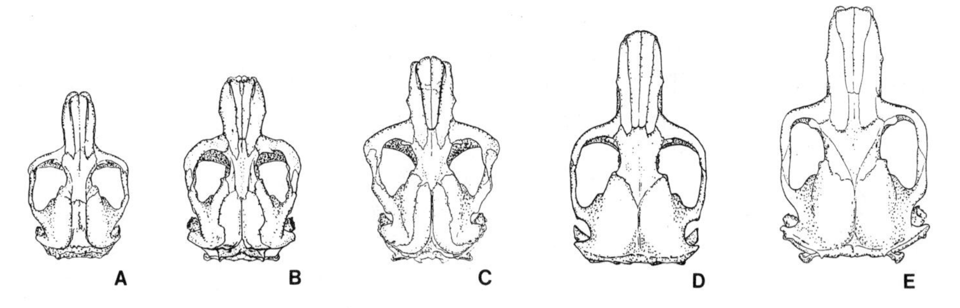

Figure 4. Representative crania among the Geomyidae. Note the general similarity in shape. (A) *Thomomys*; (B) *Geomys*; (C) *Pappogeomys*; (D) *Zygogeomys*; (E) *Orthogeomys*.

The geomyoids (literally "earth mice") also show a *Bauplan* that is highly derived in relation to the generalized scansorial condition. The geomyoids, or "pocket gophers," include five genera: *Thomomys, Geomys, Pappogeomys, Zygogeomys,* and *Orthogeomys*. In marked contrast to the heteromyids, the geomyids are remarkably similar morphologically. With the exception of size differences, all pocket gophers look basically alike (see Fig. 4).

Together, these two families form a cohesive superfamily; all members of the Geomyoidea are united by the presence of externally opening, fur-lined cheek pockets. While many other rodents have internal cheek pouches for storing food, the geomyoid cheek pockets are unique in that they are positioned outside the buccal cavity [for review, see Hill (1937), Ryan (1986)]. Long (1976) hypothesized that the cheek pockets of the geomyoids represent a radical morphological shift that may have been caused by a simple developmental change. Hence, this key innovation suggests a discontinuous, macroevolutionary origin of the entire superfamily (see also Gould, 1980).

4.2. Scenarios of Adaptation and Morphological Evolution

It is indeed a challenge to explain the evolution of these extreme and conspicuous body plans found in the Geomyoidea. Among all geomyoid rodents, the bizarre morphology of the bipedal kangaroo mice and kangaroo rats has attracted the greatest amount of attention from biologists. These forms, like their Old World counterparts (e.g., *Jaculus, Dipus, Salpingotus,* and others), are considered by many to be the epitome of desert specialization (Mares, 1983). The evolution of the enormous head, huge hindfeet, large eyes, and long tail in kangaroo mice and kangaroo rats is explained conventionally in terms of the adaptive aspects of their morphology. The common opinion among biologists is, for example, that the large head serves as a counterbalancing organ during bipedal saltation, the huge hindfeet serve as sand paddles on sandy substrates, the large eyes are extreme adaptations for nocturnal vision, and the long tail serves as a midair rudder during ricochetal locomotion [for review, see Hafner (1988)]. Virtually all of these scenarios of adaptation focus on what is termed the "antipredator morphology" of *Microdipodops* and *Dipodomys* (Kotler, 1985). Whether or not these features function in the ways suggested, the very different question remains: how did the features evolve? Because a feature functions in a certain way today, we might assume that it evolved for that purpose. However, as emphasized by Gould and Vrba (1982, p. 13), "current utility [of a feature] carries no automatic implication about [its] historical origin."

Here we focus on a causal explanation for the evolution of these features. For the sake of this discussion, we emphasize the conceptual decoupling of character evolution from present use and thereby avoid the inherently *ad hoc* nature of the scenarios in inferring character change. The causal explanation we advance to account for macroevolution in the Geomyoidea focuses on the mechanisms of heterochrony (mutations that effect changes in developmental programs) as a broad, unifying explanation to account for the evolution of the wide variety of morphological novelties that occur in this group.

4.3. Hypothesized Heterochronic Transformations

The hypothesis that morphological evolution is the result of regulatory changes in development has been promoted by many workers (e.g., de Beer, 1930, 1958; Goldschmidt, 1940; Waddington, 1957, 1962; Gould, 1977). Both Waddington (1957, 1962) and Alberch (1980) noted that morphologies do not appear in a random or continuous manner in a group, but that there is a repetition of several distinctive body plans. Such is the case in rodents in general (Fig. 2) and with geomyoid rodents in particular. An examination of both fossil and extant geomyoids reveals that virtually all forms fit into one of three general *Baupläne*: scansorial mice, hopping (kangaroo-like) rats and mice, and fossorial forms (see Fig. 2). As pointed out by Alberch (1980), epigenetic interactions may reduce the spectrum of potential novel morphologies and impose a sense of order in morphological transformations through phylogeny; as a consequence, we should expect to see developmental constraints effecting phyletic parallelism. The observation that the Geomyoidea is rife with phyletic parallelism (Wood, 1935; Hafner and Hafner, 1983) is consistent with this view.

Geomyoid rodents contain genera that are morphologically generalized (the four genera of pocket mice and spiny pocket mice), others that appear to be generally paedomorphic, or juvenilized (kangaroo mice and kangaroo rats), and still others that seem to be hypermorphic, or developed beyond the generalized condition (pocket gophers). Both paedomorphism and hypermorphism are predictable consequences of heterochronic change, which leads us to deduce that developmental heterochrony may explain macroevolutionary diversification in geomyoid rodents [for review, see Hafner, J. C., Hafner (1983), Hafner, M. S., Hafner (1984), and Hafner (1988)]. Below, we use simplified phenomenological descriptions to illustrate how regulatory changes in ontogeny may, in affecting the timing of gene action and rates of morphogenesis and growth, explain morphological transformations in geomyoid phylogeny [for review see Gould (1977), Alberch *et al.* (1979), Alberch (1980), Løvtrop (1981a,b), and Rachootin and Thomson (1981)]. We will attempt to confirm our deductive conclusion that heterochrony explains macroevolutionary diversification in the Geomyoidea by independent inductive investigations within geomyoid rodents. We will thus examine indirectly the heterochrony hypothesis of morphological change, a hypothesis that is impossible to falsify directly because it treats unique, historical events.

4.3.1. The Pocket Mouse: A Generalized Geomyoid Rodent

Pocket mice, including *Perognathus, Chaetodipus, Liomys,* and *Heteromys,* exhibit a generalized rodent *Bauplan* (Eisenberg, 1981, p. 90) and probably represent a reasonable approximation of the ancestral geomyoid condition. An animal with this ancestral morphology (i.e., a pocket mouse) is likely to have conserved the developmental patterns of the geomyoid ancestor. As the pocket mouse ages from conception, its size and shape will change, following the ancestral ontogenetic growth curve. Perturbations of "control parameters," including the onset of growth, cessation of development, and rate of growth, may deform the ancestral ontogenetic trajectory and lead to morphological transmutations [for review of this model, see Alberch *et al.* (1979)]. We hypothesize that morphological diversification in the Geomyoidae is explicable from this ontogenetic perspective. Moreover, as we shall demonstrate, the novelties that are produced as end products of heterochrony include life history attributes and morphological features that are consistent with the heterochrony model [for review, see Gould (1977)].

4.3.2. The Kangaroo Mouse: A Progenetic Rodent?

Kangaroo mice (*Microdipodops*), like kangaroo rats (*Dipodomys*), are paedomorphic, or juvenilized, rodents. We suggest that paedomorphosis, merely a gross shape phenomenon, is shared between these genera as a result of phyletic parallelism. The obvious juvenile traits shared between the genera include the large head, large brain, large eyes, and long hindfeet (Hafner, J. C., Hafner, 1983; Hafner, M. S., Hafner, 1984; Hafner, 1988). In addition, adults of both *Microdipodops* and *Dipodomys* retain the complete stapedial canal and artery (Howell, 1932; Webster and Webster, 1975; Lay, 1988); these features are lost at an early ontogenetic stage or are incompletely developed in other geomyoids. Kangaroo rats and kangaroo mice, unlike other geomyoids, also have very light, delicate skeletons whose osseous elements show a low degree of fusion; weak fusion of skeletal elements is characteristic of juvenile mammals.

The crania of both *Microdipodops* and *Dipodomys* are peculiar because of their enormous size and the presence of hypertrophied auditory bullae (Fig. 3). The general perception is that cranial enlargement is due solely to bullar inflation. It is not clear how and why the auditory bullae have become so inflated in these forms (Hafner, 1988; Lay, 1988), but bullar hypertrophy alone does not account for the huge crania of kangaroo mice and kangaroo rats. Actually, condylonasal length of the skull (a measure of skull size excluding bullar swelling) and precaudal vertebral length (a measure of body size excluding tail) are allometrically related. The allometric exponent of skull (condylonasal) length on precaudal vertebral length across the Rodentia is estimated to be 0.65 (Fig. 5; $r = 0.97$, $p \ll 0.01$). It is this general rodent trend of negative allometry that explains in part why kangaroo mice and kangaroo rats have large heads; they have short precaudal vertebral lengths and therefore have proportionately large heads. However, both of these bipedal forms have skulls that are approximately twice as large as would

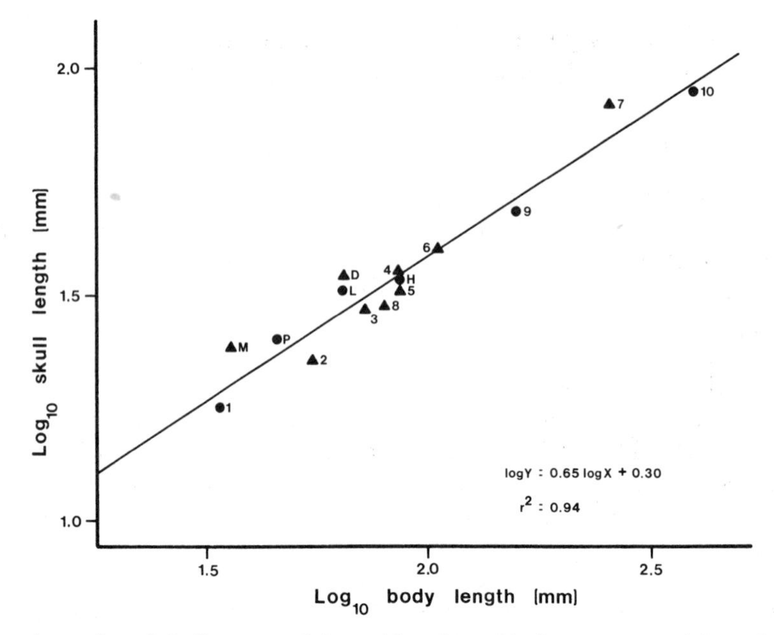

Figure 5. Relationship of skull size (condylonasal length) and body size (precaudal vertebral length) for various genera of (▲) bipedal and (●) quadrupedal rodents [data from Hatt (1932)]. (M) *Microdipodops*; (P) *Perognathus*; (D) *Dipodomys*; (L) *Liomys*; (H) *Heteromys*; (1) *Sicista*; (2) *Zapus*; (3) *Scirtopoda*; (4) *Jaculus*; (5) *Dipus*; (6) *Allactaga*; (7) *Pedetes*; (8) *Notomys*; (9) *Rattus*; (10) *Paramys*.

be predicted from their precaudal lengths alone. Functional constraints associated with bipedality do not seem to account for this disparity; indeed, Fig. 5 shows that some bipedal rodents have heads that are larger that those predicted by the regression (e.g., *Pedetes* and *Jaculus*), while other bipedal rodents have heads that are smaller than those predicted by the allometric trend (e.g., *Scirtopoda* and *Notomys*).

Despite the many paedomorphic characters shared between the bipedal heteromyids, *Microdipodops* differs from *Dipodomys* by many trenchant morphological and life history characteristics [for review, see Hatt (1932), Howell (1932), Wood (1935), and Hafner (1978, 1988)]. These differences suggest that the juvenilized morphology of kangaroo mice may have evolved in a manner that is fundamentally different from that in kangaroo rats; in short, juvenilization in *Microdipodops* may be the result of progenesis (Hafner and Hafner, 1983; Hafner, 1988). Progenesis is a heterochronic process in which somatic development is truncated because reproductive maturation is abbreviated. It involves a negative perturbation (truncation) in the ancestral ontogenetic trajectory (Alberch *et al.*, 1979) and results in a small, rapidly maturing paedomorph. Kangaroo mice, in comparison with other geomyoids, show retention of juvenile morphology and are also small.

In general, progenesis seems to be associated with animals near the *r* end of the *r–K* spectrum of life-history strategies [for review, see Gould (1977); see also McKinney (this volume)]. In comparison with other geomyoids, kangaroo mice possess many of the classical attributes of a more highly *r*-selected organism; small

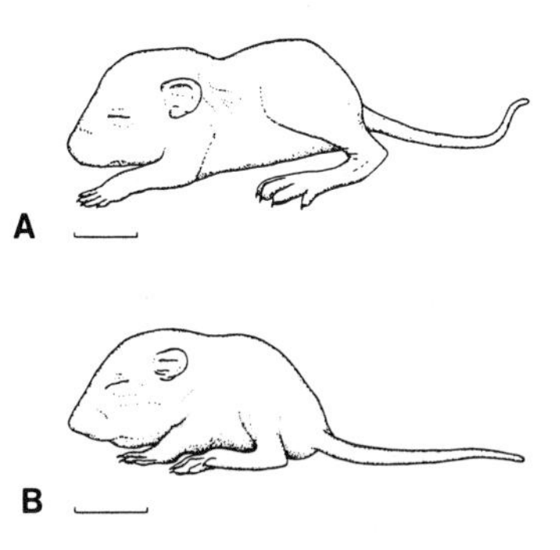

Figure 6. Comparison of (A) a kangaroo rat, age 10 days, with (B) a pocket mouse, age 9 days. Drawings modified from Lackey (1967) and Hayden and Gambino (1966). See text for discussion. Scale is 10 mm.

body size, reduced longevity (Hall and Lindsdale, 1929; Egoscue *et al.*, 1970; and personal observation), and larger litter size (Hall, 1941; O'Farrell and Blaustein, 1974; Fleming, 1977); in addition, kangaroo mice subsist on an ephemeral resource base (insects and seeds) (Hall, 1941). Kangaroo mice also inhabit ephemeral (semistabilized) sand-dune habitats in the Great Basin Desert [a cold desert, (steppe) that is characterized by an abbreviated growing season]. It follows, then, that the adaptive significance of this kind of heterochronic perturbation (progenesis) may be attributable not to the juvenilized morphology per se, but to the suite of life-history characteristics [especially small body size and rapid and precocious maturation (Wootton, 1987)] brought about by the developmental change. From this viewpoint, the juvenilized morphology of kangaroo mice may be an incidental by-product of progenesis that persists because it was not egregiously maladaptive in its initial form. Subsequent periods of natural selection, acting on developmental variability, may have refined this juvenilized morphology into the finely tuned adaptive unit we see today.

Several other features of kangaroo mouse morphology support the hypothesis of progenesis. Kangaroo mice have only 24 caudal vertebrae [the fewest number of caudal elements of all heteromyids (Hatt, 1932; J. C. Hafner, unpublished data)]. Inasmuch as ossification occurs craniocaudally in heteromyids (Van De Graaff, 1973), the reduced number of caudal elements is a predicted result of progenetic truncation of the ancestral ontogeny. In addition, certain cheek teeth of kangaroo mice contain closed roots, whereas kangaroo rats show the opaeodont (open root) condition. In accord with the progenetic hypothesis, the molar roots of kangaroo mice appear late in ontogeny and are greatly reduced in comparison with the ancestral (e.g., pocket mouse) condition (Merriam, 1891; Wood, 1935; Hall, 1941).

4.3.3. The Kangaroo Rat: A Neotenic Rodent?

The kangaroo rat and pocket mouse, although very different in body plan and body size as adults, are surprisingly similar in appearance during early stages of development. Moreover, at approximately 9 or 10 days after birth, both rodents look remarkably like *adult* kangaroo rats in that they possess large heads and large hindfeet in relation to their body size (Fig. 6). While the body proportions of the animals are basically the same at this stage (Fig. 6), the pocket mouse changes radically over time; in contrast, the gross shape of the kangaroo rat changes very little from this point onward and retains its juvenile shape. Specifically, the rate of change in shape during development seems to be retarded in the kangaroo rat in comparison with the ontogeny of the pocket mouse. Assuming that the pocket mouse retains the developmental pattern of the ancestor, then we may deduce from this comparison that a retardation of the ancestral ontogenetic shape trajectory leads to neotenic morphology. This retardation plus a longer overall growth period (delay in the termination signal of growth) leads to a descendant that is both neotenic and large. We suggest that the juvenilization of kangaroo rats occurred in this manner (see also Hafner, 1988).

It appears that neoteny is a common occurrence in both plants and animals and it is often associated with organisms near the *K* end of the *r*–*K* spectrum of life-history strategies (Gould, 1977). Kangaroo rats, which we hypothesize to be neotenic, are *k*-selected by rodent standards: they have a long life span (Egoscue *et al.*, 1970; and personal observation); show slow development (Chew and Butterworth, 1959; Butterworth, 1961; Eisenberg and Isaac, 1963; Hayden and Gambino, 1966; Lackey, 1967; Fleming 1977); have long gestation periods (Eisenberg and Isaac, 1963; Fleming 1977); have enlarged brains (Hafner and Hafner, 1984); and have small litters (Hall, 1946; Butterworth, 1960; Eisenberg and Isaac, 1963; Fleming, 1977).

According to the retardation model, features that appear late in the ancestral ontogeny would be expected to be absent or reduced in size (and retain the juvenile shape) in the neotenic descendant. There is evidence that both the distal phalanges (Van De Graaff, 1973) and the roots of molars (Zakrzewski, 1981) are features that appear late in ontogeny. Thus, the reduction and/or loss of the hallux in adult kangaroo rats and the generally opaeodont (open-rooted) condition of the molars argue forcefully for retardation. Conversely, two characters appear discordant with the neoteny hypothesis: enlarged auditory bullae and elongated tail. Kangaroo rats, like kangaroo mice, have hypertrophied bullae and elongated tails. In both rodents, bullar inflation may be due to peramorphosis (*sensu* Alberch *et al.*, 1979; McNamara, 1986); that is, the bullae may have developed beyond the ancestral condition. The high number of caudal vertebrae in kangaroo rats (Hatt, 1932) also seem to argue that tail elongation in kangaroo rats (but not kangaroo mice; see above) may result from peramorphosis. Hence, tail elongation in kangaroo rats and kangaroo mice may have been achieved in two different ways: additional caudal vertebrae in the former, and longer, yet fewer, caudal vertebrae in the latter. As such, these rodents may represent a natural parallel to the replicate selection experiments with mice (*Mus*) by Rutledge *et al.* (1975). Thus, kangaroo rats, although generally juvenilized in appearance, may be the end product of different heterochronic processes (i.e., paedomorphosis and peramorphosis) that acted concurrently. The simultaneous action of different heterochronic processes was also described for an Eocene echinoid (McKinney, 1984) and for human evolution (Gould, 1977).

4.3.4. The Pocket Gopher: A Hypermorphic Rodent?

The adult pocket gopher, unlike other geomyoids, possesses a heavily ossified skeleton; the skull, in particular, is exceptionally rugose and shows a high degree of fusion among cranial elements. These features (skeletal rugosity and fusion) are characteristic of later stages of mammalian ontogeny. This, coupled with the

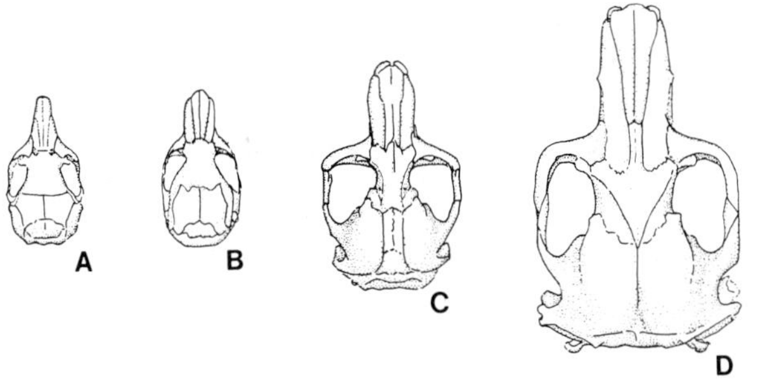

Figure 7. Comparison of (B) a neonatal pocket gopher (*Thomomys*) with (A) an adult pocket mouse (*Chaetodipus*) and (C,D) adult pocket gophers of different genera (*Thomomys* and *Orthogeomys*, respectively). Note the similarity between the juvenile pocket gopher (B) and the adult pocket mouse (A).

fact that neonatal pocket gophers look remarkably like mature pocket mice (Fig. 7), leads us to suggest that hypermorphosis is involved in the morphological evolution of pocket gophers.

Hypermorphosis is a process wherein there is a positive perturbation (delay) in the signal for cessation of growth in the ancestral ontogeny; the hypermorphic descendant is termed a peramorph (Alberch *et al.*, 1979). This developmental perturbation allows prolongation of the development of somatic features and permits pocket gophers to surpass or extend the ancestral ontogenetic trajectory; this is recapitulation through hypermorphosis. Hence, juvenile levels of morphological differentiation in the pocket gopher may be adult features of the ancestor. In the developing pocket gopher, the "pocket mouse" morphotype is attained at the same size of the adult ancestor, but the pocket gopher is still juvenile at this size; adult pocket gophers are large geomyoids. It is for this reason that we reject recapitulation by acceleration as the process leading to peramorphosis in pocket gophers.

4.3.5. Conclusions on Geomyoid Heterochrony

We have hypothesized that heterochronic disturbances during development have resulted in morphological evolution in the rodent superfamily Geomyoidea. Heterochronic changes in the generalized (ancestral) pocket mouse ontogeny may have effected both recapitulation and reverse recapitulation in geomyoids. Recapitulation, as seen in pocket gophers, may have been achieved through hypermorphosis. Although kangaroo rats and kangaroo mice are both paedomorphic, reverse recapitulation appears to have resulted via different heterochronic processes: neoteny in kangaroo rats and progenesis in kangaroo mice. If we are correct, the kangaroo mouse [or "dwarf kangaroo rat," as it was originally called by Merriam (1891)] is not merely a scaled-down version of the neotenic kangaroo rat, but is a separate, unique progenetic form.

It should be emphasized that the epigenetic interactions effecting these morphological transformations are probably much more complex than the simple heterochronic changes that we have outlined. For example, the kangaroo mouse appears to be mainly progenetic, but its enlarged auditory bullae may be the result of acceleration. Similarly, the gross morphology of the kangaroo rat is hypothesized to be mainly a result of neoteny and a longer overall growth period, but it is possible that the hypertrophied auditory bullae and elongated tail resulted from hypermorphosis [see above and Lay (1988)]. Morphogenesis is a complex process and "pure" examples of, say, neoteny unaccompanied by other heterochronic processes are probably very uncommon [for discussion, see Gould (1977), Alberch *et al.* (1979), Fink (1982), and McKinney (1984)]. Before one can understand fully morphological differentiation in the Geomyoidea, much more complete data are needed on growth and development, longevity, age at maturation, litter sizes, food habits, and many other crucial natural history parameters.

Over the past decade or so, it has become increasingly clear that evolutionary changes in shape and size arise from evolutionary change in developmental programs. For the geomyoids, we propose that their disparate morphological modifications evolved through heterochronic shifts in ontogeny. Our emphasis here on developmental heterochrony stems from the recent realization that a single heterochronic perturbation (i.e., a point mutation affecting development) may cause sweeping morphological changes when its effects are amplified by the myriad of pleiotropic and epigenetic effects that occur during development. All developmental processes that are genetically correlated with those that are directly affected by the heterochronic event will, in turn, be altered [for discussion, see Atchley (1987)]. What, then, is the role of natural selection in geomyoid macroevolution? It is most parsimonious to postulate that natural selection acts on developmentally integrated character complexes that are affected by heterochronic changes in ontogeny. This explanation is much simpler than one that views natural selection as acting on each morphological feature independently. The evolution of each novel morphological trait, then, might not be attributable to the direct, "fine-tuning" action of natural selection, but may occur largely as an inseparable and highly interdependent subunit of a developmentally integrated character complex that is, in turn, governed by various hierarchical developmental processes. The study of heterochrony seems to be an epistemologically sound way of probing the causal underpinnings of phylogenetic transmutation, or macroevolution (*sensu* Simpson, 1944, 1953), in this group. Moreover, the trends noted here may explain not just geomyoid patterns of differentiation, but the distinct body plans seen in other rodent groups as well.

5. Selective and Developmental Constraints

5.1. Adaptive Significance

As was stated at the outset, the intent of this chapter is to focus on patterns of morphological differentiation in rodents in an attempt to gain insight into the causal evolutionary processes that may be responsible for effecting macroevolution. In any discussion of macroevolution, however, it is difficult to maintain a proper perspective on the subject of adaptation. While our emphasis is on the evolution and not the present adaptive significance of the rodent *Baupläne*, it is important here to remark briefly on the relevance of the adaptationist program to the study of rodent biology.

The adaptationist program, when correctly implemented, is a profitable method of scientific research because it is of heuristic value [for review, see Mayr

[1983)]. However, Gould and Lewontin (1979) are correct in criticizing those studies of adaptation that ignore the integrated aspect of the phenotype and atomize the individual in an attempt to determine the adaptive significance of a specific trait. A more holistic approach, advocated by Gould and Lewontin (1979), and the classical adaptationist approach are both important in studies of adaptation. Actually, there is no reason why both approaches cannot be pursued simultaneously; the research programs are not mutually exclusive.

To date, the adaptationist approach has dominated the study of rodent morphological evolution. Indeed, consider the generally accepted scenarios of adaptation for each of the three novel body plans in rodents: a "flying" squirrel has gliding membranes because gliding is adaptive in predator avoidance; a fossorial rodent has a heavy, excessively rugose skull because it is advantageous for an animal burrowing (often chewing) its way through the soil; and a bipedal rodent has long hindlimbs because rapid bipedal locomotion aids in predator avoidance. These adaptationist explanations may or may not be true; this is not the issue here. The point is, once a convincing explanation has been proffered, it quickly becomes dogma. Once dogma is firmly entrenched, competing hypotheses (adaptationist or otherwise) are rarely considered and few, if any, attempts are made actually to test hypotheses about the functional descriptions of characters (Hafner, 1988). On the whole, the adaptationist program has not produced testable functional hypotheses for most of the novel morphologies seen in rodents. Accordingly, it now seems appropriate to consider more holistic approaches to the study of adaptation. In so doing, it might be profitable to investigate the adaptive significance of the *Bauplan* in its entirety (including life-history features), as opposed to the conventional method of focusing on the adaptiveness of individual morphological traits.

5.2. Selective or Developmental Constraints?

It is important to recognize the distinction between developmental and selective constraints in the study of morphological evolution [for review, see Maynard Smith *et al.* (1985)]. However, at any particular instance, it may be exceedingly difficult to resolve the relative contributions of developmental and selective constraints in shaping macroevolutionary patterns. Although Maynard Smith *et al.* (1985) suggest several ways of distinguishing between selective and developmental constraints, they also note that no method is absolutely foolproof.

Ultimately, it may be impossible to identify unambiguously the constraints that are responsible for producing the evolutionary novelties we see in rodents. The main problem is that the production of an evolutionary novelty itself is a unique event that is unlikely to be duplicated; hence, macroevolution is generally not amenable to direct investigation. In rodents, however, if heterochrony can be shown to be the underlying mechanism responsible for the production of similar novel body plans in several unrelated phyletic groups (see below), then it would be proper to conclude that developmental constraints are of at least some importance in effecting macroevolutionary patterns. We hasten to add, however, that even if heterochrony is implicated in the morphological differentiation of the major rodent body plans, the nonrandom association between environment and the body plans (see Section 2.3) makes it impossible to factor out environmental considerations in the course of macroevolution. With these problems in mind, and in consideration of our present level of understanding of the role of development in evolution, it seems that our best hope is to attempt to identify what mechanism or combination of mechanisms is most likely responsible for the origin of morphological novelties.

6. Prospectus and Directions for Future Study

Our hypotheses concerning heterochrony are based largely on the patterns of morphological differentiation seen in adult specimens of geomyoid rodents. These hypotheses predict that developmental programs of the derived forms (e.g., kangaroo rats, kangaroo mice, and pocket gophers) are altered in a specific fashion relative to the generalized (ancestral) condition. Moreover, these heterochronic hypotheses may have relevance beyond the taxonomic boundary of the Geomyoidea; we propose that much of the morphological differentiation seen in the rodents may be explained by heterochronic perturbations in development. In particular, we suggest that morphological convergence associated with two of the major rodent body plans, bipedality and fossoriality, is explicable by the heterochrony model.

Our hypotheses concerning the morphological transformations in geomyoid rodents are testable with the accumulation of new, comparative data on growth and development. We predict, for example, that the growth and development of the neotenic kangaroo rat is retarded relative to that of the generalized (pocket mouse) condition; existing data, although scanty, seem to support this. Our hypothesis also predicts developmental truncation in kangaroo mice; however, nothing is known about growth and development of kangaroo mice. If future work demonstrates that the ontogeny of the kangaroo mouse is *not* truncated relative to the generalized pattern, then our hypothesis of the progenetic origin of the kangaroo mouse is falsified. Similarly, our hypothesis concerning hypermorphosis in pocket gophers is amenable to direct falsification by comparative data on growth and development.

We predict that the convergent evolutionary trends toward bipedality and fossoriality seen in rodents may have been caused largely by heterochronic changes in development. For example, bipedality in the dipodid and murid rodents may be associated with paedomorphosis. In this regard, we hypothesize that the dipodid genus *Salpingotus* (dwarf jerboa), the ecological equivalent to the kangaroo mouse, is a progenetic form and should show truncated development; nothing is known about the growth and development of *Salpingotus*. We also predict that the Old World counterparts to the kangaroo rat (e.g., jerboas, including *Dipus* and *Jaculus*) are neotenic. Interestingly, Kirmiz (1962) reports that the development of *Jaculus* is retarded relative to that of the rat (*Rattus*); in this case, the developmental information is consistent with our heterochrony model. Other striking examples of convergence are provided by the fossorial rodents; hypermorphosis may be involved in the trend toward fossoriality. Comparative developmental studies should reveal whether forms such as the tucutuco (*Ctenomys*) or the various kinds of mole rats (e.g., *Spalax*, *Tachyoryctes*, and *Bathyergus*) are hypermorphic relative to the generalized condition.

Future work exploring the evolution of ontogenies may prove to be a profitable way of understanding morphological evolution in rodents. Future workers interested in macroevolution in rodents should seek to gather the kinds of data on growth and development that can be brought to bear on these hypotheses of heterochrony. In this regard, Atchley's (1987) developmental quantitative genetics model stressing genetic variance–covariance structure seems most appropriate. Basic descriptive data (e.g., postnatal growth curves) are currently lacking for most species of rodents. Accordingly, we encourage other workers to investigate the embryological and postnatal ontogeny of rodents, as was done recently in neotomine–peromyscine rodents (Creighton and Strauss, 1986). Manipulative embryological studies (e.g., DuBrul and Laskin, 1961) and studies of developmental integration (character correlation) are also needed to evaluate fully the role of developmental perturbation in morphological evolution. While the omnipresent force of natural selection serves ultimately as the arbitrator in determining the success or failure of evolutionary novelties, we should not overlook the possible importance of developmental heterochrony in the evolution of morphological trends in the Rodentia.

ACKNOWLEDGMENTS. Appreciation is extended to D. J. Hafner and R. W. Thorington, Jr. for their critical comments on early drafts of this material and for invaluable discussion during the course of this work. We thank S. Warschaw and S. Muth for their clerical assistance. This research was supported in part by research grants from the National Science Foundation BSR-8600644 (J.C.H.) and BSR-8607223 (M.S.H.).

References

Alberch, P., 1980, Ontogenesis and morphological diversification, *Am. Zool.* **20**:653–667.

Alberch, P., Gould, S. J., Oster, G. F., and Wake, D. B., 1979, Size and shape in ontogeny and phylogeny, *Paleobiology* **5**:296–317.

Atchley, W. R., 1987, Developmental quantitative genetics and the evolution of ontogenies, *Evolution* **41**:316–330.

Bartholomew, G. A., Jr., and Caswell, H. H., Jr., 1951, Locomotion in kangaroo rats and its adaptive significance, *J. Mammal.* **32**:155–169.

Brookfield, J. F. Y., 1982, Adaptation and functional explanation in biology, *Evol. Theor.* **5**:281–290.

Butterworth, B. B., 1960, A comparative study of sexual behavior and reproduction in the kangaroo rats *Dipodomys deserti* Stephens and *D. merriama* Mearns, Ph.D. dissertation, University of Southern California.

Butterworth, B. B., 1961, A comparative study of growth and development of the kangaroo rats, *Dipodomys deserti* Stephens and *Dipodomys merriami* Mearns, *Growth* **25**:127–139.

Chew, R. M., and Butterworth, B. B., 1959, Growth and development of Merriam's kangaroo rat, *Dipodomys merriami*, *Growth* **23**:75–95.

Creighton, G. K., and Strauss, R. E., 1986, Comparative patterns of growth and development in cricetine rodents and the evolution of ontogeny, *Evolution* **40**:94–106.

Darlington, P. J., 1957, *Zoogeography: The Geographical Distribution of Animals*, Wiley, New York.

De Beer, G. R., 1930, *Embryology and Evolution*, Clarendon Press, Oxford.

De Beer, G. R., 1958, *Embryos and Ancestors*, Clarendon Press, Oxford.

DuBrul, E. L., and Laskin, D. M., 1961, Preadaptive potentialities of the mammalian skull: An experiment in growth and form, *Am. J. Anat.* **109:**117–132.

Egoscue, H. J., Bittmenn, J. G., and Petrovich, J. A., 1970, Some fecundity and longevity records for captive small mammals, *J. Mammal.* **51:**622–623.

Eisenberg, J. F., 1975, The behavior patterns of desert rodents, in: *Rodents in Desert Environments* (I. Prakash and P. K. Ghosh, eds.), pp. 189–224, Junk, The Hague.

Eisenberg, J. F., 1981, *The Mammalian Radiations, An Analysis of Trends in Evolution, Adaptation, and Behavior,* University of Chicago Press, Chicago.

Eisenberg, J. F., and Isaac, D. E., 1963, The reproduction of heteromyid rodents in captivity, *J. Mammal.* **44:**61–66.

Fink, W. L., 1982, The conceptual relationship between ontogeny and phylogeny, *Paleobiology* **8:**254–264.

Fleming, T. H., 1977, Growth and development of two species of tropical heteromyid rodents, *Am. Midl. Nat.* **98:**109–123.

Goldschmidt, R., 1940, *The Material Basis of Evolution,* Yale University Press, New Haven.

Gould, S. J., 1977, *Ontogeny and Phylogeny,* Harvard University Press, Cambridge.

Gould, S. J., 1980, *The Panda's Thumb,* Norton, New York.

Gould, S. J., and Lewontin, R. C., 1979, The spandrels of San Marco and the Panglossian paradigm: A critique of the adaptationist programme, *Proc. R. Soc. Lond.* **B 205:**581–598.

Gould, S. J., and Vrba, E. S., 1982, Exaptation—A missing term in the science of form, *Paleobiology* **8:**4–15.

Hafner, J. C., 1978, Evolutionary relationships of kangaroo mice, genus *Microdipodops, J. Mammal.* **59:**354–366.

Hafner, J. C., 1988, Macroevolutionary diversification in heteromyid rodents: Heterochrony and adaptation in phylogeny, in: *Biology of the Family Heteromyidae* (H. H. Genoways and J. H. Brown, eds.), American Society of Mammalogists (in press).

Hafner, J. C., and Hafner, M. S., 1983, Evolutionary relationships of heteromyid rodents, *Great Basin Nat. Mem.* **7:**3–29.

Hafner, M. S., and Hafner, J. C., 1984, Brain size, adaptation and heterochrony in geomyoid rodents, *Evolution* **38:**1088–1098.

Hall, E. R., 1941, Revision of the rodent genus *Microdipodops, Field Mus. Nat. Hist. Zool. Ser.* **27:**233–277.

Hall, E. R., 1946, *Mammals of Nevada,* University of California Press, Berkeley.

Hall, E. R., and Linsdale, J. M., 1929, Notes on the life history of the kangaroo mouse (*Microdipodops*), *J. Mammal.* **10:**298–305.

Hatt, R. T., 1932, The vertebral columns of richochetal rodents, *Bull. Am. Mus. Nat. Hist.* **63:**599–738.

Hayden, P., and Gambino, J. J., 1966, Growth and development of the little pocket mouse, *Perognathus longimembris, Growth* **30:**187–197.

Hill, J. E., 1937, Morphology of the pocket gopher mammalian genus *Thomomys, Univ. Calif. Publ. Zool.* **42:**81–172.

Howell, A. B., 1932, The saltatorial rodent *Dipodomys:* The functional and comparative anatomy of its muscular and osseous systems, *Proc. Am. Acad. Arts Sci.* **67:**377–536.

Keast, A., 1972, Comparisons of contemporary mammal faunas of southern continents, in: *Evolution, Mammals, and Southern Continents* (A. Keast, F. C. Erk, and B. Gloss, eds.), State University of New York Press, Albany.

Kirmiz, J. P., 1962, *Adaptation of Desert Environment,* Butterworth, London.

Kotler, B. P., 1985, Owl predation on desert rodents which differ in morphology and behavior, *J. Mammal.* **66:**824–828.

Lackey, J. A., 1967, Growth and development of *Dipodomys stephensi, J. Mammal.* **48:**624–632.

Lay, D. M., 1988, Anatomy of the heteromyid ear, in: *Biology of the Family Heteromyidae* (H. H. Genoways and J. H. Brown, eds.), American Society of Mammalogists (in press).

Long, C. A., 1976, Evolution of mammalian cheek pouches and a possibly discontinuous origin of a higher taxon (Geomyoidea), *Am. Nat.* **110:**1093–1097.

Løvtrup, S., 1981a, Introduction to evolutionary epigenetics, in: *Evolution Today* (G. G. E. Scudder and J. L. Reveal, eds.), pp. 139–144, Hunt Institute for Botanical Documentation, Carnegie-Mellon University, Pittsburgh.

Løvtrup, S., 1981b, The epigenetic utilization of the genomic message, in: *Evolution Today* (G. G. E.

Scudder and J. L. Reveal, eds.), pp. 145–161, Hunt Institute for Botanical Documentation, Carnegie-Mellon University, Pittsburgh.

Mares, M. A., 1983, Desert rodent adaptation and community structure, *Great Basin Nat. Mem.* **7**:30–43.

Maynard Smith, J., Burian, R., Kauffman, S., Alberch, P., Campbell, J., Goodwin, B., Lande, R., Raup, D., and Wolpert, L., 1985, Developmental constraints and evolution, *Q. Rev. Biol.* **60**:265–287.

Mayr, E., 1983, How to carry out the adaptationist program?, *Am. Nat.* **12**:324–334.

McKinney, M. L., 1984, Allometry and heterochrony in an Eocene echinoid lineage: Morphological change as a by-product of size selection, *Paleobiology* **10**:407–419.

McNamara, K. J., 1986, A guide to a nomenclature of heterochrony, *J. Paleontol.* **60**:4–13.

Merriam, C. H., 1891, Description of a new genus and species dwarf kangaroo rat from Nevada (*Microdipodops magacephalus*), *N. Am. Fauna* **5**:115–117.

Nowak, R. M., and Paradiso, J. L., 1983, *Walker's Mammals of the World*, 4th ed., Johns Hopkins University Press, Baltimore.

O'Farrell, M. J., and Blaustein, A. R., 1974, *Microdipodops magacephalus*, *Mamm. Species* **46**:1–3.

Rachootin, S. P., and Thomson, K. S., 1981, Epigenetics, paleontology, and evolution, in: *Evolution Today* (G. G. E. Scudder and J. L. Reveal, eds.), pp. 181–193, Hunt Institute for Botanical Documentation, Carnegie-Mellon University, Pittsburgh.

Romer, A. S., 1966, *Vertebrate Paleontology*, 3rd ed., University of Chicago Press, Chicago.

Rutledge, J. J., Eisen, E. J., and Legates, J. E., 1975, Correlated response in skeletal traits and replicate variation in selected lines of mice, *Theor. Appl. Genet.* **46**:26–31.

Ryan, J. M., 1986, Comparative morphology and evolution of cheek pouches in rodents, *J. Morphol.* **190**:27–41.

Simpson, G. G., 1944, *Tempo and Mode in Evolution*, Columbia University Press, New York.

Simpson, G. G., 1953, *The Major Features of Evolution*, Columbia University Press, New York.

Van De Graaff, K. M., 1973, Comparative development osteology in three species of desert rodents, *Peromyscus eremicus*, *Perognathus intermedius*, and *Dipodomys merriami*, *J. Mammal.* **54**:729–741.

Waddington, C. H., 1957, *The Strategy of the Genes*, Allen and Unwin, London.

Waddington, C. H., 1962, *New Patterns in Genetics and Development*, Columbia University Press, New York.

Webster, D. B., and Webster, M., 1975, Auditory systems of Heteromyidae: Functional morphology and evolution of the middle ear, *J. Morphol.* **146**:343–376.

Wood, A. E., 1935, Evolution and relationship of the heteromyid rodents with new forms from the territory of western Northern America, *Ann. Carnegie Mus.* **24**:73–262.

Wootton, J. T., 1987, The effects of body mass, phylogeny, habitat, and trophic level on mammalian age at first reproduction, *Evolution* **41**:732–749.

Zakrzewski, R. J., 1981, Kangaroo rats from the Borchers Local Fauna, Blancan, Meade County, Kansas, *Trans. Kansas Acad. Sci.* **84**:78–88.

Chapter 13

Heterochrony in Primates

BRIAN T. SHEA

1. Introduction and Background

Until recently the study of heterochrony in primate evolution has focused primarily on arguments concerning human neoteny. Swiss zoologist Julius Kollman (1905), who introduced the term neoteny in the late 1800s, suggested that early humans could be traced to pygmy groups that had developed from anthropoid apes via juvenilization and neoteny. De Beer (1930), building on Bolk's (1926, 1929) important work, made human evolution one of the central examples in his classic work *Embryology and Evolution*, which established the study of heterochrony within the modern synthesis. Continued emphasis is evidenced in the work of Abbie (1952, 1958, 1964), Montagu (1962, 1981), and Gould (1977).

One significant factor in the relative lack of work on heterochrony in the evolution of nonhuman primates was the reluctance of A. H. Schultz to utilize concepts of allometric growth and heterochrony. Schultz was a great comparative anatomist whose life's work (Schultz, 1969) in part involved the documentation

BRIAN T. SHEA • Departments of Anthropology and Cell Biology and Anatomy, Northwestern University, Chicago, Illinois 60611.

of the ontogenetic patterns underlying differences in adult morphology among primates. Although clearly interested in the relationships between ontogeny and phylogeny, Schultz did not analyze his data in terms of relative growth or heterochrony. This undoubtedly hindered the study of heterochrony in primate evolution until quite recent times.

Gould's (1977) masterful synthesis provided an important foundation and impetus for studies of heterochrony in the evolution of primates, as it did for so many other groups. The present review will focus principally on primatological research carried out subsequent to the publication of Gould's book. I will attempt to utilize this body of work to illustrate points of theoretical and methodological significance, as well as to highlight areas of future research in primate heterochrony.

2. Methodology

A number of recent papers have dealt with aspects of theory and method in heterochronic analyses (e.g., Alberch *et al.*, 1979; Bonner and Horn, 1982; Fink, 1982; Shea, 1983c; McNamara, 1986; McKinney, 1984). In this chapter I will discuss interspecific morphological transformations in terms of shifts in the commonly used parameters of size, shape, and timing. The comparative analysis of bivariate patterns of growth allometry figure prominently in this approach, since they enable us to determine when shape changes are correlated with overall size change (however these might be related to developmental timing), and when size/shape interrelationships are dissociated to produce new morphological configurations at comparable sizes. *Ontogenetic scaling* (Gould, 1975; Shea, 1981), or the differential truncation or extension of common patterns of growth allometry, is used here to refer to cases of interspecific allometric shape change. Such allometric shape change may be the simple result of differences in overall size. Developmental timing is based on traditional absolute (e.g, age in months, years) or relative (e.g., dental eruption, epiphyseal fusion) criteria.

Gould's (1977) clock models will be used schematically to depict changes in these size, shape, and timing parameters (see Fig. 1). I prefer these clocks because of the ease with which one can simultaneously represent shifts in all three factors. More precise quantification is not necessary, given the presentation of bivariate allometries elsewhere.

3. Size and Allometric Shape Change

Recent work has uncovered a number of cases among primates where significant interspecific adult shape variation can be directly related to differences in either overall body size or the total size of a particular body region. Ontogenetic scaling of growth allometries characterizes these comparisons. My own work on some of the anthropoid primates provides many of these examples. Figures 1a–1d illustrate this type of heterochrony change using Gould's clock model.

The African apes are three relatively large, knuckle-walking species, which

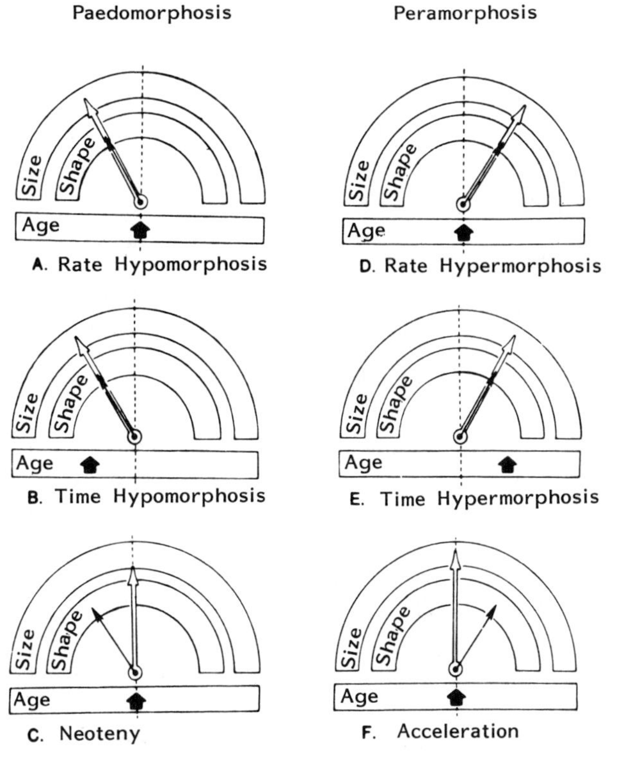

Figure 1. A modification of Gould's (1977) clock models used to illustrate changes in size, shape, and timing between ancestors and descendants. Ancestral values are represented by the dashed line, and the shift in position of the clock's hands document heterochronic change. When the size and shape vectors remain together, ancestors and descendants share common growth allometries. Note that a paedomorphic or peramorphic transformation may result from at least three different processes. See text, Gould (1977), Alberch *et al.* (1979), and Shea (1983d) for additional discussion.

exhibit a progressive increase in overall size and degree of sexual dimorphism: pygmy chimpanzees (*Pan paniscus*) average around 40 kg; common chimpanzees (*Pan troglodytes*) average around 50 kg; and gorillas (*Gorilla gorilla*) weigh approximately 90 kg for females and 160 kg for males. Many of the pervasive shape differences within the cranium and postcranium among these apes result from ontogenetic scaling of proportions. Figure 2 provides representative examples [see Shea (1981, 1983a–d, 1984, 1985a, 1986a) for additional details]. For these cases, gorillas may be described as peramorphic chimpanzees, and common chimpanzees are in part similarly related to pygmy chimpanzees (although see Section 4).* The ecological basis of the size shifts among these apes probably relates to dietary differences. Large body size permits the gorilla to select more fibrous food items not heavily utilized by the smaller chimpanzees (Clutton-Brock and Harvey, 1979a), which concentrate on scattered and highly nutritious food sources (Ghig-

* For the sake of simplicity here I am assuming a particular direction of heterochronic transformation. Of course, the chimpanzee may represent a paedomorphic gorilla, or both taxa may result via some such transformation from a third, unknown group. Fink (1982, and this volume) has discussed some of difficulties involved in determining the precise direction of heterochronic transformations. Since I am primarily interested in elucidating the morphological differences among these extant taxa in terms of heterochrony, I am less concerned with identifying specific character states and phylogenetic relationships, although this is certainly necessary for a full understanding of the morphological transformation, particularly in an ecological context.

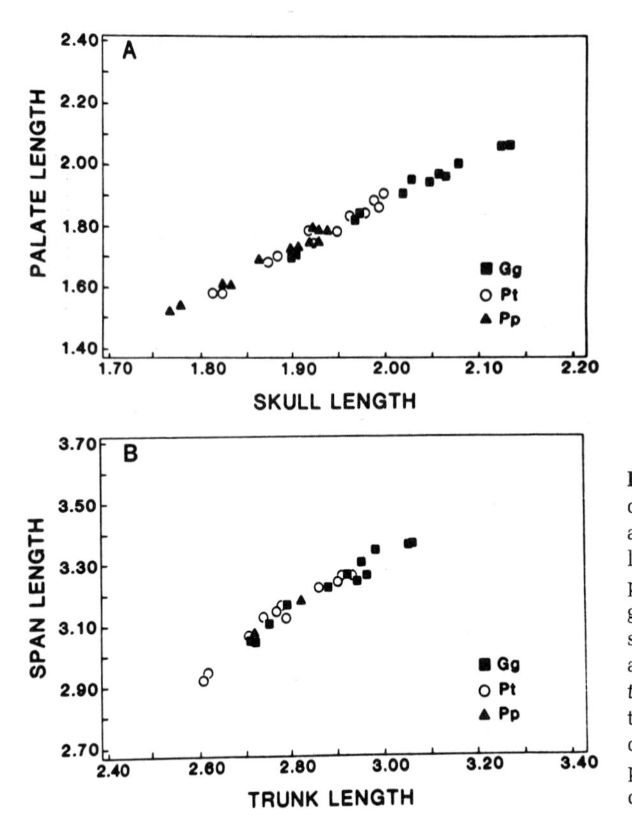

Figure 2. Ontogenetic allometric plots of (A) palate length vs. skull length and (B) arm span length vs. trunk length (head-to-fork length) in African pongids. These are cross-sectional growth data, with each point representing the mean value for a different age group. (■) *Gorilla gorilla;* (○) *Pan troglodytes;* (▲) *Pan paniscus.* Note that the three species share common ontogenetic allometries for these comparisons. See text for additional discussion.

lieri, 1984; see also Tutin and Fernandez, 1985). In fact, it has been suggested that the smaller body size and reduced sexual dimorphism of chimpanzees is related to their need for increased mobility in food procurement (Rodman, 1984; Ghiglieri, 1984).

The brachiating lesser apes of Asia include the small, frugivorous gibbons (6–8 kg) and the larger, more folivorous siamangs (12–14 kg). These groups exhibit a number of differences in cranial and postcranial shape in addition to their overall size differences (Schultz, 1933; Corruccini, 1981; Creel and Preuschoft, 1976). Yet, at least for the skull, most of the shape differences between the gibbons and siamangs appear to result from ontogenetic scaling (Shea, B. T., unpublished results) (see Figs. 3C,D) for representative examples), and siamangs may be described as peramorphic gibbons. The ecological basis of the size (and correlated shape) differentiation between these lesser apes may mirror the case for the chimpanzees and gorillas, since the larger siamang is also the more folivorous of the two (Raemakers, 1984).

The talapoin monkey (*Cercopithecus [Miopithecus] talapoin*) of West Africa is the smallest of the Old World anthropoids, weighing approximately 1250 g. In addition to this size difference, talapoins exhibit a number of shape differences in comparison to their larger relatives of the genus *Cercopithecus*, such as gracile, juvenilized crania (Verheyen, 1962) and different interlimb proportions. As illustrated in Figs. 3A,B, however, recent work indicates that these shape differences

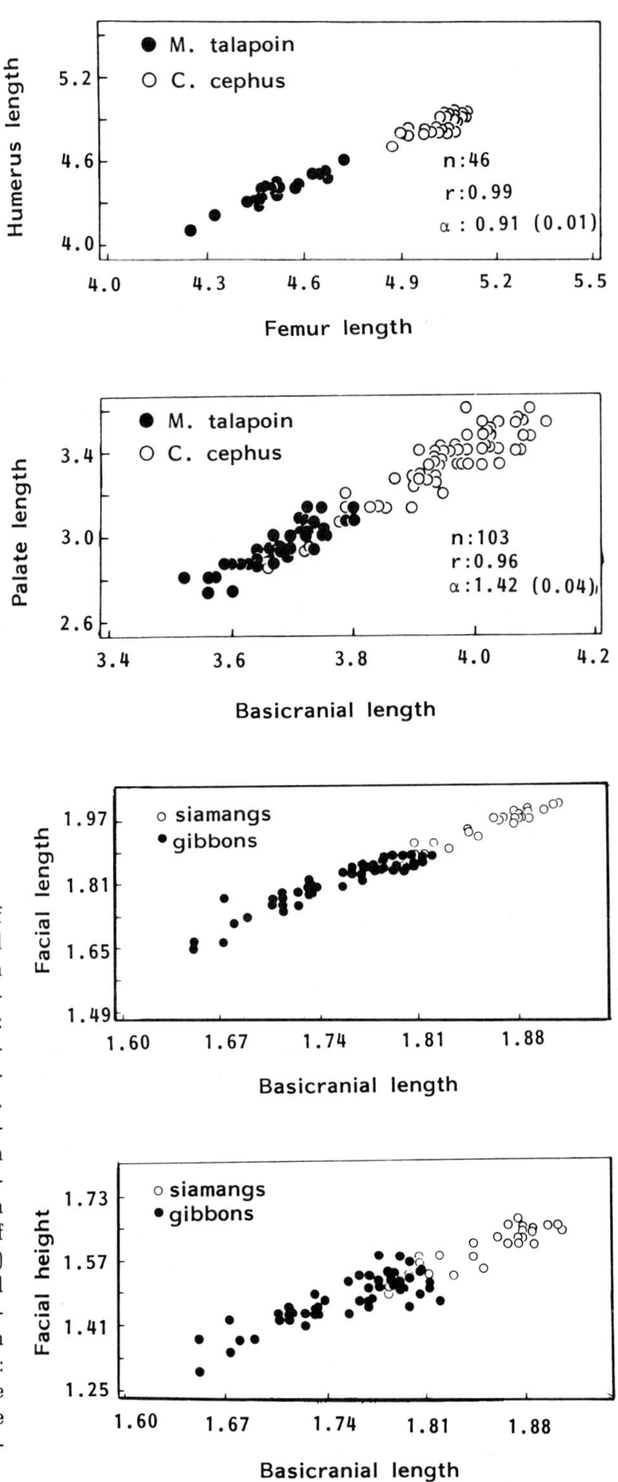

Figure 3. (A,B) Allometric plots of humerus length vs. femur length and palate length vs. basicranial length in talapoin and moustached (*Cercopithecus cephus*) monkeys. Key: *n*, sample size; *r*, correlation coefficient; α, coefficient of growth allometry (standard error of the slope). Note that the shape differences between the adult endpoints in both cases appear to be the result of differential extension of a common growth allometry. (C,D) Two plots of facial dimensions (length, height) vs. overall skull size (basicranial length) in cross-sectional growth series of gibbons and siamangs. Both trajectories are positively allometric: facial proportions in siamangs are allometric enlargements of those seen in gibbons. See text for additional discussion.

are primarily allometric correlates of the small size of the talapoins (Shea, B. T., unpublished results). It seems most likely that this small size is derived (Verheyen, 1962; Delson and Andrews, 1975), but there is no evidence to indicate that the talapoin monkey matures any earlier than related *Cercopithecus* species (Gautier-Hion and Gautier, 1976). Small body size in talapoin monkeys is likely related to their ecological specialization on the relatively unexploited food source of insects in the gallery forests of West Africa (Gautier-Hion, 1973).

The above cases provide examples where considerable shape differentiation appears to be the correlated result of selection for size change in particular ecological contexts. Only a comparative ontogenetic and heterochronic approach can determine this. Our knowledge of these ontogenetic bases of interspecific differentiation enables us to avoid the assumption or conclusion that adult shape differences reflect other, "special" (or nonallometric) adaptive changes, such as adaptations to differing dietary regimes in the case of skull form in the African pongids (Shea, 1983b). Ontogenetic scaling may be viewed as an example of historical factors (Lauder, 1981; Wake, 1982) in evolutionary change, since allometric trajectories channel transformations when overall size changes rapidly for adaptive or nonadaptive reasons (Huxley, 1932; Simpson, 1953; Gould and Lewontin, 1979; Shea, 1985a; Wayne, 1986). The heterochronic approach also allows us to elucidate the ways in which entire morphologies are intercorrelated and integrated.

4. Size/Shape Dissociations

The primates present many cases where ontogenetic allometries are dissociated, presumably in response to selection for specific novel proportions (as opposed to simple size change or growth duration). Figures 1e and 1f illustrate this type of heterochronic change using Gould's clock model. A great many of the ontogenetic modifications discussed by Schultz (1969) represent dissociations that yield functionally important adult specializations, e.g., the derived positive allometries of human hindlimbs and gibbon forelimbs. But, as noted previously, a heterochronic perspective is only particularly informative when developmental interrelationships among a number of features are involved.

The evolution of skull form within the chimpanzees provides an interesting such case. Cranial form in *Pan paniscus* is strongly paedomorphic, as demonstrated in a number of studies (Coolidge, 1933; Weidenreich, 1941; Fenart and Deblock, 1974; Cramer, 1977; Shea, 1983b, 1982, 1984; Laitman and Heimbuch, 1984). Although debate continues over the degree or even presence of overall body size differences between the two chimpanzee species [see Shea (1984) and Jungers and Susman (1984) for a review], it is well established that the markedly dwarfed and paedomorphic skull of the pygmy chimpanzee is produced developmentally by a dissociation and retardation of particularly facial growth in these apes (Shea, 1983c). This provides an example of paedomorphosis via neoteny (see Fig. 4), assuming that the proportions observed in *Pan troglodytes* are in fact primitive. I have noted that the neotenic evolution of skull form in pygmy chim-

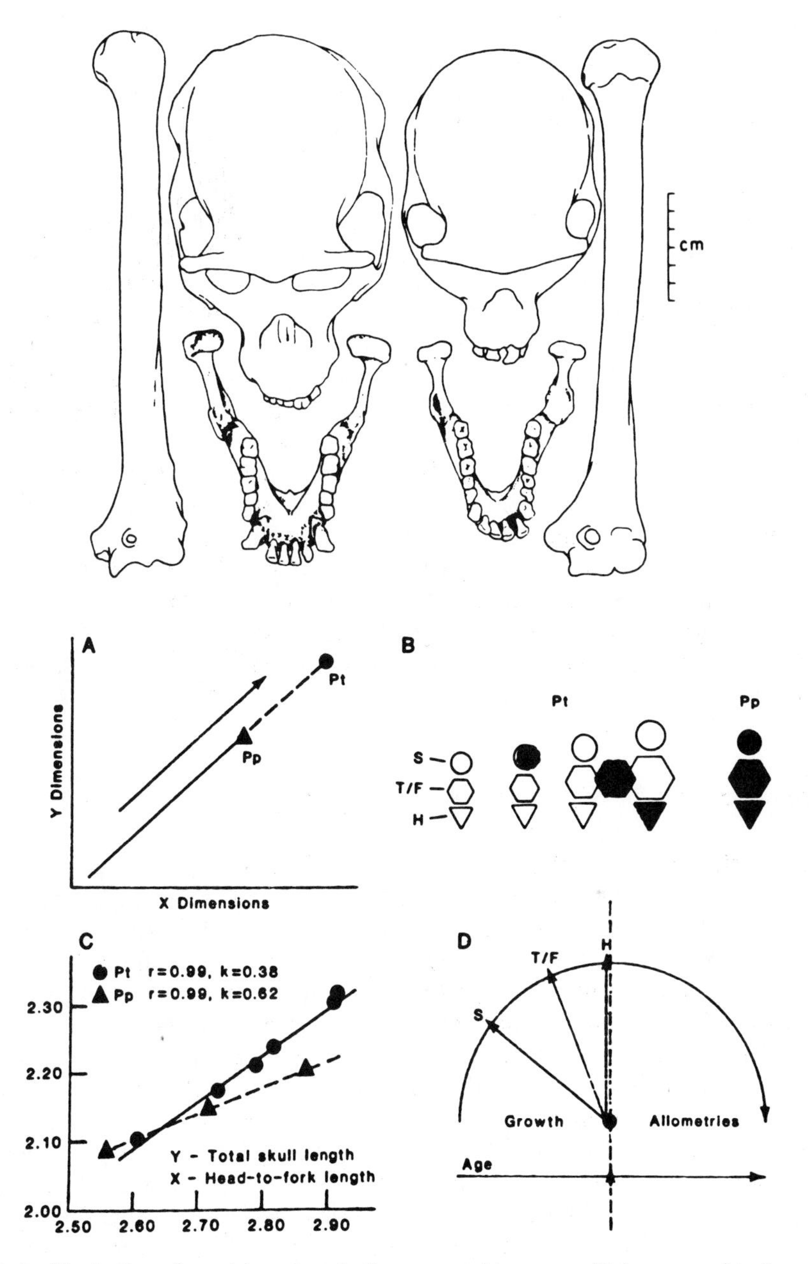

Figure 4. An illustration of cranial neoteny in the pygmy chimpanzee (Pp) compared to the common chimpanzee (Pt). The top half is from McHenry and Corruccini (1981) and illustrates the dwarfed overall skull size of *Pan paniscus* compared to other portions of the body, here represented by the humerus. The bottom half of the figure is from Shea (1983c), and illustrates the fact that (A) within body regions, including the skull, common and pygmy chimpanzees share common growth allometries, (B) the skull is differentially dwarfed and paedomorphic in Pp compared to the rest of the body, (C) growth allometries of skull vs. body size have significantly lower slopes in Pp, and (D) the various body regions of Pp exhibit a gradient of differential dwarfing and paedomorphosis, with the skull most strongly dwarfed and the hindlimbs not at all.

panzees yields reduced dentofacial sexual dimorphism (Shea, 1983c, 1984). This relative lack of dimorphism correlates well with behavioral differences between pygmy and common chimpanzees emerging from field and laboratory observations. For example, *Pan paniscus* exhibits higher male/female affinity and less sexual differentiation in social structure than seen in *Pan troglodytes* (Kuroda, 1980, 1986; Kano, 1980). In fact, Kuroda (1980) describes some of the behavioral characteristics of pygmy chimpanzees as paedomorphic. Recently, Dahl (1986) has shown that adult *Pan paniscus* resemble subadult *Pan troglodytes* in aspects of their external genital anatomy and patterns of perineal swellings and estrus cycling. It appears that neoteny may have played an important role in the differentiation of a variety of features in chimpanzee evolution, and this case provides an excellent example of how ontogeny offers a storehouse of morphologies that may be recast in novel combinations (presumably with a minimum of genetic and developmental alterations) when the need arises (Gould, 1977).

Manley-Buser (1986) has recently completed a detailed morphometric study of hominoid foot bones using heterochronic analyses. Among other things, she finds that many of the most significant shape differences distinguishing pedal morphology in modern *Homo sapiens* from the great apes result from a global dissociation of ontogenetic trajectories, such that humans are usually transposed below great apes in bivariate plots of pedal dimensions vs. overall size. Manley-Buser thus concludes that neoteny has played an important role in the evolution of our distinctive pedal morphology.

5. Ontogenetic Scaling and Adaptation

The principles of ontogenetic scaling and size/shape dissociations as discussed in the preceding sections can be productively utilized to identify morphological configurations that likely were selected for in response to new functional requirements. This essentially entails the use of ontogenetic scaling as a "criterion of subtraction" (Gould, 1966, 1975; Shea, 1985a) when comparing close relatives of differing overall size and shape. Interspecific shape differences resulting from the differential extension of common patterns of growth allometry may simply be the correlated result of selection for overall size, and not the result of selection for the specific shape changes themselves.* Parenthetically, this does not necessarily mean that such interspecific shape differences are nonadaptive and without functional significance, since we know that allometric growth is frequently related to changing functional requirements during ontogeny and growth (Goss, 1964; Clutton-Brock and Harvey, 1979b; Frazzetta, 1975; Shea, 1985a, 1986a). Nevertheless, a fundamental dissociation of (presumably ancestral) growth allometries to produce new size/shape configurations is good evidence for the selection of novel proportions. Gould (1977) provides many examples of such adaptive dissociations and discusses their possible ecological contexts.

* One must also acknowledge the possibility that specific shapes that result from allometries may be the selective target and increased or decreased size may be the correlated by-product. Clutton-Brock and Harvey (1979b) discuss this cogently.

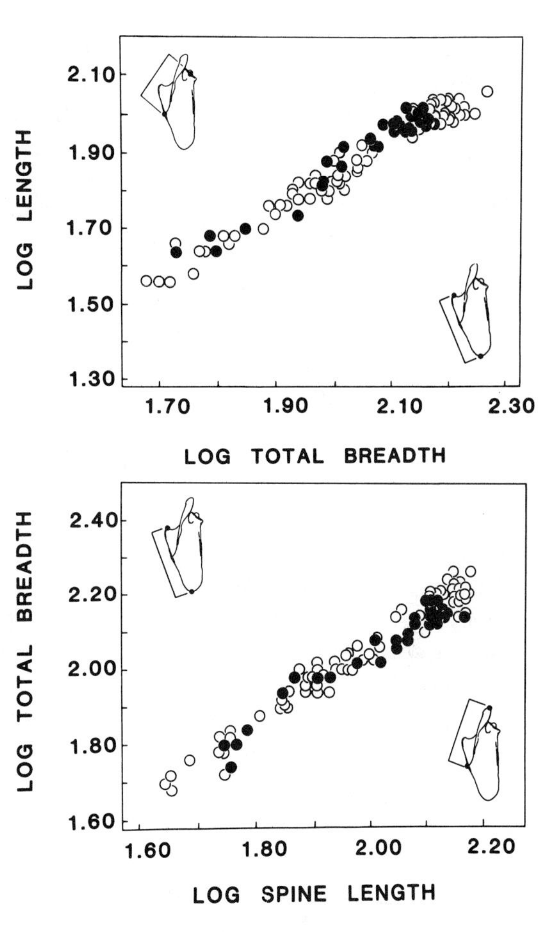

Figure 5. Cross-sectional growth allometric plots of scapula shape in (●) pygmy and (○) common chimpanzees. Scapulae of adult *Pan paniscus* resemble those of subadult *Pan troglodytes* in shape. See text for additional discussion.

Here I simply wish to stress once again the value of comparative ontogenetic and heterochronic perspectives in elucidating aspects of functional biology and adaptation of specific body regions (Wake, 1982), with examples from studies of primate morphology. I have used comparative ontogenetic analyses to address shape differences in scapula form between pygmy and common chimpanzees (Shea, 1986a). A number of authors had previously suggested that scapula morphology in pygmy chimpanzees differed from that of common chimpanzees, more closely approximating that of the brachiating gibbons, and therefore indicating increased reliance on suspensory locomotion in the smaller chimpanzees (Coolidge, 1933; Horn, 1975; Susman *et al.*, 1980). My results (Shea, 1986a) confirmed the interspecific shape differences, but demonstrated ontogenetic scaling of all proportions examined (Fig. 5); thus, adult pygmy chimpanzees had scapula proportions similar to those of common chimpanzees of comparable overall size. Since there is no evidence to suggest that subadult and adult common chimpanzees locomote as differently as has been suggested for adult pygmy and common chimpanzees, this finding does not support the conclusion that the interspecific shape differences are indicative of substantially different locomotor behaviors and adaptations.

Levitch's (1986) work on heterochrony and limb proportions in New World monkeys provides an example where fundamental dissociations do occur with size change and thus suggest selection for new proportions. She found that in comparative ontogenetic analyses, forelimb (and especially radius) lengths of one type of callitrichid (*Leontopithecus*) were transposed above those of other callitrichids when plotted against various size dimensions. Other aspects of limb proportions were in fact ontogenetically scaled in the New World monkeys (see below). These comparative analyses highlight those fundamental dissociations that are not simple size correlates [at least in the sense of ontogenetic scaling; see Shea (1981, 1985a) for discussion] by distinguishing them from cases of ontogenetic scaling. The elongate distal forelimbs of this callitrichid may be related to its feeding adaptation of probing within narrow confines for insects (Hershkovitz, 1977), or perhaps may reflect some differences in locomotor specialization.

My studies on limb proportions within the great apes provide an additional example. Here comparative ontogenetic analyses demonstrated that the increasing intermembral index (ratio of forelimb to hindlimb) from pygmy chimpanzees to common chimpanzees to gorillas is not a simple result of ontogenetic scaling, and thus requires an explanation based on the development of novel proportions (Shea, 1981). This is in contrast to the case of talapoin monkeys discussed above, where intermembral index differences result from simple ontogenetic scaling.

Many more examples could be given, but these suffice to illustrate the point that a heterochronic approach permits us to focus more productively on the novel shape changes that most likely reflect new adaptations to changing behaviors and environments.

6. Shifts in Growth Timing or Duration

It is surprising and perhaps significant that we can cite very few examples where considerable interrelated morphological change between primate species results from either precocious maturation (time hypomorphosis or progenesis) or extended duration of growth (time hypermorphosis). One likely candidate may be the small callitrichids among platyrrhine primates (New World monkeys), which a number of authorities have viewed as phyletic dwarfs (Leutenegger, 1980; Ford, 1980). Marmosets and tamarins develop precociously when compared to close relatives such as squirrel (*Saimiri*) and *Cebus* monkeys (Herschkovitz, 1977). Levitch (1986) has recently shown that a number of differences in limb proportions, particularly those relating distal and proximal elements within each limb, result from ontogenetic scaling when callitrichids are compared to larger hypothetical "ancestors" such as squirrel monkeys (*Saimiri*) or owl monkeys (*Aotus*). These shape differences represent examples (from a perhaps much larger set of as yet basically unexamined features) that are related in part to the precocious maturation of these small monkeys. These interspecific shape differences result primarily from time hypomorphosis (Fig. 1b), and a heterochronic perspective strengthens the notion that such shape transformations and the small overall size of the callitrichids may be secondary consequences of selection for truncated growth duration in relation to the need to increase reproductive turnover [Leutenegger (1980) argues that twinning in callitrichids is a derived correlate of their small size and developmental schedule].

It is appropriate at this point to reiterate that we cannot normally read information about the *duration* of growth from allometric (or any size/shape) plots (Shea, 1983d). There are many cases where not just paleontologists (who, after all, usually have no direct way of gauging time to maturity or growth duration) but also neontologists have assumed that truncated or extended growth allometries reflect shortened or lengthened periods of growth, respectively. But size is not a valid proxy for time, particularly for between-species comparisons. For this reason, I have suggested an addition to, and revision of, previous terminology of heterochronic processes (Shea, 1983d) (see Fig. 1). As illustrated in Fig. 1, the extension or truncation of growth allometries *may or may not* be accompanied by lengthening or shortening of the duration of growth (the traditional classification assumes this to be true). Altered rates of overall weight growth, while maintaining ancestral allometries and maturation times, will result in heterochronic changes in morphology and the extension or truncation of size/shape patterns. I distinguish these cases as rate vs. time hypo- and hypermorphosis. All the cases I discussed in Section 3 provide examples of rate hypo- or hypermorphosis. Under the traditional classification we would conclude from the allometric plots that talapoins have truncated, and gorillas have prolonged, growth *periods* relative to their morphological neighbors—which they emphatically do not [see Shea (1983d, 1986b) for gorillas and Gautier-Hion and Gautier (1976) and Shea (1986b) for talapoins].

The reason for emphasizing these distinctions is the same as that so eloquently argued by Gould (1977), i.e., we need to distinguish results from underlying processes, because it is variation among the latter that suggests fundamentally different selective targets and scenarios. For example, ecological correlates of small size and paedomorphosis will presumably be quite different in the case of reduced growth rate as opposed to truncated growth time or duration. Assuming for the sake of argument that Old World talapoin monkeys and New World callitrichids are paedomorphic versions of their larger relatives, in the first case selection was probably on reduced growth rate and smaller adult size in order to reduce feeding competition and exploit insects as a dietary resource, while in the second case selection appears to have been (at least in part) on truncated growth *duration*, perhaps as a means of increasing reproductive turnover.

This issue of distinguishing between rate- and time-based simple heterochronies has significant and problematic ramifications for the identification of heterochrony in the fossil record. Quite simply, except in the unusual case where we have some size-independent measure of chronological age (e.g., Charles *et al.*, 1986; Bromage and Dean, 1985; Smith, 1986; McKinney, this volume; Jones, this volume), it will not be possible to distinguish rate from time processes. One of the classic examples from the allometry and heterochrony literature, the case of the peramorphic Irish Elk (*Megalocerus*), has been claimed to be a product of (time) hypermophosis (Gould, 1974, 1977). But we cannot assume that the Irish Elk took longer to reach maturity and final size than its ancestral relatives; it may have simply grown much bigger in the same amount of time (as in gorillas vs. chimpanzees), and thus been a product of rate rather than time hypermorphosis (Shea, 1983d). This general argument would also apply to another classic case from the literature on allometry and heterochrony, body size and horn size in titanothere evolution (McKinney and Schoch, 1985). In fact, Geist (1986) has recently suggested an interesting scenario of antler evolution in the Irish Elk that is not based on growth prolongation, and that differs in a number of respects from the traditional allometric argument (Gould, 1974).

7. Sexual Dimorphism

A number of authors have recently stressed the importance of elucidating the ontogenetic bases of sexual differences in adult morphology as a means to understanding how patterns of sexual dimorphism relate to social, ecological, and nutritional factors (Schultz, 1969; Wiley, 1974; Ralls, 1977; Fedigan, 1982; Jarman, 1983; Shea, 1985b, 1986a). In particular, Jarman (1983) has noted that similar patterns of adult dimorphism may be achieved via diverse ontogenetic pathways, and variation among these may reflect fundamental differences in social structure or ecological factors. I generalized these issues in terms of heterochrony and the parameters of size, shape, and timing, and applied this analysis to selected cases of sexual dimorphism among anthropoid primates (Shea, 1986b).

Heterochrony provides the means to distinguish among the intersexual shape differences that result from simple size differences (i.e., allometric or ontogenetically scaled shape differences) and those that reflect fundamental dissociations of growth allometries. In humans and nonhuman primates, many intersexual limb proportion differences are simple allometric correlates (Wood, 1975), while pelvic shape divergences result from fundamental dissociations of growth allometries (Mobb and Wood, 1977; Leutenegger and Larson, 1985). We also wish to know the rate of growth of particular structures and the time taken to reach a given size. Any of the methodologies currently utilized for interspecific comparisons may of course be adapted for intersexual (or interindividual) comparisons. I have adapted Gould's clock models for use in male vs. female comparisons (here the "ancestral" setting represents the female and the size, shape, and timing hands are shifted to reflect appropriate changes for the male), as well as for between-species comparisons of patterns of dimorphism [where an extra set of "hands" is added to the basic clock to depict size, shape, and timing values for the second species; see Fig. 6 and Shea (1986b) for additional details].

Figure 7 illustrates only three of numerous ways in which a given degree of sexual dimorphism in overall weight may be produced. A preliminary analysis of weight dimorphism in anthropoids has shown that in certain groups a marked increase in dimorphism is not accompanied by increased sexual *bimaturism* (Wiley, 1974), or difference in male/female patterns of growth timing and duration. A contrast of the very dimorphic *Cercopithecus neglectus* with the smaller and less dimorphic *Cercopithecus talapoin* fits such a pattern, based on the data of Gautier-Hion and Gautier (1976). On the other hand, the increased size dimorphism of species of *Papio* as compared to *Macaca* is accompanied by significantly increased levels of sexual bimaturism. This also appears to be true for a comparison of the very dimorphic *Gorilla* with the smaller and less dimorphic chimpanzees (Shea, 1985b, 1986b). Once again, the rate and time distinctions noted

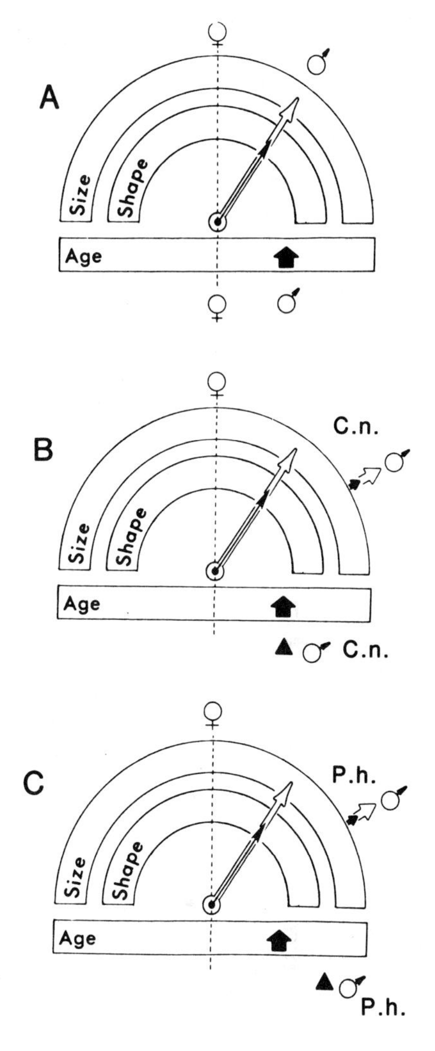

Figure 6. Use of Gould's clock models in the analysis of sexual dimorphism. (A) Comparison of males and females of a single species, with the clock hands representing the male pattern relative to the female baseline (dashed). (B) An extra set of hands is added, and two dimorphic pairs are contrasted. Here, we see that *Cercopithecus neglectus* (C.n.) is more dimorphic than the talapoin baseline, and both species exhibit comparable degrees of sexual bimaturism. (C) Comparison of *Papio hamadryas* (P.h.) with a baseline of *Macaca* (various species). Note that the increased dimorphism of the hamadryas baboon is accompanied by an increased amount of sexual bimaturism. See text for further discussion.

above become important here, for increased dimorphism may be produced by increased rate differentiation, time differentiation (bimaturism), or a combination of the two.

The ecological and/or social correlates of the different ontogenetic processes yielding a given degree of dimorphism are potentially the most important insight resulting from such a heterochronic analysis. Jarman (1983) has attempted to tie the rate vs. time distinction among large, terrestrial herbivores to whether male strategy involves attempting polygyny by using spatially stable territories and permanently accompanying a group of females, or by other means. Such factors are yet to be elucidated for primates, but certainly the ecological inputs and selective pressures favoring *accelerated* growth in males may be quite different from those favoring *prolonged* male growth, even when the final size and morphology attained are comparable.

A heterochronic approach to sexual dimorphism also permits us to identify

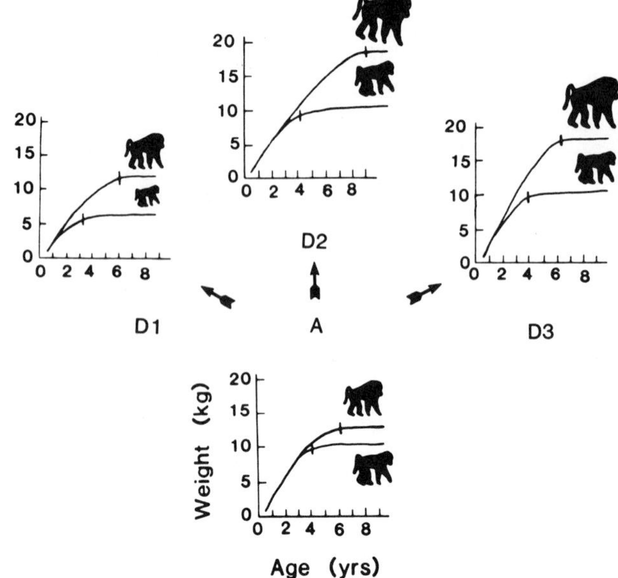

Figure 7. A schematic representation of how a given degree of body weight dimorphism may be produced by various rate and timing processes. The female/male weight ratio in the hypothetical ancestor A is 10/12 or 83%; in D1–D3 this ratio is 6.5/12, 10/18, and 10/18, respectively, yielding a comparable degree of dimorphism (but not terminal sizes) in each case, approximately 55%. The vertical marks on each growth curve indicate the approximate size and age at which physical maturity is reached and growth has basically ceased. In D1, increased weight dimorphism is produced by the earlier maturation of the female, and the growth of the male remains unchanged. In D2, the male attains larger size than the ancestral condition primarily by extending the duration of growth, thus increasing the degree of bimaturism. Female growth remains the same as in the ancestral condition. In D3, the overall size and degree of dimorphism are the same as in D2, but the timing of maturation for both sexes remains unaltered from the ancestral condition, so bimaturism does not change. See text for additional discussion.

whether one or both sexes have altered the ontogenetic processes yielding final adult outcomes. For example, the patas monkey (*Cercopithecus patas*) appears to attain its very marked weight dimorphism via a time truncation (precocious maturity) of growth in the female relative to the male. Although male patas monkeys become sexually and physically mature at around 5 years of age (roughly the same time as males of closely related species), female patas appear to reach maturity at about 2.5 years of age, the earliest yet recorded for a cercopithecine monkey. Rowell (1977) has discussed the possible ecological basis of precocious female maturation in patas monkeys. Without a comparative ontogenetic perspective we would simply focus on the static male–female terminal size differences and miss some fundamentally important variation in the processes that yield adult dimorphisms.

Leutenegger and Larson (1985) examined sexual dimorphism in postcranial bones of four genera of New World monkeys (*Callithrix, Saguinus, Saimiri*, and *Cebus*) in terms of heterochronic shifts. They found a peramorphic pattern between the sexes of *Cebus* and *Saimiri*, particularly in dimensions of the limb bones and pectoral girdle; these allometric extensions resulted primarily from time hypermorphosis, or prolonged duration of growth in the male. In contrast, sexual differences in pelvic shape in *Cebus* and *Saimiri* result from allometric dissociations and presumably relate to the need for a large bony birth canal in females. More detailed analyses of sexual dimorphism from this perspective are needed for most mammalian groups, and I believe that this may be one of the most important areas for the application of heterochrony in the near future.

8. Subspecific Differentiation

Heterochronic processes have been implicated in the differentiation of various subspecies in certain primates. One of the major foci of Bolk's (1929) work on neoteny related to the morphological differences among human races, although this application was inconsistent and misleading in many ways. Perhaps one of the clearest examples of simple developmental shifts causing a host of interrelated and marked size and shape changes among human groups is presented by the pygmies. In addition to being the smallest extant humans known, the African pygmies differ from their larger relatives in a variety of body proportions, such as relatively elongate distal extremities and relatively short hindlimbs (Marquer, 1972). These shape differences have been taken as evidence of substantial distinctiveness or even differing adaptations (Marquer, 1972; Hiernaux, 1977). However, preliminary work on anthropometric and skeletal data (Shea and Pagezy, 1988) indicates that the major proportion differences may be the simple result of ontogenetic scaling (Fig. 8). Interestingly, the only endocrine studies carried out in the field indicate decreased levels of the production of insulinlike growth factor I (Merimee *et al.*, 1982), a primary systemic growth hormone. Based on animal models of endocrine growth abnormalities (see Section 10), this is what we might have predicted. Although the overall size decrease might be adaptive in these Central Africans (Hiernaux, 1977), these preliminary studies suggest that we probably should not seek specific adaptive explanations for many of the altered proportions, since they appear to be simple allometric correlates of the size change. Data on growth timing for the pygmies are hard to come by, since date of birth

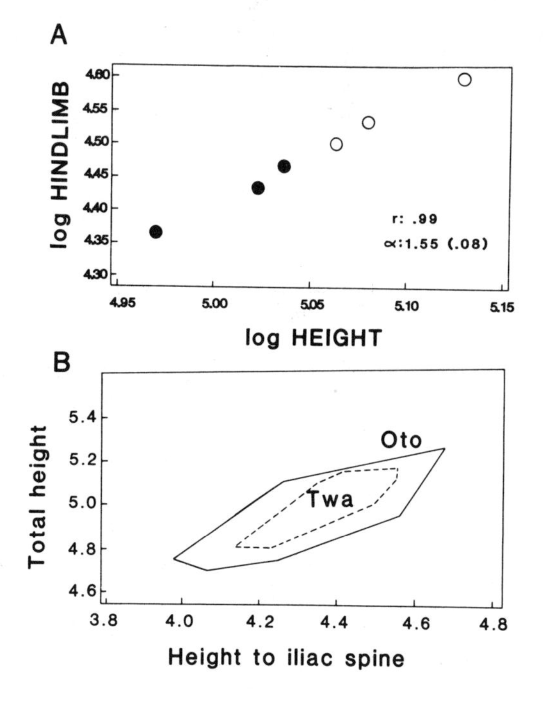

Figure 8. A plot of hindlimb length vs. body height in (●) three groups of African pygmies and (○) three groups of neighboring Africans of larger size. The points represent male adult means; data are taken from Hiernaux (1966). Note the strong positive allometry of hindlimb length and the close approximation to a single allometric trajectory. (B) A plot of total height vs. height to the anterior superior iliac spine (approximating lower limb length) in the Twa pygmies and the larger Oto villagers in Central Africa (unpublished data provided by Helene Pagezy). The polygons enclose the cross-sectional ontogenetic point scatter for both groups. Note the sharing of a common growth allometry; regression statistics reveal strong positive allometry of hindlimb length in this comparison also. See text for additional discussion.

is not recorded. Van de Koppel and Hewett (1986) argue on the basis of growth data collected among the West African Aka pygmies that the small adult size results from early growth cessation and the relative lack of a pubertal growth spurt. However, their data are cross-sectional and age estimates are very tenuous, since "the Aka are not concerned about their age and are consequently unaware of it" (van de Koppel and Hewlett, 1986, p. 95). We require longitudinal data for individuals of known age before this question can be settled. In any case, many of the morphological differences between pygmies and larger Africans appear to be the result of ontogenetic scaling, which may reflect a combination of time and rate hypomorphosis, assuming their small size is derived, as is almost certainly the case (Hiernaux, 1977). Additional work in progress should further clarify the developmental bases of the pygmies' characteristic body proportions, but suffice it to say that this provides a good example of how heterochrony can be involved in the production of intraspecific variation in humans.

A second interesting case involves subspecific variation within the common chimpanzee, *Pan troglodytes*, which is traditionally divided into three geographically distinct groups, *P. t. versus*, *P. t. troglodytes*, and *P. t. schweinfurthii* (ranging from west to east across Central Africa). Analyses of the skull (Shea and Groves, 1987) and the postcranium (Jungers and Susman, 1984) have demonstrated that shape differences among these subspecies are primarily the result of ontogenetic scaling. But Jungers and Susman (1984) have argued that the eastern variant of common chimpanzees is significantly smaller in overall size than at least *P. t. troglodytes* and probably *P. t. versus* as well. In fact, they suggest that *P. t. schweinfurthii* and pygmy chimpanzees (*Pan paniscus*) do not differ significantly in overall weight. If this turns out to be true, then *P. t. schweinfurthii* could be derived from *P. t. troglodytes* via the process of acceleration (Fig. 1f), since as adults, *schweinfurthii* attains the same within-skeleton size and shape as *troglodytes*, but at a smaller overall body weight. Alternatively, the proposed weight differences among the subspecies of common chimpanzees may be the result of small sample sizes, seasonal weight variation, or some other factor. Additional work on other species of primates should reveal further possible cases where significant intraspecific differentiation results from heterochronic processes.

9. Life History and Brain Scaling

Patterns of brain/body allometry and their ontogenetic bases have traditionally been of central concern to primatologists. This is in part due to the fact that our own order is generally a highly encephalized one (Jerison, 1979), as well as our understandable interest in the development of the highly derived and brainy *Homo sapiens*.

It is well established that the marked increase in brain size during human evolution was developmentally produced by relatively simple heterochronic change. Humans follow a fundamental mammalian curve of brain/body allometry, but extend the duration of the (primarily fetal) period of high relative growth (Fig. 9) (Count 1947; Gould, 1977). Our relatively large brains therefore result

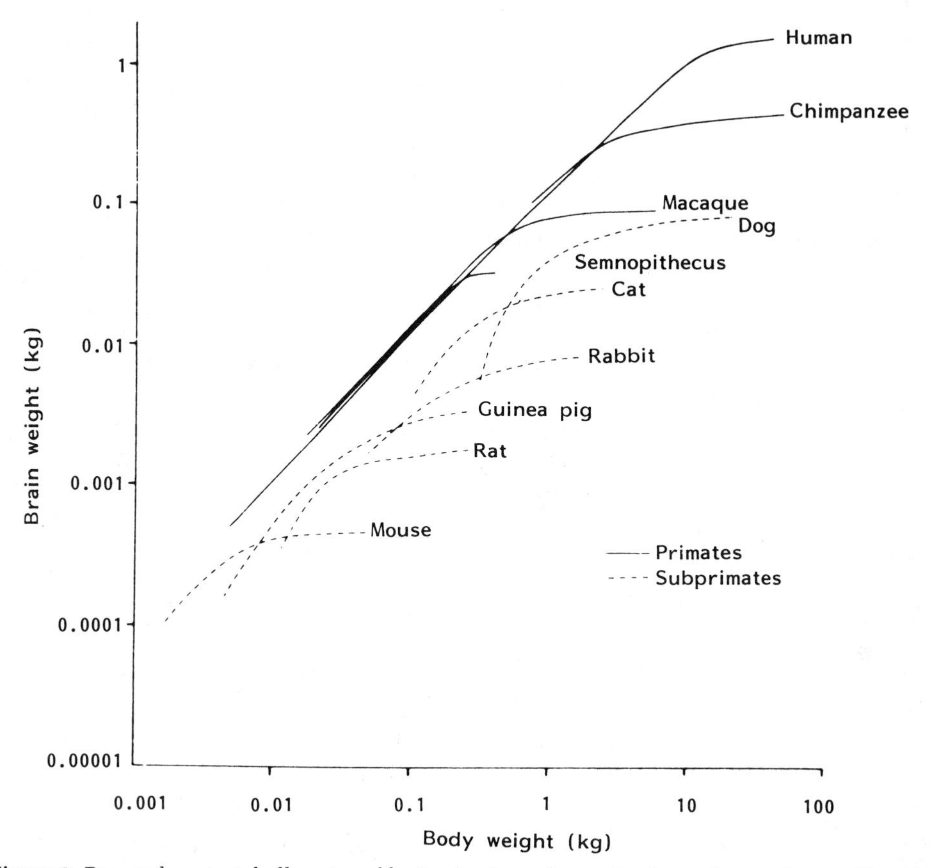

Figure 9. Pre- and postnatal allometry of brain size in various primates and nonprimates. Note that humans share a common anthropoid pattern, but extend the period of strong brain growth. From Holt *et al.* (1975).

from *time hypermorphosis*, and they yield a high brain/body ratio that is paedomorphic, given the general postnatal negative allometry of brain/body growth. A very important (and largely unanswered) question relevant to notions of human neoteny is whether any other significant morphological changes are the direct and correlated result of this heterochronic modification of life history schedules (Shea, in press). Furthermore, as noted in Section 5, optimally we would like to distinguish between cases where selection for timing changes (e.g., prolonged growth and delayed maturation) results in morphological change, and those where selection for a specific morphological configuration required alterations in the timing of growth (Bonner and Horn, 1982).

Other, more general aspects of brain/body allometry have also been elucidated by heterochronic approaches. The fact that closely related species differing in body size exhibit allometric coefficients of brain/body scaling resembling static adult intraspecific levels (0.2–0.4), rather than typical interspecific levels (0.66–−0.75), has been a vexing problem (Gould, 1975). Gould (1975) hinted at a possible solution and a focus for additional work when he suggested that the various coefficients of brain/body allometry reflected different developmental and evolutionary pathways to size increase. Lande (1979) used quantitative genetics to show that a slope around 0.3 would be expected, given selection for increased body size alone, based on genetic patterns of covariation of brain size and body size in rodents. In 1983, I suggested that the lowered interspecific slope often observed within genera resulted from the developmental basis of size increase among closely related species (Shea, 1983a). Noting Leutenegger's (1977) conclusion that slopes of neonatal/maternal weight scaling are lower in congeneric comparisons than those at higher taxonomic levels, I argued that body size variation among closely related species is produced primarily via differential concentration on postnatal growth stages, when the brain exhibits very little growth. This is also the case for size variation among individuals within a species, and thus we see very low brain/body slopes here as well. Therefore, the lowered relative brain size of male gorillas compared to female conspecifics does not reflect differences in adaptive strategies or behavioral capacities, but rather the developmental pathways utilized to produce larger body size in the males.* Further work may reveal that relative brain, tooth, and neonatal size exhibit progressively lowered allometric slopes as the degree of intersexual or congeneric interspecific size differentiation increases (Shea, 1983a).

More recently, Atchley and colleagues (Riska and Atchley, 1985; Atchley *et al.*, 1984) have further generalized aspects of these arguments and suggested a genetic and hormonal basis for slope variations in brain/body allometry at different taxonomic levels. They emphasize that the hormonal factors controlling early growth [such as insulinlike growth factor II (Hintz, 1985)] produce correlated growth in both brain and body size, while other factors [such as insulinlike growth factor I (Hall *et al.*, 1981)] underlie later growth in overall size and do not affect brain growth directly. Additional research on the covariation of patterns of size, shape, and age will be central to our understanding of heterochrony (see Section 10). Atchley (1987) has provided some important theoretical and methodological beginnings for such a project of research.

10. Genetics and Epigenetics

Much of the theory underlying heterochronic changes in morphology draws either implicitly or explicitly on the covariation, and potential dissociation of such, among developmental parameters such as size, shape, and timing. Yet we admittedly have little direct evidence for most groups, and this is particularly true of the primates, since their long generation times do not lend them to genetic studies or experimental developmental approaches. Most of the important data that we have on these issues come from work on rodents, amphibians, or other groups (e.g., Atchley *et al.*, 1984; Cheverud *et al.*, 1983a,b; Lande, 1982; Alberch and Gale, 1985; Hall, 1983, 1984; Hanken, 1982; Wake, 1980).

* In addition to a consideration of brain/body allometry, this approach may help us understand why a number of features scale at different rates, depending on the taxonomic level examined. The negative allometry of the postcanine dentition among closely related species may be due to the fact that the complex factors controlling tooth size act early in development and ontogeny and thus are less affected by variation in later postnatal growth.

Paleontologists and neontologists using heterochronic approaches typically work on the phenotypic level, "read" the results of morphological transformations from their comparative analyses, and construct scenarios relating global or regional change to selection for developmental shifts in particular ecological contexts (see Section 3 for examples of this approach). But the argument that selection for overall size change has resulted in allometrically correlated shape change via ontogenetic scaling assumes that patterns of genetic correlation closely resemble the patterns of phenotypic correlation we directly examine (Lande, 1979). Or, as Cheverud (1982) notes, explanations of rate hypo- or hypermorphosis based on ontogenetic scaling assume that there is no genetic correlation between the lengths of ontogenetic vectors and their slopes. Arguments concerning the dissociation of duration of growth from the underlying allometric controls (as in cases of time hypo- and hypermorphosis) assume a similar lack of correlation between the timing and size/shape parameters of growth.

In addition to knowledge of genetic correlations among the features we examine, it is also highly desirable to have information on epigenetic factors, or the diverse mechanisms by which genes express their phenotypic effects (Hall, 1983). The developmental factors that control growth, such as tissue interactions or hormones, are of fundamental importance in studies of heterochrony (Hall, 1983, 1984; Bryant and Simpson, 1984). Indeed, such factors in part produce the genetic correlations that are examined by the quantitative geneticist (Atchley *et al.*, 1984; Atchley, 1987; Riska, 1986). Considering only size/shape associations as an example, we desperately need information on the developmental bases of changes in ontogenetic allometries (Bonner and Horn, 1982). What are the developmental bases of ontogenetic extrapolations, of slope divergences, and of transpositions? Are such developmental controls easily manipulated, so that global dissociations can be easily produced by shifts in hormones, as hypothesized by Gould (1971) in his considerations of evolution via geometric similarity?

New information on these important issues of the genetic and epigenetic bases of heterochronic change is emerging, although space allows only a few examples. J. M. Cheverud (in press) has recently compared genetic and phenotypic correlation matrices drawn from studies in the literature, and he finds a strong correspondence, as suggested by Lande (1979) in a less exhaustive study. This provides some confidence to those working on the phenotypic level, though one must always be aware of assumptions being made in this regard [for example, Atchley *et al.* (1981) demonstrate a strong correspondence between phenotypic and genetic covariance patterns in skull dimensions of rats, but a weaker correspondence for similar measures in mice]. In a selection experiment where time of sexual maturation was selected for and altered significantly in six generations of mice, Drickamer (1981) reported a dissociation from overall size and growth rate. Mice that matured earlier and later did so at sizes smaller and larger than the original control stock (specific size/shape patterns of the skeleton were not examined). This is the sort of information needed to assess more reasonably heterochronic hypotheses such as the argument that selection for precocious maturation will result in smaller size (time hypomorphosis or progenesis), or that prolonging certain growth periods will yield correlated dissociation of size/shape relations and a retardation of shape change [neoteny as envisioned by Gould (1977, p. 344)].

As a final example, the use of giant transgenic mice (Palmiter *et al.*, 1982)

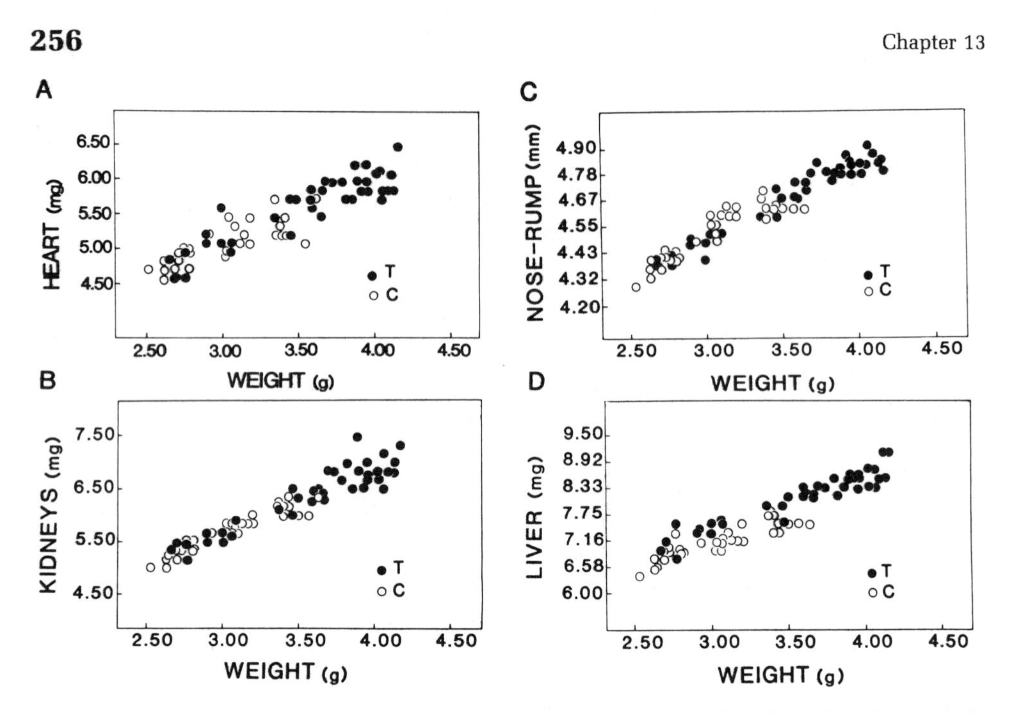

Figure 10. Plots of (A) heart weight, (B) kidneys weight, (C) nose–rump length, and (D) liver weight against body weight in (●, T) giant transgenic and (○, C) control mice. In A–C, the two groups share common growth allometries; in D, a divergence of trajectories and acceleration of liver growth in the transgenics are apparent. See text and Shea *et al.* (1987) for additional discussion.

and various mutants presenting endocrine abnormalities (Bartke, 1979) in studies of relative growth have increased our understanding of the control of ontogenetic allometries (Shea *et al.*, 1987, and in preparation). Results here support the notion that changes in levels of growth hormone and insulinlike growth factor I primarily yield simple extensions or truncations of patterns of growth allometry that are not otherwise fundamentally altered (Fig. 10). This suggests a possible mechanism underlying the frequent observation of ontogenetic scaling in closely related series differing in overall size; a simple change in one of the extrinsic controls can produce correlated but allometrically regulated change in many features. Shifts in the intrinsic controls, such as initial size of stem cell populations, rates of cell division per unit of control substance, or number of receptors, are probably involved in more fundamental changes in growth allometries, such as slope divergences or vertical transpositions (Katz, 1980; Bryant and Simpson, 1984).

A program of research such as that proposed by Hall (1984) or Atchley (1987) would address many of these fundamental issues. This is also likely to be one of the most productive areas of research into questions of heterochrony in the next decade.

11. Primate Fossil Record

There are few cases where heterochrony has been used to interpret morphological transformations among fossil primates, because of the generally poor state of preservation of most primate lineages. Pilbeam and Gould (1974) considered variation in skull form between gracile (*Australopithecus africanus*) and robust (*A. robustus* and *A. boisei*) australopiths as a reflection of size change and allometric extrapolation (peramorphosis via hypermorphosis in heterochronic terminology). More recently, I argued that an allometric comparison of facial morphology in the gracile and robust australopithecines suggests some fundamental dissociation, rather than simple ontogenetic scaling, of proportions (Shea, 1985a). This finding supports the traditional explanation (e.g., Robinson, 1972) that facial shape differences reflect divergent dietary adaptations. The lowered levels of interspecific brain/body scaling among the australopiths noted by Pilbeam and Gould (1974) probably do relate to size increase via the ontogenetic mechanisms discussed previously in Section 9.

Another example of a fossil primate that exhibits low relative brain size is the early catarrhine *Aegyptopithecus* from Oligocene deposits of Africa. This genus was also quite large in overall size for primates of that time, and its small relative brain size may be a reflection of this rapid size increase or phyletic gigantism (Radinsky, 1973; Gould, 1975; Shea, 1983a). Alternatively, the low relative brain size of *Aegyptopithecus* is viewed as evidence of its primitive status and behavioral capabilities (Fleagle and Kay, 1983). Once again this provides an example of how heterochronic approaches can play an important role in whether we assess shape differences as resulting from size-related or non-size-related changes. The debate over relative brain size in the early Miocene hominoid *Proconsul africanus* [compare Walker *et al.* (1983) with Leutenegger (1984)] provides one example with significant adaptational and systematic implications.

Most of the discussion of heterochrony in the primate fossil record has dealt with neoteny of skull form in human evolution. Because of the dearth of well-preserved skulls, let alone adequate samples of subadults, these studies have not involved detailed and quantified comparisons. Normally, they note the general paedomorphosis of the *Homo* lineage resulting from enlargement of the brain and dimunition of the facial mass, and place various fossil taxa such as australopiths, *Homo erectus*, or *H. sapiens neandertalensis* on phyletic diagrams (e.g., de Beer, 1930; Abbie, 1952).

A series of French scientists has recently used heterochrony to interpret morphological changes during hominid evolution (Berge *et al.*, 1988; Chaline *et al.*, 1986). They focus on morphological developments that have traditionally been of central concern to anthropologists, particularly the restructuring of pelvic anatomy in relation to bipedal locomotion, and the explosive growth of brain size in the lineage leading to *Homo sapiens*. The shift in pelvic morphology is viewed as a result of the heterochronic process of *acceleration*, while increased encephalization results from "a combination of *neoteny* plus *hypermorphosis*" (Chaline *et al.*, 1986). Figure 11 illustrates their view of hominid evolution based on heterochronic changes in "punctuated" and "gradualist" modes. It remains to be demonstrated that this approach to labeling the transformation of particular (and in all probability reasonably unrelated) features will appreciably increase our understanding of the processes of hominization.

12. Human Paedomorphosis and Neoteny

There is no need to review the historical background or morphological context of the debate over human neoteny, since this has been done recently and effec-

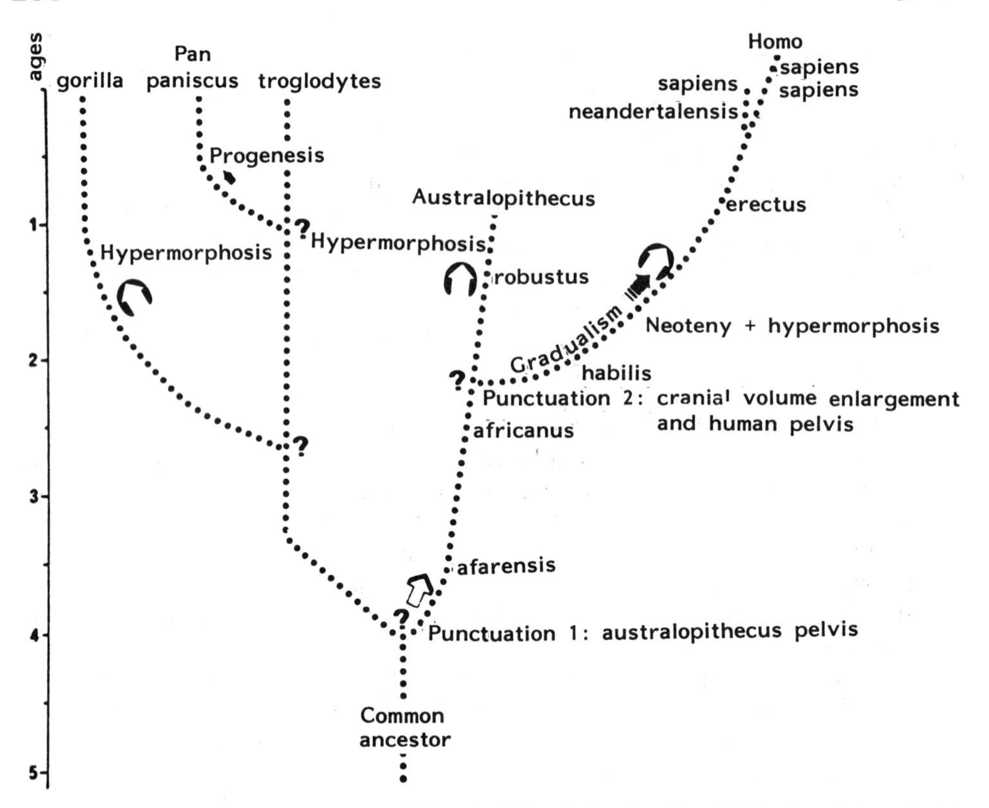

Figure 11. A speculative reconstruction of hominoid evolution based on heterochronic transformations. The primary morphological features plotted on the chart are pelvic shape and brain size. The authors' assessments as to whether these heterochronic transformations are punctuated or gradual are also indicated. See text for additional discussion. From Chaline *et al.* (1986).

tively by both Gould (1977) and Montagu (1981). Furthermore, space limitations prohibit a detailed exposition here of my own views on the pros and cons of the neoteny hypothesis of human evolution. I would like to stress, however, that although human neoteny has in some sense become the *cause célèbre* of heterochrony, it is in fact a rather "messy" and largely unconvincing case.

It is a given that modern human morphology has evolved from that of some form of primitive hominoid and early members of our own lineage via heterochronic processes (how could it be otherwise?). The central issue of the neoteny hypothesis and debate is whether a substantial portion of our morphology represents an *intercorrelated* set of features that can be attributed to predominantly one or two heterochronic processes. An additional fundamental problem is whether our *morphological* characteristics can be viewed as the *correlated consequence* of changes in the *duration* of various growth periods.

I believe that certain of our salient features are the result of neotenic processes, although I do not accept that the bulk of them form such an interrelated complex. There are many incorrect, inconsistent, and contradictory aspects of the neoteny hypothesis as it has been articulated. I list and briefly describe below some of the most important problematic areas requiring additional consideration and new research (Shea, in press).

12.1. Nonpaedomorphic Features

A significant problem with a neotenic theory of human origins (or later anagenetic transformations within hominid lineages) is that many of our characteristic features are simply not paedomorphic, as Schultz (1969) and others have shown. Among these are almost the entire suite of features associated with bipedal adaptations, perhaps the defining attribute of the hominid clade. This does not preclude neoteny from having played a prominent role in the development of other human features, but it does mean that neoteny is not particularly relevant to one of our most important character complexes.

12.2. Size-Related Features

Many of the features offered as primary evidence for the role of neoteny in differentiating the sexes and races of modern humans [particularly by Bolk (1926, 1929) and later Montagu (1981)] are in fact simply correlates of changes in overall body size. Gracile skeletons, narrower joint surfaces, relatively larger brains, and smaller facial masses may be paedomorphic, but they do not result from neoteny (rather rate and/or time hypomorphosis as in Fig. 1). If females were neotenic males, they would develop their characteristic adult female "shapes" at the same overall size as the (larger) male, and not at sizes characteristic of subadult males.

12.3. A "Superficial Juvenilization"?

While our overall cranial morphology resembles that of infant or fetal higher primates in having a relatively large brain and small face, this superficial similarity overlies a number of new features of adult human skull form that cannot be explained by invoking neotenic processes (although other heterochronic processes may be involved, but once again apparently not in an integrated package). For example, Moss *et al.* (1982) demonstrate that new morphological features of the human skullbase cannot be derived from the fetal or subadult morphologies of nonhuman primates via the developmental processes of retardation and neoteny. There is considerable evidence that the adult human skullbase and upper respiratory tract exhibit a number of specializations related to the development of our speech-producing apparatus (e.g., Laitman *et al.*, 1978; Laitman and Crelin, 1980; Laitman and Heimbuch, 1982) [see Lieberman (1984) for discussion]. In this area, human infants resemble adult apes, rather than the converse as predicted by neoteny. In fact, Laitman and Heimbuch (1984) demonstrate that in terms of basicranial flexion and growth, adult pygmy chimpanzees depart less from their juvenile morphologies than do adult humans.

Recent work on the detailed processes of facial growth by Bromage (1985) has demonstrated that human facial morphology results from fundamentally different, as opposed to merely retarded, patterns of deposition, resorption, and remodeling. This suggests that the "facial shortening" that adult humans and subadult nonhuman primates share is in fact a nonhomologous similarity. Other specific arguments concerning human cranial form and the neoteny hypothesis have been discussed by Starck (1962), Gould (1977), Dullemeijer (1975), and Shea (in press). In short, I believe that many of the "paedomorphic" features of human skull form are superficial resemblances that in fact result not from the retardation of common developmental processes, but rather from important and novel human specializations of skull form and development.

12.4. Growth Duration and Morphological Relationships

A fundamental inconsistency in many previous considerations of neoteny is the confusion of the length of various growth periods with the morphological shape changes observed within and across these periods. Neoteny refers to decreased rates of morphological shape change, *not an increased duration of growth in time*. The fact that humans are frequently referred to as "slow-growing" mammals primarily involves the absolute *duration* of our various life periods, infant through adult—thus, a better description would be "long-growing." But it certainly has not been demonstrated that dissociation and retardation of ancestral patterns of shape change are an ineluctable developmental consequence of this lengthening. We know that in many other groups where such timing changes occur in the length of the growth periods, no general paedomorphosis emerges. For examples, we need look no further than our close relatives. The great apes exhibit prolonged developmental periods relative to monkeys (much as humans do relative to the apes themselves), and yet they are certainly not paedomorphic simians.

Many of the difficulties with a hypothesis of human neoteny (at least for the skull) can be appreciated by reference to a reasonably simple case among our close relatives, the chimpanzees. Pygmy chimpanzees exhibit general cranial paedomorphosis relative to common chimpanzees (Coolidge, 1933; Weidenreich, 1941; Fenart and Deblock, 1973; Cramer, 1977; Shea, 1983b,c, 1984; Laitman and Heimbuch, 1984), and this juvenilization reflects a neotenic dissociation of rates of overall skull (particularly facial) growth relative to total size (Shea, 1983c, 1984). Metric and qualitative studies provide no reason for questioning that this retardation involves homologous growth processes. Pygmy chimpanzees exhibit no novel specializations of skull form not seen in common chimpanzee subadults or adults. As noted above, in aspects of basicranial morphology, adult pygmy chimpanzees diverge less from their juvenile morphology than do adult humans from their own.

The neoteny hypothesis is badly in need of new data, particularly from areas such as genetics, developmental biology, and experimental endocrinology. We need to know more about the interrelationships among various components of growth, such as how the duration of a particular period is related to rates of shape change within that period. We need information on whether the genetic and physiological factors that control early growth are also responsible for a retardation of overall differentiation and shape change relative to those controlling later growth.

13. Summary

I have attempted to demonstrate in this review that heterochronic analyses have increased our understanding of the evolution of the primates, and that, in turn, primatological investigations have contributed to our general knowledge of the importance of heterochrony in evolution. Examples discussed illustrate the significance of allometric factors, timing variations, historical factors, and the use of size/shape associations and dissociations in the interpretation of primate morphology. The need to distinguish adequately between rate and time shifts underlying allometric truncations and extrapolations is also stressed, and a heterochronic approach to sexual dimorphism is offered.

A relatively simple heterochronic basis for significant and interrelated morphological change is suggested for a number of primate species and subspecies comparisons. The well-publicized case for human neoteny is briefly but critically reviewed, and I consider several possible avenues for future work on this important topic in heterochrony.

Three major areas are likely to play a prominent and productive role in future investigations of heterochrony, in primates as well as other groups. One is in the analysis of the ontogenetic bases (and the socioecological correlates thereof) of patterns of sexual dimorphism. A second is greater focus on determining the genetic and epigenetic bases of morphological variations among individuals and groups—an understanding of the intrinsic and extrinsic controls of growth (Bryant and Simpson, 1984). Finally, a topic only touched upon in the above review, we require multidisciplinary investigations that combine traditional comparative approaches with (1) laboratory and experimental studies of the factors controlling growth and development and (2) field studies of the detailed socioecological contexts of various developmental and life-history trajectories. Such an integrated and multilevel approach will better enable us to elucidate the "ecological control of development" (Gould, 1977).

ACKNOWLEDGMENTS. I would like to thank Mike McKinney for inviting me to participate in the symposium on heterochrony and evolution at the 1986 meetings of the Geological Society of America, and for his help with this chapter. I also thank L. Levitch and K. Manley-Buser for providing me with unpublished manuscripts describing their work on primates. I am indebted to the National Science Foundation (grants BNS 7810558 and 8608950) and the National Institutes of Health (grants DE 07597 and DK 36463) for providing financial support for my studies of growth and form in primates and other mammals.

References

Abbie, A. A., 1952, A new approach to the problem of human evolution, *Trans. R. Soc. S. Aust.* **75**:70–88.

Abbie, A. A., 1958, Timing in human evolution, *Proc. Linn. Soc. N. S. Wales* **88**:197–213.

Abbie, A. A., 1964, The factor of timing in the emergence of distinctively human characters, *Pap. Proc. R. Soc. Tasmania* **98**:63–71.

Alberch, P., and Gale, E. A., 1985, A developmental analysis of an evolutionary trend: Digital reduction in amphibians, *Evolution* **39**:8–23.

Alberch, P., Gould, S. J., Oster, G. F., and Wake, D. B., 1979, Size and shape in ontogeny and phylogeny, *Paleobiology* **5**:296–317.
Atchley, W. R., 1987, Developmental quantitative genetics and the evolution of ontogenies. *Evolution* **41**:316–330.
Atchley, W. R., Rutledge, J. J., and Cowley, D. E., 1981, Genetic components of size and shape. II. Multivariate covariance patterns in the rat and mouse skull, *Evolution* **35**:1037–1055.
Atchley, W. R., Riska, B., Kohn, L. A. P., Plummer, A. A., and Rutledge, J. J., 1984, A quantitative genetic analysis of brain and body size associations, their origin and ontogeny: data from mice, *Evolution* **38**:1165–1179.
Bartke, A., 1979, Genetic models in the study of anterior pituitary hormones, in: *Genetic Variation in Hormone Systems* (J. G. M. Shire, ed.), pp. 113–126, CRC Press, Boca Raton, Florida.
Berge, C., Didier, M., Chaline, J., and Dommergues, J.-L., 1988, La bipèdie des hominides: Comparisons des itinéraires ontogénétiques des dimensions fémoropelviennes des pongides, australopithèques et hommes, Paper presented at the Colloque International: Ontogenèse et Evolution, Dijon, France, 1986, (in press).
Bolk, L., 1926, On the problem of anthropogenesis, *Proc. Section Sci. Kon. Akad. Wetens. Amsterdam* **29**:465–475.
Bolk, L., 1929, Origin of racial characteristics in man, *Am. J. Phys. Anthropol.* **13**:1–28.
Bonner, J. T., and Horn, H. S., 1982, Selection for size, shape, and developmental timing, in: *Evolution and Development* (J. T. Bonner, ed), pp. 259–276, Springer-Verlag, Berlin.
Bromage, T. G., 1985, Taung facial remodeling: A growth and development study, in: *Hominid Evolution: Past, Present, and Future* (P. V. Tobias, ed.), pp. 239–246, Liss, New York,
Bromage, T. G., and Dean, M. C., 1985, Re-evaluation of the age at death of immature fossil hominids, *Nature* **317**:525–527.
Bryant, P. J., and Simpson, P., 1984, Intrinsic and extrinsic control of growth in developing organs, *Q. Rev. Biol.* **59**:387–415.
Chaline, J., Marchand, D., and Berge, C., 1986, L'évolution de l'homme: Un modele gradualiste ou ponctualiste?, *Bull. Soc. R. Belge Anthropol. Prehist.* **97**:77–97.
Charles, D. K., Condon, K., Cheverud, J. M., and Buikstra, J. E., 1986, Cementum annulation and age determination in *Homo sapiens*. I. Tooth variability and observer error, *Am. J. Phys. Anthropol.* **71**:311–321.
Cheverud, J. M., 1982, Relationships among ontogenetic, static, and evolutionary allometry, *Am. J. Phys. Anthropol.* **59**:139–149.
Cheverud, J. M., 1988, A comparison of phenotypic and genetic patterns of correlation, *Evolution* (in press).
Cheverud, J. M., Rutledge, J. J., and Atchley, W. R., 1983a, Quantitative genetics of development: Genetic correlations among age-specific trait values and the evolution of ontogeny, *Evolution* **37**:895–905.
Cheverud, J. M., Leamy, L. J., Atchley, W. R., and Rutledge, J. J., 1983b, Quantitative genetics and the evolution of ontogeny. I. Ontogenetic changes in quantitative genetic variance components in randombred mice, *Genet. Res.* **42**:65–75.
Clutton-Brock, T. H., and Harvey, P. H., 1979a, Home range size, population density and phylogeny in primates, in: *Primate Ecology and Human Origins* (I. S. Bernstein and E. O. Smith, eds.), pp. 201–214, Garland Press, New York.
Clutton-Brock, T. H., and Harvey, P. H., 1979b, Comparison and adaptation, *Proc. R. Soc. Lond. B* **205**:547–565.
Coolidge, H. J., Jr., 1933, *Pan paniscus*: Pygmy chimpanzee from south of the Congo River, *Am. J. Phys. Anthropol.* **18**:1–57.
Corruccini, R. S., 1981, Analytical techniques for Cartesian coordinate data with reference to the relationship between *Hylobates* and *Symphalangus* (Hylobatidae; Hominoidea), *Syst. Zool.* **30**:32–40.
Count, E. W., 1947, Brain and body weight in man: Their antecedents in growth and evolution, *Ann. N. Y. Acad. Sci.* **46**:993–1122.
Cramer, D. L., 1977, *Craniofacial Morphology of Pan paniscus*, Karger, Basel.
Creel, N., and Preuschoft, H., 1976, Cranial morphology of the lesser apes: A multivariate statistical study, *Gibbon Siamang* **4**:219–303.
Dahl, J. F., 1986, Cyclic perineal swelling during the intermenstrual intervals of captive female pygmy chimpanzees (*Pan paniscus*), *J. Hum Evol.* **15**:369–386.

De Beer, G. R., 1930, *Embryology and Evolution*, Clarendon, Oxford.

Delson, E., and Andrews, P., 1975, Evolution and interrelationships of the catarrhine primates, in: *Phylogeny of the Primates* (W. P. Luckett and F. S. Szalay, eds.), pp. 405–446, Plenum Press, New York.

Drickamer, L. C., 1981, Selection for age of sexual maturation in mice and the consequences for population regulation, *Behav. Neural Biol.* **31**:82–89.

Dullemeijer, P., 1975, Bolk's foetalization theory, *Acta Morphol. Neerl. Scand* **13**:77–86.

Fedigan, L. M., 1982, *Primate Paradigms: Sex Roles and Social Bonds*, Eden Press, Montreal.

Fenart, R., and Deblock, R., 1973, *Pan paniscus* et *Pan troglodytes* Craniometrie, Étude Comparative et Ontogénetique selon les methodes classiques et vestibulaire, *Mus. Roy. Afrique Central, Ann. Ser. in 8°, Sci. Zool.*, **204**:1–473.

Fink, W. L., 1982, The conceptual relationship between ontogeny and phylogeny, *Paleobiology* **8**:254–264.

Fleagle, J. G., and Kay, R. F., 1983, New interpretations of the phyletic position of Oligocene hominoids, in: *New Interpretations of Ape and Human Ancestry* (R. L. Ciochon and R. S. Corruccini, eds.), pp. 181–210, Plenum Press, New York.

Ford, S., 1980, Callithricids as phyletic dwarfs, and the place of the Callithricidae in the Platyrrhini, *Primates* **21**:31–43.

Frazzetta, T. H., 1975, *Complex Adaptations in Evolving Populations*, Sinnauer, Sunderland, Massachusetts.

Gautier-Hion, A., 1973, Social and ecological features of Talapoin monkeys: Comparisons with other cercopithecines, in: *Comparative Ecology and Behavior of Primates* (R. P. Michael and J. H. Crook, eds.), pp. 148–170, Academic Press, London.

Gautier-Hion, A., and Gautier, J.-P., 1976, Croissance, maturite sexuelle et sociale, et reproduction chez les cercopithecines forestiers africains, *Folia Primatol.* **26**:165–184.

Geist, V., 1986, The paradox of the great Irish stags, *Nat. Hist.* **5**(3):54–65.

Ghiglieri, M. P., 1984, *The Chimpanzees of Kibale Forest*, Columbia University Press, New York.

Goss, R. J., 1964, *Adaptive Growth*, Academic Press, New York.

Gould, S. J., 1966, Allometry and size in ontogeny and phylogeny, *Biol. Rev.* **41**:587–640.

Gould, S. J., 1971, Geometric similarity in allometric growth: A contribution to the problem of scaling in the evolution of size, *Am. Nat.* **105**:113–136.

Gould, S. J., 1974, The evolutionary significance of "bizarre" structures: Antler size and skull size in the "Irish Elk," *Megaloceros giganteus*, *Evolution* **28**:191–220.

Gould, S. J., 1975, Allometry in primates, with emphasis on scaling and the evolution of the brain, in: *Approaches to Primate Paleobiology* (F. S. Szalay, ed.), pp. 244–292, Karger, Basel.

Gould, S. J., 1977, *Ontogeny and Phylogeny*, Harvard University Press, Cambridge.

Gould, S. J., and Lewontin, R. C., 1979, The spandrels of San Marco and the Panglossian paradigm: A critique of the adaptationist programme, *Proc. R. Soc. Lond. B* **205**:481–598.

Hall, B. K., 1983, Epigenetic control in development and evolution, in: *Development and Evolution* (B. C. Goodwin, N. Holder, and C. G. Wylie, eds.), pp. 353–379, Cambridge University Press, Cambridge.

Hall, B. K., 1984, Developmental processes underlying heterochrony as an evolutionary mechanism, *Can. J. Zool.* **62**:1–7.

Hall, K., Sara, V. R., Enberg, G., and Ritzen, E. M., 1981, Somatomedins and postnatal growth, in: *Biology of Normal Human Growth* (E. M. Ritzen, K. Hall, A. Zetterberg, A. Aperia, A. Larsson, and R. Zetterstrom, eds.), pp. 275–283, Raven Press, New York.

Hanken, J., 1982, Appendicular skeletal morphology in minute salamaders, genus *Thorius* (Amphibia: Plethodontidae): Growth regulation, adult size determination, and natural variation, *J. Morphol.* **174**:57–77.

Herschkovitz, P., 1977, *New World Monkeys (Platyrrhini)*, Vol. 1, University of Chicago Press, Chicago.

Hiernaux, J., 1968, *La Diversité Humaine en Afrique Subsaharienne*, Recherches Biologiques, Institut de Sociologie de l'Université Libre de Bruxelles, Brussels.

Hiernaux, J., 1977, Long-term biological effects of human migration from the African savana to the equatorial forest: A case study of human adaptation to a hot and wet climate, in: *Population Structure and Human Variation* (G. A. Harrison, ed.), pp. 187–217, Cambridge University Press, London.

Hintz, R. L., 1985, Control mechanisms of prenatal bone growth, in: *Normal and Abnormal Bone Growth* (A. Dixon and B. G. Sarnat, eds.), pp. 25–34, Liss, New York.

Holt, A. B., Cheek, D. B., Mellitus, E. D., and Hill, D. G., 1975, Brain size and the relation of the primate to the nonprimate, in: *Fetal and Postnatal Cellular Growth: Hormones and Nutrition* (D. B. Cheek, ed.), pp. 23–44, John Wiley, New York.

Horn, A. D., 1975, Adaptations of the pygmy chimpanzee (*Pan paniscus*) to the forests of the Zaire Basin, *Am. J. Phys. Anthropol.* **42:**307.

Huxley, J. S., 1932, *Problems of Relative Growth*, MacVeagh, London.

Jarman, P., 1983, Mating system and sexual dimorphism in large, terrestrial, mammalian herbivores, *Biol. Rev.* **58:**485–520.

Jerison, H. J., 1979, Brain, body and encephalization in early primates, *J. Hum. Evol.* **8:**615–635.

Jungers, W. L., and Susman, R. L., 1984, Body size and skeletal allometry in African apes, in: *The Pygmy Chimpanzee: Evolutionary Biology and Behavior* (R. L. Susman, ed.), pp. 131–178, Plenum Press, New York.

Kano, T., 1980, The social group of pygmy chimpanzees (*Pan paniscus*) of Wamba, *Primates* **23:**171–188.

Katz, M. J., 1980, Allometry formula: A cellular model, *Growth* **44:**89–96.

Kollman, J., 1905, Neue Gedanken uber das alter Problem von der Abstammung des Menschen, *Corresp.-Bl. Deutsch. Ges. Anthropol. Ethnol. Urges.* **36:**9–20.

Kuroda, S., 1980, Social behavior of the pygmy chimpanzees, *Primates* **20:**161–183.

Kuroda, S., 1986, Developmental retardation and behavioral characteristics in the pygmy chimpanzees, Paper presented at the symposium, Understanding Chimpanzees, November 1986, Chicago Academy of Sciences.

Laitman, J. T., and Crelin, E. S., 1980, Developmental change in the upper respiratory system of human infants, *Perinatol. Neonatol.* **4:**15–22.

Laitman, J. T., and Heimbuch, R. C., 1982, The basicranium of Plio-Pleistocene hominids as an indicator of their upper respiratory systems, *Am. J. Phys. Anthropol.* **59:**323–343.

Laitman, J. T., and Heimbuch, R. C., 1984, A measure of basicranial flexion in *Pan paniscus*, the pygmy chimpanzee, in: *The Pygmy Chimpanzee* (R. L. Susman, ed.), pp. 49–64, Plenum Press, New York.

Laitman, J. T., Heimbuch, R. C., and Crelin, E. S., 1978, Developmental change in a basicranial line and its relationship to the upper respiratory system in living primates, *Am. J. Anat.* **152:**467–482.

Lande, R., 1979, Quantitative genetic analysis of multivariate evolution, applied to brain:body allometry, *Evolution* **33:**402–416.

Lande, R., 1982, A quantitative genetic theory of life history evolution, *Ecology* **63:**607–615.

Lauder, G. V., 1981, Form and function: Structural analysis in evolutionary morphology, *Paleobiology* **7:**430–442.

Leutenegger, W., 1977, Neonatal–maternal weight relationship in macaques: An example of intrageneric scaling, *Folia Primatol.* **27:**153–159.

Leutenegger, W., 1980, Monogamy in callitrichids: A consequence of phyletic dwarfism? *Int. J. Primatol.* **11:**95–98.

Leutenegger, W., 1984, Encephalization in *Proconsul africanus*, *Nature* **309:**287.

Leutenegger, W., and Larson, S., 1985, Sexual development of the postcranial skeleton of New World monkeys, *Folia Primatol.* **44:**82–95.

Levitch, L. C., 1986, Ontogenetic allometry of small-bodied platyrrhines, *Am. J. Phys. Anthropol.* **69:**230.

Lieberman, P., 1984, *The Biology and Evolution of Language*, Harvard University Press, Cambridge.

Manley-Buser, K. A., 1986, A heterochronic study of the human foot. *Am. J. Phys. Anthropol.* **69:**235.

Marquer, P., 1972, Nouvelle contribution a l'étude du squelette des pygmées occidentaux du centre Africain comparé à celui des pygmées orientaux, *Mem. Nat. Hist. A* **72:**1–122.

McHenry, M., and Corruccini, R. S., 1981, *Pan paniscus* and human evolution, *Am. J. Phys. Anthrop.* **54:**355–367.

McKinney, M. L., 1984, Allometry and heterochrony in an Eocene echinoid lineage: Morphological change as a byproduct of size selection, *Paleobiology* **10:**407–419.

McKinney, M. L., and Schoch, R. M., 1985, Titanothere allometry, heterochrony, and biomechanics: Revising an evolutionary classic, *Evolution* **39:**1352–1363.

McNamara, K. J., 1986, A guide to the nomenclature of heterochrony, *J. Paleontol.* **60:**4–13.

Merimee, T. J., Zapf, J., and Froesch, E. R., 1982, Insulin-like growth factors (IGFs) in pygmies and subjects with the pygmy trait: Characterization of the metabolic actions of IGFI and IGFII in man, *J. Clin. Endocrinol. Metab.* **55:**1081–1088.

Mobb, G. E., and Wood, B. A., 1977, Allometry and sexual dimorphism in the primate innominate bone, *Am. J. Anat.* **150:**531–538.

Montagu, M. F. A., 1962, Time, morphology, and neoteny in the evolution of man, in: *Culture and the Evolution of Man* (M. F. A. Montagu, ed.), pp. 324–342, Oxford University Press, New York.

Montagu, M. F. A., 1981, *Growing Young*, McGraw-Hill, New York.

Moss, M. L., Moss-Salentijn, L., Vilmann, H., and Newell-Morriss, L., 1982, Neuro-skeletal topology of the primate basicranium: its implications for the "fetalization hypothesis," *Gegenbaurs morph Jahrb., Leipzig* **128:**58–67.

Palmiter, R. D., Norstedt, G., Gelinas, R. E., Hammer, R. E., and Brinster, R. L., 1982, Metallothionein-human GH fusion genes stimulate growth in mice, *Science* **222:**809–814.

Pilbeam, D. R., and Gould, S. J., 1974, Size and scaling in human evolution, *Science* **186:**892–901.

Radinsky, L., 1973, *Aegyptopithecus* endocasts: oldest record of a pongid brain, *Am. J. Phys. Anthrop.* **39:**239–248.

Raemakers, J., 1984, Large versus small gibbons: Relative roles of bioenergetics and competition in their ecological segregation in sympatry, in: *The Lesser Apes* (H. Preuschoft, D. J. Chivers, W. Y. Brockelman, and N. Creel, eds.), pp. 209–218, Edinburgh University Press, Edinburgh.

Ralls, K., 1977, Sexual dimorphism in mammals: Avian models and unanswered questions, *Am. Nat.* **111:**917–938.

Riska, B., 1986, Some models for development, growth, and morphometric correlation, *Evolution* **40:**1303–1311.

Riska, B., and Atchley, W. R., 1985, Genetics of growth predict patterns of brain-size evolution, *Science* **229:**668–671.

Robinson, J. T., 1972, *Early Hominid Posture and Locomotion*, University of Chicago Press, Chicago.

Rodman, P. S., 1984, Foraging and social systems of orangutans and chimpanzees, in: *Adaptations for Foraging in Nonhuman Primates* (P. S. Rodman and J. G. H. Cant, eds.), pp. 161–194, Columbia University Press, New York.

Rowell, T. E., 1977, Variation in age at puberty in monkeys, *Folia Primatol.* **27:**284–296.

Schultz, A. H., 1933, Observations on the growth, classification, and evolutionary specializations of gibbons and siamangs, *Hum. Biol.* **5:**212–255.

Schultz, A. H., 1969, *The Life of Primates*, Universe Books, New York.

Shea, B. T., 1981, Relative growth of the limbs and trunk in the African apes, *Am. J. Phys. Anthropol.* **56:**179–202.

Shea, B. T., 1982, Growth and size allometry in the African Pongidae: Cranial and postcranial analyses, Ph.D. dissertation, Duke University, Durham, North Carolina.

Shea, B. T., 1983a, Phyletic size change and brain/body scaling: A consideration based on the African pongids and other primates, *Int. J. Primatol.* **4:**33–62.

Shea, B. T., 1983b, Size and diet in the evolution of African ape craniodental form, *Folia Primatol.* **40:**32–68.

Shea, B. T., 1983c, Paedomorphosis and neoteny in the pygmy chimpanzee, *Science* **222:**521–522.

Shea, B. T., 1983d, Allometry and heterochrony in the African apes, *Am. J. Phys. Anthropol.* **62:**275–289.

Shea, B. T., 1984, An allometric perspective on the morphological and evolutionary relationships between pygmy (*Pan paniscus*) and common (*Pan troglodytes*) chimpanzees, in: *The Pygmy Chimpanzee: Evolutionary Biology and Behavior* (R. L. Susman, ed.), pp. 89–130, Plenum Press, New York.

Shea, B. T., 1985a, Ontogenetic allometry and scaling: A discussion based on the growth and form of the skull in African apes, in: *Size and Scaling in Primate Biology* (W. L. Jungers, ed.), pp. 175–206, Plenum Press, New York.

Shea, B. T., 1985b, The ontogeny of sexual dimorphism in the African apes, *Am. J. Primatol.* **8:**183–188.

Shea, B. T., 1986a, Scapula form and locomotion in chimpanzee evolution, *Am. J. Phys. Anthropol.* **70:**475–488.

Shea, B. T., 1986b, Ontogenetic approaches to sexual dimorphism in anthropoids, *Hum. Evol.* **1:**97–110.

Shea, B. T., in press, Neoteny and heterochrony in human evolution, in: *The Cambridge Encyclopedia of the Human Species* (R. D. Martin, D. Pilbeam, and S. Jones, eds.), Cambridge University Press, Cambridge.

Shea, B. T., and Groves, C. P., 1987, Evolutionary implications of size and shape variation in the genus *Pan*, *Am. J. Phys. Anthropol.* **72:**253.
Shea, B. T., and Pagezy, H., 1988, Allometric analyses of body form in Central African pygmies, *Amer. J. Phys. Anthrop.* **75:**269–270 (Abstract).
Shea, B. T., Hammer, R. E., and Brinster, R. L., 1987, Growth allometry of the organs in giant transgenic mice, *Endocrinology* **121:**1–7.
Shea, B. T., Hammer, R. E., and Brinster, R. L., in preparation, Cranial and postcranial skeletal growth allometries in giant transgenic mice.
Simpson, G. G., 1953, *The Major Features of Evolution*, Columbia University Press, New York.
Smith, B. H., 1986, Dental development in *Australopithecus* and early *Homo*, *Nature* **323:**327–330.
Starck, D., 1962, *Der heutige Stand des Fetalisations-problems*, Paul Parey, Hamburg.
Susman, R. L., Badrian, N. L., and Badrian, A. I., 1980, Locomotor behavior of *Pan paniscus* in Zaire, *Am. J. Phys. Anthropol.* **53:**69–80.
Tutin, C. E. G., and Fernandez, M., 1985, Foods consumed by sympatric populations of *Gorilla g. gorilla* and *Pan t. troglodytes* in Gabon: Some preliminary data, *Int. J. Primatol.* **6:**27–43.
Van de Koppel, J. M. H., and Hewlett, B. S., 1986, Growth of Aka pygmies and Bagandus of the Central African Republic, in: *African Pygmies* (L. L. Cavalli-Sforza, ed.), pp. 95–102, Academic Press, New York.
Verheyen, W. N., 1962, Contribution a la craniologie comparée des primates, *Mus. R. Afr. Cent. Tervuren Belg. Ann. Ser. Octav. Ser. Zool.* **105:**1–247.
Wake, D. B., 1980, Evidence of heterochronic evolution: A nasal bone in the Olympic salamander, *Rhyacotriton olympicus*, *J. Herpetol.* **14:**292–295.
Wake, D. B., 1982, Functional and evolutionary morphology, *Perspect. Biol. Med.* **25:**603–620.
Walker, A., Falk, D., Smith, R., and Pickford, M., 1983, The skull of *Proconsul africanus*: Reconstruction and cranial capacity, *Nature* **305:**525–527.
Wayne, R. K., 1986, Cranial morphology of domestic and wild canids: The influence of development on morphological change, *Evolution* **40:**243–261.
Wiedenreich, F., 1941, The brain and its role in the phylogenetic transformation of the human skull, *Trans. Am. Philos. Soc.* **31:**321–442.
Wiley, R. H., 1974, Evolution of social organization and life-history patterns among grouse, *Q. Rev. Biol.* **49:**201–227.

III

Cause, Abundance, and Implications of Heterochrony

Chapter 14

Genetic Basis for Heterochronic Variation

VICTOR AMBROS

1. Introduction

1.1. Heterochrony from a Cell Biological and Genetic Point of View

Genes encode the functional and structural components of cells and hence directly or indirectly define the behavior of individual cells and groups of cells in a developing system. In this sense the genome of an organism defines and controls

VICTOR AMBROS • Biological Laboratories, Harvard University, Cambridge, Massachusetts 02138.

the developmental processes that generate the final form of the organism. To examine the feasibility of rapid heterochronic change, it is important to determine how many genes must be mutated to cause heterochrony and the mechanisms by which those genes control developmental timing.

The genetic study of heterochrony provides two potential benefits: first, genetic analysis addresses important questions concerning the number and kinds of mutations required to introduce certain kinds of heterochronic developmental changes and thus bears on the frequency and ease with which heterochrony can occur; second, mutations that cause heterochrony will generally identify genes that play critical roles in controlling or regulating temporal patterns of development in the particular group of organisms under study. Once so identified, such genes can be subjected to more detailed genetic, molecular biological, and biochemical analysis so that the mechanisms of their effects on normal development can be investigated. Thus, the study of the genetic basis of heterochrony will simultaneously illuminate both evolutionary and developmental processes. The major focus of this chapter will be studies of nematode heterochronic mutants (Ambros and Horvitz, 1984). Laboratory mutagenesis of the species *Caenorhabditis elegans* has resulted in mutants with abnormal timing of certain developmental events. The heterochronic mutants of *C. elegans* provide an opportunity to use genetic techniques to probe mechanisms of temporal regulation and heterochronic change in nematodes. These nematode findings will be discussed in their own right and in the context of heterochrony in other organisms where genetic and molecular factors regulating temporal patterns of development have been revealed.

1.2. Experimental Approaches

One apparent property of heterochrony is that simple changes in the relative timing of developmental events can often have radical effects on final morphology, behavior, and/or life history (Alberch *et al.*, 1979; Gould, 1977). For this reason, heterochrony is thought to be a major source of radical developmental changes in evolution. Genetic analysis can further assess the ease with which such radical developmental changes can occur by testing whether heterochrony can arise from simple genetic changes. Genetic studies of heterochrony have employed two kinds of experimental approaches. In one approach, naturally occurring strains that are capable of interbreeding, and that show a heterochronic difference in their development, are crossed and the heterochronic developmental difference is analyzed as a standard genetic trait. The segregation patterns of the heterochronic developmental traits among the progeny of such crosses reveals whether the heterochrony is a dominant or recessive trait and whether it is caused by a single gene or multiple genes. The second approach is similar to the first, except that laboratory-induced developmental mutants ("heterochronic mutants") of a single species (instead of natural strains) are crossed and the genetic properties of their heterochronic defects are analyzed (Ambros and Horvitz, 1984).

The advantage of studying heterochronic mutants in an organism readily amenable to developmental and genetic analysis is that multiple mutations of a specific gene can be isolated independently and analyzed so that the full spectrum of mutations that is possible for a single gene may be examined. Furthermore, different genes affecting the timing of the same developmental event can be identified and studied. In principle, all genes that can be mutated individually to affect the relative timing of particular events can be identified. By combining mutations of different genes within the same individual, functional interactions among the products of these genes can be examined. This type of detailed genetic analysis has been possible for nematode heterochronic mutants (Ambros and Horvitz, 1987; Ambros, V., unpublished results). This is a significant advantage over the genetic analysis of naturally occurring strains, where one is limited to the experimental material provided by "nature."

A disadvantage of the use of laboratory-derived heterochronic mutants to study evolutionary heterochrony is that considerable uncertainty can exist as to whether the same heterochronic differences induced by mutation in the laboratory actually occur naturally and, if so, whether the precise genes identified by mutations are actually responsible for natural heterochrony. A demonstration of the feasibility of heterochronic change in one species through mutation of a given gene is not sufficient to prove that that gene *actually* causes heterochrony of evolutionary significance. More concrete support for an evolutionary role for a particular gene would require transfer of that gene from one species to another. Techniques for introduction of foreign genes into the germ line of whole organisms exist for plants (de Block *et al.*, 1984), mammals (Constantini and Lacy, 1981), insects (Rubin and Spradling, 1983), and nematodes (Stinchcomb *et al.*, 1985; Fire, 1985). Thus, the genetic analysis of heterochronic mutants will undoubtedly progress to the point where defined genes will be directly tested for evolutionary roles by intraspecific gene transfer.

Despite the limitations of interpreting developmental genetics in an evolutionary context, the heterochronic mutants of nematodes provide the opportunity to address, for a simple biological system, several relevant questions. First, one can test whether significant heterochrony can be caused by simple mutations of single genes. The results of the work described below provide further evidence that single-gene mutations can cause heterochrony and that heterochrony like that induced genetically in the laboratory does exist among living nematode species collected from the wild. Second, through the detailed analysis of the heterochronic genes of nematodes, an understanding is emerging of how these genes actually function, i.e., what kinds of products they encode, and how and when during development those products act to specify the proper timing of developmental events.

2. Nematode Development

2.1. Cell Lineages

Nematode development is commonly considered to consist of two primary phases: embryogenesis, which begins at fertilization and is completed by the time the newly hatched larva emerges, and postembryonic development, during which the larva develops through a series of stages (usually four) to the adult. The patterns of early cell divisions that occur during embryonic development may be highly conserved among nematodes (Sulston *et al.*, 1983). Also, the newly hatched larvae of different nematode species appear nearly identical in cellular anatomy, suggesting that later stages of nematode embryonic development might be highly conserved as well (Sulston and Horvitz, 1977; Sternberg and Horvitz, 1981, 1982; Sulston *et al.*, 1983; Ambros and Fixsen, 1987; Fixsen, 1985). These observations suggest that many of the developmental changes between nematodes are modifications of postembryonic development.

Nematodes offer an excellent opportunity to study the roles of genes in both developmental control and in evolutionary change. The ease of nematode culture and genetic manipulation (Brenner, 1974) has facilitated the isolation of large numbers of developmental mutants of *Caenorhabditis elegans* (Sulston and Horvitz, 1981; Ambros and Horvitz, 1984; Ferguson and Horvitz, 1985). The simple anatomy and rapid life cycle of the nematode *C. elegans* has allowed a detailed analysis of its wild-type development. As a result, the complete cell lineage of *C. elegans* has been elucidated (Sulston and Horvitz, 1977; Kimble and Hirsh, 1979; Sulston *et al.*, 1983). This lineage is nearly invariant in the wild type; each cell is formed after a defined lineage history and at a specific time during development. The postembryonic cell lineages of other nematode species can be determined and compared with each other and to *C. elegans* (Sternberg and Horvitz, 1981, 1982; Fixsen, 1985; Ambros and Fixsen, 1987). In this way, developmental differences among nematodes, particularly differences in the relative timing of events, can be characterized on the level of individual cell divisions and the differentiation of particular cells. Intraspecific comparisons of nematode cell lineages will be discussed further in a later section.

2.2. Temporal Patterns of Postembryonic Development in *Caenorhabditis elegans*

Caenorhabditis elegans postembryonic development consists of four larval stages (L1–L4). Each of these stages is characterized by stage-specific patterns of cell division and differentiation, and/or the formation of stage-specific cuticles (Sulston and Horvitz, 1977; Cox *et al.*, 1981). By following the development of living mutant animals, perturbations in this normal temporal pattern of postembryonic events can be easily recognized and characterized. Because of the ease with which abnormalities in the timing of developmental events can be identified and the sophisticated genetic manipulations that are possible (Brenner, 1974), *C. elegans* is a convenient animal in which to study the genetic programming of temporal patterning of development. Mutations that affect temporal patterns of development in *C. elegans* have defined "heterochronic" genes (Chalfie *et al.*, 1981; Ambros and Horvitz, 1984). These genes and their possible roles in evolutionary change will be discussed in more detail below.

The studies of nematode heterochrony discussed here focus on events in the development of the hypodermis of *C. elegans*. Hypodermal cells form the cuticle at each of the four larval molts and also function as blast cells. These blast cells, called "seam" cells, include seven initial cells named V1–V6 and T and certain of their progeny. Seam cells divide at approximately the time of each larval molt (e.g., see Figs. 1a, 2a, and 3a) (Sulston and Horvitz, 1977). These cell divisions cease after the fourth larval stage (Figs. 2a and 3a). To describe the behavior of specified cells at a particular time in development, the term "cell fate" is used. The "fate" of a cell is simply *what it does* in some stated temporal context. We would say that as each lateral hypodermal cell lineage proceeds, seam cells express a series of stage-specific cell fates. For example, the fate of the "T" cell in the *first larval stage* is to generate a set of specific cell types (a set of neuronal cells, one of which dies, and a blast cell) by a specific cell division pattern. The fate expressed by a granddaughter of T, (T.ap), in the *second larval stage* is to divide and generate a different cell division pattern and a different set of specific cell types (Fig. 1a). Thus, the two fates expressed by T-lineage blast cells in the L1 and L2 stages, respectively, are distinct, based on the pattern, number, and types of progeny cells that they produce. Similarly, the V-cell lineages contain stage-specific division patterns (Fig. 2a). Biochemical and electron microscopic studies have revealed that the cuticles formed at the L1, L2, and L3 molts are distinct in composition and architecture from the adult cuticle formed at the L4 molt (Cox *et al.*, 1981). Thus, the final larval stage marks a significant switch in hypodermal cell fate from "larval" (cell divisions, larval cuticle formation) to "adult" (no division, adult cuticle formation). Thus, the fates expressed by seam cells at various stages of *C. elegans* development are (1) formation of larval cuticles and generation of stage-specific patterns, numbers, and types of progeny, and (2) formation of adult cuticle and terminal differentiation. *Caenorhabditis elegans* heterochronic mutants affect the timing of all these events, i.e., stage-specific cell division patterns, cuticle formation, and cessation of molting.

3. Heterochronic Mutants of *Caenorhabditis elegans*

3.1. General Description

Heterochronic mutants of *C. elegans* have been isolated based on their morphological defects after mutagenesis of wild type (with chemicals or radiation) (Ambros and Horvitz, 1984). These defects include abnormalities in cuticle morphology and molting behavior, as described below. Since these defects are easily detectable by visual inspection of living worm cultures by low-magnification microscopy, heterochronic mutants have been recovered with relative ease. The precise nature of the defects of these mutants, including their cell lineage abnormalities, have been characterized by high-magnification Nomarski differential interference microscopy (Ambros and Horvitz, 1984).

Six genes have been identified that can be mutated to cause heterochrony. These are *lin-4* (Chalfie *et al.*, 1981), *lin-14, lin-28,* and *lin-29* (Ambros and Horvitz, 1984), *lin-41, and lin-42* (V. Ambros, unpublished results). The defects caused by mutations in these genes vary from gene to gene. Mutations in some genes, such as *lin-4* and *lin-14*, cause extensive heterochrony, where the timing of events in nearly all nongonadal cell lineages is altered with respect to gonadal development and the molting cycle. Other genes, such as *lin-29*, are more specialized, affecting the timing of one specific event of hypodermal development (Ambros

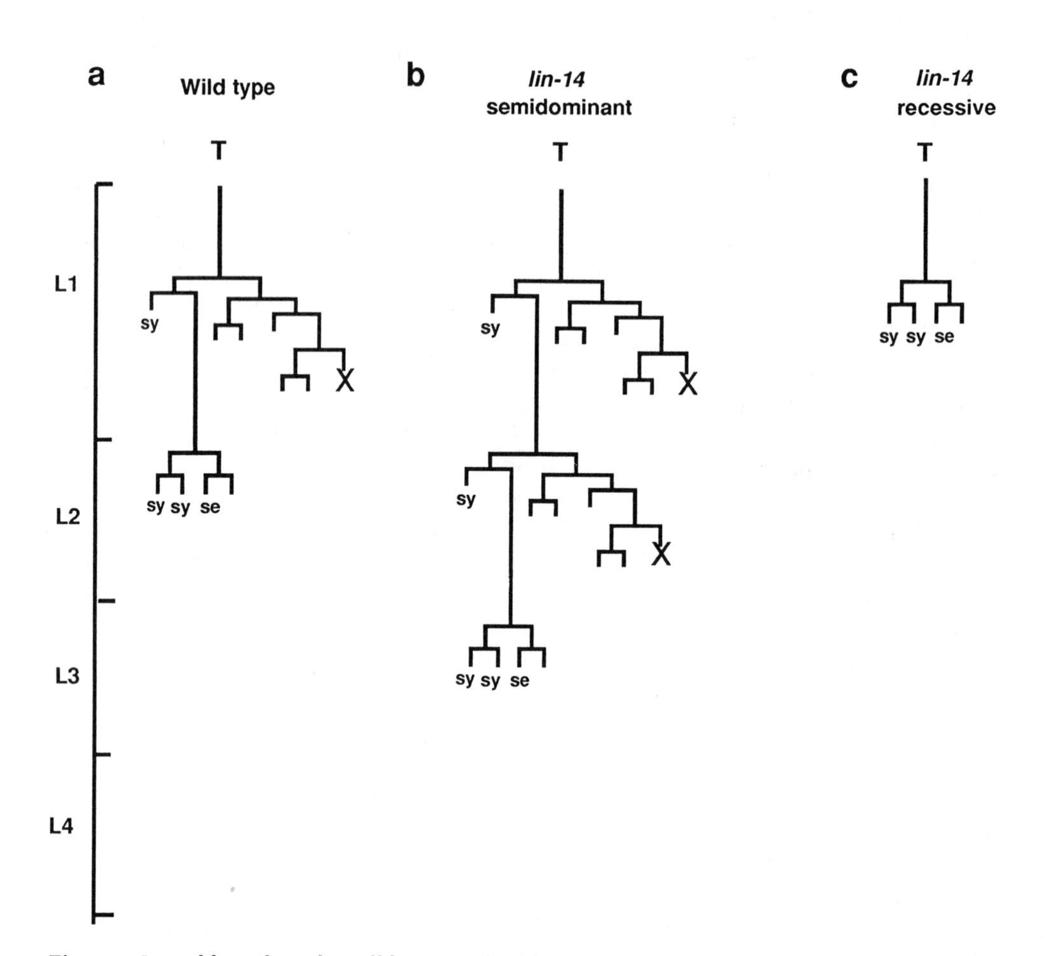

Figure 1. Lateral hypodermal T-cell lineages of wild-type and two opposite classes of *lin-14* mutants. Cell divisions were followed in living nematodes (Sulston and Horvitz, 1977). Cell lineages are diagrammed and cells are named according to Sulston and Horvitz (1977). The vertical scale to the left indicates postembryonic developmental stage, with the time of hatching and each postembryonic molt marked by a horizontal line. L1–L4 are the four larval stages; supernumerary stages of the *lin-14* semidominant mutant are not shown (see Fig. 3). Unless otherwise noted in this and other figures, all cell divisions were along an anterior–posterior axis in the animal. Anterior is to the left in the diagram. "Seam cells" (se) are morphologically distinguishable by Nomarski microscopy from hypodermal syncytial nuclei (sy) and neuronal cells (unlabeled cells). × denotes cells undergoing programmed cell death. The semidominant mutant (b) displays retarded development in the T lineage. In the L2 stage, one granddaughter of T generates a cell division pattern and descendant cell types very similar to those normally generated only during the L1 by the wild-type T cell (a). Conversely, in the recessive mutant (c), the T cell does not generate the normal L1-specific cell division pattern, but instead precociously generates a cell division pattern normally generated during the L2. The four progeny generated during the L1 by T in the recessive mutant appear by Nomarski microscopy (Sulston and Horvitz, 1977) morphologically similar to the four corresponding progeny generated in the L2 in the wild type.

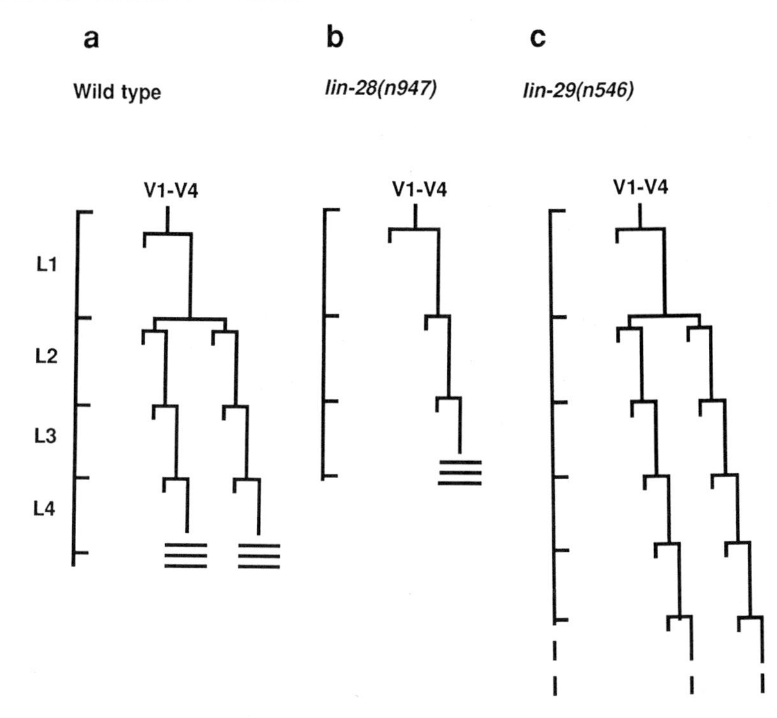

Figure 2. Heterochronic lateral hypodermal cell lineage patterns of *lin-28* and *lin-29* mutants (see also Ambros and Horvitz, 1984). The times that seam cells cease division and form adult-specific cuticular structures known as lateral alae are shown by triple horizontal bars. Adult alae formation occurs only at the L4 molt in the wild type (a), but occurs at the L3 molt in *lin-28* mutants (b) and does not occur in *lin-29* mutants (c). The *lin-28* mutants undergo only three molts, while *lin-29* mutants undergo extra molts. In *lin-29* animals, seam cells divide once at each extra molt. The schedule of gonadal development appears unaffected in these mutants.

and Horvitz, 1984) (Fig. 2c). The general properties that mutations in all these genes share are: (1) they effect only postembryonic development (i.e., only events after hatching of the larva from the egg); (2) they all seem to affect only *nongonadal* cell lineages; and (3) they all lead to heterochrony by causing apparent temporal transformations in cell fates. I will expand on these points below and use as examples mutations of three genes, *lin-14*, *lin-28*, and *lin-29*.

3.2. Retarded Mutants Reiterate Larval Cell Lineage Patterns and Undergo Supernumerary Larval Molts

Certain alleles of the gene *lin-14* and recessive alleles of the gene *lin-29* lead to "retarded" development in *C. elegans*, in which developmental events normally specific for early stages of development occur at later stages (Figs. 1b, 2c, and 3b) (Ambros and Horvitz, 1984). These retarded defects result from apparent temporal transformations in cell fates; cells at later stages do not express their normal fates, but instead express the fates normally specific to cells at earlier

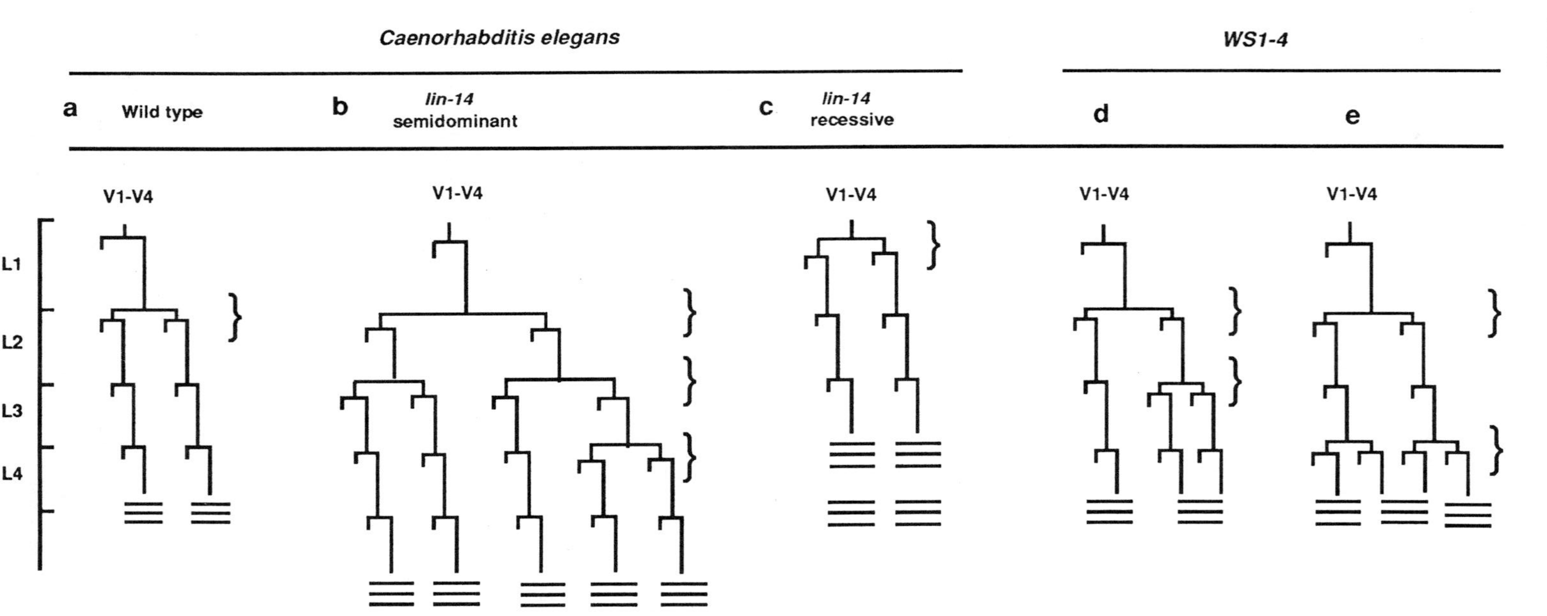

Figure 3. Lateral hypodermal cell (V1–V4) lineage patterns of wild-type *C. elegans*, two different phenotypic classes of *C. elegans lin-14* mutants (see also Ambros and Horvitz, 1984) and the wild type of another Rhabditidae species *WS1-4*. The *WS1-4* was isolated from the wild by placing soil samples on agar seeded with *Escherichia coli* and containing nematode culture medium (Brenner, 1974). Animals that left the soil and moved into the bacteria were individually picked and their progeny were cultured under conditions previously described for *C. elegans* (Brenner, 1974). The newly hatched L1 larvae of *WS1-4* are essentially identical to those of *C. elegans*. Because of the similarities between newly hatched larvae of different species in the number, position, and type of all cells, we have named the postembryonic blast cells of these species using the nomenclature developed for *C. elegans* (Sulston and Horvitz, 1977). The times that seam cells cease cell division and form adult-specific cuticular structures known as lateral alae are shown by triple horizontal bars. Adult alae formation occurs only at the L4 molt in the wild type, and at later (panel b) or earlier (panel c) stages in *lin-14* mutants. The "retarded" mutant shown in panel b underwent an extra molt beyond the normal four. A distinctive cell division pattern (indicated by a bracket) occurs in the L2 of the wild type and at later (panel b) or earlier (panel c) stages of *lin-14* mutants. In another Rhabditidae species, *WS1-4*, a similar cell division pattern is reiterated at both the L2 and L3 stages (panel d) or at the L2 and L4 stages (panel e). [The two alternative cell lineages shown in panels d and e are variably expressed in *Ws1-4* (Ambros and Fixsen, 1987).] Despite the apparent cell lineage reiteration in *Ws1-4*, these animals undergo four larval stages, similar to wild-type *C. elegans*.

stages of the same cell lineage. As a result of these temporal transformations, these mutants reiterate aspects of larval development. For example, in animals carrying a *lin-14* semidominant mutation, a T-lineage blast cell in the L2 does not divide in the manner that it would in wild-type animals, but instead reiterates an L1-specific pattern (Fig. 1b). In addition, the lateral hypodermal cells of *lin-14* and *lin-29* retarded mutants do not form adult-specific cuticle at the proper time (the fourth larval molt), but form a larval cuticle instead; Figs 2c and 3b). These animals undergo extra larval stages beyond the normal four. Hypodermal cells in retarded mutants appear to have failed to switch from their larval fates to their adult fate at the normal time (Ambros and Horvitz, 1984). Cell lineage reiterations are common in heterochronic mutants of *C. elegans*, and, as will be described below, also seem to occur between nematode species.

3.3. Precocious Mutants Express Adult Characteristics at Larval Stages

In mutants carrying recessive alleles of *lin-14* or *lin-28*, developmental events can occur "precociously," i.e., earlier than normal (Figs. 1c, 2b, and 3c). These defects are caused by temporal transformations in cell fates opposite to those observed in retarded mutants. For example, in animals carrying *lin-14* recessive mutations, normally L2-specific cell division patterns occur precociously in the L1, and the normal L1 division patterns are deleted (Figs. 1c and 3c). In certain precocious mutants, hypodermal cells at the L3 molt do not express their normal fates (to divide and form larval cuticle), but instead cease cell division and form adult cuticle (Figs. 2b and 3c). Thus, the fates of these cells in *lin-14* and *lin-28* precocious mutants appear to be transformed into the fates normally specific to cells later in the same lineage (Ambros and Horvitz, 1984). The *lin-28* mutants also cease the molting cycle precociously by one stage (after only three molts). This molting defect is opposite to the extra molts displayed by *lin-14* or *lin-29* retarded mutants.

3.4. Simple Mutations of Single Genes Can Cause Heterochrony in *Caenorhabditis elegans*

Genetic analysis of heterochronic mutants indicates that the mutations that cause heterochrony are not necessarily complex changes in gene structure and function. One piece of evidence that these mutations are likely simple genetic changes is the frequency with which heterochronic mutants have been recovered after mutagenesis. In *C. elegans*, mutations that cause loss of function of an "average" gene through a simple change in the DNA of that gene (i.e., an alteration in one base pair or a deletion of one or more adjacent base pairs) occur at a frequency of approximately 5×10^{-4} (Brenner, 1974). Mutations in *lin-14*, *lin-28*, and *lin-29* were recovered at approximately this frequency, indicating that they are probably straightforward mutations causing loss of function of single genes (Ambros and Horvitz, 1984).

Genetic and molecular mapping experiments further suggest that heterochronic mutants carry simple mutations. Recombination analysis of heterochronic mutations has shown that if multiple mutations are required to cause heterochrony in these cases, then they must be located exceedingly close to one another (Ambros and Horvitz, 1984). Furthermore, the DNA sequences of *lin-14* have been isolated from the wild-type *C. elegans* genome and from mutants (Ruvkun *et al.*, 1988). Sequence analysis of this *lin-14* DNA suggests that two *lin-14* mutations that cause retarded development are each small deletions of several hundred base pairs. Other *lin-14* mutations seem to be "point" mutations.

4. Developmental Roles of Heterochronic Genes in *Caenorhabditis elegans*

4.1. Control of Cell Fates and Developmental Patterns

Genetic analysis of *C. elegans* developmental mutants has identified genes that play crucial roles in specifying and expressing the normal *C. elegans* developmental program (Sulston and Horvitz, 1981; Greenwald *et al.*, 1983; Ambros and Horvitz, 1984; Ferguson and Horvitz, 1985; Fixsen *et al.*, 1985). These developmental control genes can be mutated to cause transformations of the fates of cells, such that certain cells adopt the fates normally specific to other cells.

Three general classes of cell fate transformations occur in *C. elegans* developmental mutants: (1) in heterochronic mutants, temporal transformations occur where cells at one stage in development express fates normally specific for cells at another time in development (Chalfie *et al.*, 1981; Ambros and Horvitz, 1984); in "homeotic" mutants, spatial transformations occur where cells adopt the fates normally expressed by cells at another position in the animal (Sulston and Horvitz, 1981; Greenwald *et al.*, 1983; Ferguson and Horvitz, 1985); in sex transformer mutants, sexual transformations occur where cells express fates normally expressed by analogous cells in the other sex (Hodgkin and Brenner, 1977). Since mutations in these genes cause cells to forsake their normal fates and adopt the fates of other cells, the normal roles of these genes are to specify that certain cells select one fate as opposed to another. These genes seem to be involved in specifying patterns of expression of diverse cell fates: in the case of homeotic genes, spatial patterns; in the case of "sex determination genes," sexual patterns; or, in the case of heterochronic genes, temporal patterns. Thus, the heterochronic genes may be members of a general class of developmental control genes that function to control the patterns of expression of specific cell fates during development.

4.2. A Temporal Gradient of *lin-14* Activity

The genetic properties of *lin-14* mutations have suggested a model for how *lin-14* controls temporal patterns of development (Ambros and Horvitz, 1984). One class of *lin-14* mutations seems to elevate *lin-14* activity and results in retarded expression of larval cell fates. Conversely, mutations that reduce *lin-14* activity cause precocious expression of adult fates. These basic observations, together with other genetic data, have led to the proposal that cells with multiple potential fates are generated at successive stages of the wild-type lateral hypodermal lineages and that for each of these cells, the level of *lin-14* activity selects which fate is expressed. The normal sequence in which specific fates are expressed in each hypodermal lineage is proposed to be caused by a decrease in the level of *lin-14* activity from high at earlier stages to low at later stages. In *lin-14* mutants, abnormal levels of *lin-14* activity at certain stages cause the retarded or precocious expression of cell fates at those stages (Ambros and Horvitz, 1987).

4.3. A System of Temporal Information in *Caenorhabditis elegans*

The *lin-14* mutations cause temporal transformations of cell fates in a wide variety of cell types and at several stages of development. The timing of events in the lateral and ventral hypodermis, muscle cell lineages, and intestine can be altered, while notably, the gonad is unaffected in its development (Ambros and Horvitz, 1984). Thus, *lin-14* has an extensive yet sharply delineated realm of action. For those cells affected by *lin-14* mutations, it appears that *lin-14* has some fundamental role in making stage-specific developmental choices. The above model, whereby *lin-14* diminishes during development and thereby causes various cell fates to be expressed in defined temporal sequences, suggests that *lin-14* may encode a general temporal signal—a kind of "clock." Perhaps those heterochronic genes with broad effects on the timing of events in many tissues (e.g., *lin-14*), are involved in conveying temporal information to diverse cell types. Genes with more tissue- or stage-specific effects (e.g., *lin-29*) may be required for the reception or utilization of that temporal information for cell-type-specific or developmental stage-specific determinative choices. Analysis of doubly mutant strains containing mutations in both *lin-14* and *lin-29* has supported this model (V. Ambros, in preparation). The regulatory activity of the *lin-14* product seems to require or act via the *lin-29* product. This is consistent with a signal–response hierarchical relationship between these two genes.

4.4. Molecular Mechanisms

Thus far, no definitive conclusions can be drawn concerning the molecular mechanisms by which heterochronic genes of *C. elegans* may act. The persistence of larval features in sexually mature animals of *lin-14* and *lin-29* mutants is reminiscent of the neoteny introduced by decreased sensitivity to or decreased levels of thyroid hormone in amphibians [reviewed by Gould (1977)]. Another example of heterochrony resulting from manipulation of extracellular hormone level is the occurrence of extra larval stages or, conversely, precocious differentiation of adult features in insects following experimental perturbation of juvenile hormone levels (Wigglesworth, 1940). Despite these analogies, there is no evidence that any of the *C. elegans* genes encode products that act cell-extrinsically, in a hormone-like manner. It is possible that all these genes act cell-intrinsically. Examples of developmental control genes that seem to act cell-intrinsically are the homeotic genes of insects [reviewed by Morata and Lawrence (1977)]. Homeotic genes regulate the pattern of segment-specific morphogenesis and differentiation. The products of these genes seem to reside in the nuclei of cells and their products have structural features characteristic of DNA-binding proteins, suggesting a direct role in the regulation of segment-specific gene expression (Scott and Weiner, 1984). By analogy, at least some of the heterochronic genes may act cell-intrinsically and regulate stage-specific gene products.

5. Heterochrony among Nematode Species

5.1. Interspecific Cell Lineage Comparisons

The role of heterochrony in nematode evolution has not been well-documented. Nematodes are not easily fossilized, and the anatomical features detectable in fixed specimens do not clearly reflect temporal patterns of growth and development. Until recently, only the more gross aspects of life history could be observed in living nematodes. Nonetheless, evidence for heterochrony has been gathered in this way. For example, several species of *Seinara* were found to vary in the number of postembryonic molts and in the time of hatching relative to the first molt (Hechler and Taylor, 1966). More recently, Nomarski differential interference microscopy has been used for extended cell lineage analysis by direct observation of living nematodes (Sulston and Horvitz, 1977). This technique allows the precise characterization of temporal as well as spatial patterns of cell division, differentiation, and morphogenesis in various cultured nematode species.

Certain nematode species, especially free-living (nonparasitic) varieties, can be isolated from the wild and cultured in the laboratory. The relatively simple cellular anatomy of the free-living nematodes in general allows the direct comparison of cell lineages between different species on the level of individual cells and cell divisions. Using Nomarski microscopy of living specimens, it is relatively easy to determine at what stages in development specific cell lineages begin to differ between *C. elegans* and other species and what the precise nature of these differences are (Sternberg and Horvitz, 1981, 1982; Ambros and Fixsen, 1987; Fixsen, 1985). If one assumes that genes like those regulating development in *C. elegans* play evolutionary roles, then developmental differences between species might be similar or identical to mutationally induced developmental changes in *C. elegans*.

Studies of the cell lineage differences among species of four nematode families have revealed that many developmental differences between species of the same family and between families can be characterized as apparent transformations of cell fates (Sternberg and Horvitz, 1981, 1982; Fixsen, 1985; Ambros and Fixsen, 1987). In particular, heterochronic differences between species of the same family seem to be caused by temporal transformations of cell fate (Ambros and Fixsen, 1987).

5.2. Cell Lineage Reiterations

Certain kinds of heterochronic changes caused by *lin-14* mutations can also be observed between nematode species. Of particular interest here are apparent cases of heterochronic differences in cell lineage pattern. The lateral hypodermal lineages of the free-living *Rhabditidae* species *WS1-4* differ from the same lineages of *C. elegans* (another *Rhabditidae* species) in a manner that could be interpreted as heterochrony (Fig. 3d and 3e). The species *WS1-4* seems to variably express two types of lateral hypodermal lineages: Although the L1 and L2 patterns appear identical to *C. elegans*, a pattern like that expressed in the L2 can be repeated at either the L3 or L4 stages (Figs. 3d and 3e). (In another *Rhabditidae* species, *WS7-4*, only the L4 stage repeat of the L2 pattern occurs (Ambros and Fixsen, 1987). In *C. elegans*, the reiteration of a normally L2-specific cell lineage pattern at later stages is caused by mutations that elevate *lin-14* activity (Ambros and Horvitz, 1984). Thus, it is possible that these differences between the cell lineages in *C. elegans* and *WS1-4* (and *WS7-4*) could be primarily caused by a simple change in the level of activity of a gene like *lin-14*.

Reiterative and nonreiterative lateral hypodermal lineages have been observed in two separate families: the three species *C. elegans*, *WS1-4*, and *WS7-4* belong to the *Rhabditidae* family based on pharyngeal and male tail morphology (Goodey, 1963); the lateral hypodermal lineages of the latter two species reiterate during the L3 and L4 features specific to the L2 in *C. elegans*. Similarly, comparisons of the lateral hypodermal lineages of two members of the family *Panagrolaimidae* have shown that in one species these lineages reiterate at later stages certain events that occur only at the L2 stage early in the other species (Fixsen, 1985).

6. Possible Significance of *Caenorhabditis elegans* Heterochronic Mutants to Nematode Evolution

6.1. Simple Genetic Mutations Can Cause Diverse Kinds of Heterochrony in a Nematode

Caenorhabditis elegans heterochronic mutants display two major complementary classes of heterochrony (Gould, 1977; Alberch *et al.*, 1979). In retarded mutants, sexually mature individuals express normally juvenile characteristics. On the other hand, in precocious mutants, sexually immature individuals express normally adult characteristics. These two general classes of defects displayed by heterochronic mutants include a wide variety of specific kinds of changes in the timing of developmental events. These changes also result in significant alterations in anatomy, morphology, and life history, including (1) changes in the timing of stage-specific division patterns, resulting in overproduction of specific cell types and/or omission of other cell types; (2) changes in the timing of stage-specific cuticle formation; (3) altered number of larval molts, either one fewer than normal (e.g., *lin-28*, Fig. 2b) or greater than normal (e.g., *lin-29*, Fig. 2c). The specific set of developmental events affected in a particular heterochronic mutant can depend on which gene is mutated and also can vary significantly for different alleles of a single gene. For *lin-14* in particular, defects can be quite allele-specific (Ambros and Horvitz, 1987). In contrast to *lin-14* mutations, which can have broad effects in many lineages and at several stages of development, mutations of certain genes, such as *lin-29* (Fig. 2c) (see also Ambros and Horvitz, 1984) affect the fate of only one cell type at only one or a few stages of development. The broad range of changes in the temporal patterns of development observed in heterochronic mutants suggests that mutation of these genes, either singly or in combination, could give rise to diverse kinds and varying severities of heterochronic changes among nematode species.

Although the specific heterochronic phenotypes displayed by *C. elegans* mutants may identify the limits or constraints on the possible changes in developmental timing in nematodes, they also suggest a considerable flexibility in the potential evolutionary change that might result from mutations of these genes. For example, depending upon the state of the *lin-14* locus, adult cuticle synthesis can occur at any one of three stages, the L3, L4, or L5 molts (Ambros and Horvitz, 1987) (Figs. 3a–3c). In these same mutants, the fates expressed by many of the cells at earlier stages seems to be independent of the stage of adult cuticle formation (Ambros and Horvitz, 1987). This kind of independent regulation of the timing of various events within the same lineage would provide a potentially rich spectrum of possible heterochronic changes for nematodes.

Since the heterochronic mutants that thus far have been characterized in *C. elegans* display altered stage-specificity of nongonadal developmental events while gonad development proceeds according to the normal schedule, the timing of events in gonadal and nongonadal cell lineages may be controlled separately. It will be interesting to determine whether genes can be identified in *C. elegans* that affect the timing of gonadal development alone, or both gonadal and nongonadal lineages.

6.2. Cell Lineage Reiterations and Transformations of Cell Fate in Nematode Evolution

Heterochrony in *lin-14*, *lin-28*, and *lin-29* mutants as well as in mutants of other genes seems to be caused by temporal transformations of cell fates. This suggests that heterochrony among nematodes may be commonly caused by such cell fate transformations. Observations of cell lineage differences between nematode species has confirmed the plausibility of this hypothesis (Fixsen, 1985; Ambros and Fixsen, 1987).

Cell lineage reiterations appear to underlie certain differences in development between nematode species. Furthermore, reiterations have been observed in two separate families of nematodes. Such common cell lineage differences could reflect evolutionary changes that occur with relative ease, and therefore might result from relatively few mutational events. An example is described below. A clearer picture of just how common cell lineage reiterations are will require a more extensive survey, ideally involving species from most existing nematode families. In general, if heterochrony in *C. elegans* mutants is caused by mutations in genes common to all nematodes, then the types of developmental changes observed in these mutants could be nearly ubiquitous. It is important to note that genetic evidence from *C. elegans* indicates that a change from nonreiterative to reiterative development might be easily reversible. Specifically, the reiterative development of the mutant *lin-4(e912)* (Chalfie *et al.*, 1981) can be suppressed by simple recessive mutations in *lin-14* (Ambros and Horvitz, 1984).

7. Genetics of Heterochrony in Other Organisms

Besides the heterochronic mutants of *C. elegans*, there are several other cases, including birds, fish, amphibians, and plants, where heterochrony has been shown to be caused by one or a few genetic loci. Thus, these *C. elegans* mutants supplement a body of evidence for rapid evolutionary change in temporal patterns of development by mutation of single genes. The kinds of developmental changes represented by these examples of genetically induced heterochrony include two major classes: (1) alterations in the relative rates of growth of various aspects of the organisms and (2) changes in the relative timing of events, without alterations in growth rates.

"Rate genes" were found to underlie differences in pigmentation in gypsy moth caterpillars (Goldschmidt, 1981). The rates of pigmentation deposition, controlled by a defined set of genes, affect the ultimate pattern of markings characteristic of certain moth races. Further examples are described in an extensive review of the genetic basis of morphological evolution in plants (Gottlieb, 1984). For example, Sinnott (1958) studied two races of the squash, *Lagenaria*, that have different fruit shapes. These shape differences were known to arise from different relative rates of fruit elongation. After crossing these two races, Sinnott found that the pattern of segregation of parental allometric constants among the F_1 and F_2 progeny indicated a single gene difference.

The heterochronic genes of *C. elegans* seem primarily to control the relative timing of events, and not relative growth rates. As was mentioned in a previous section, paedomorphosis in axolotl (*Ambystoma mexicanum*), where retarded expression of larval characteristics in the adult seems to be caused by homozygosity for a single recessive allele (Humphrey, reviewed by Tompkins, 1978), is a case of genetically induced heterochrony analogous to the retarded *lin-14* mutants. Similarly, a sex-linked gene seems to control the time of onset of sexual maturation in the platyfish, *Xiphorus maculatus* (Kallman and Schreibman, 1973). In chickens, the creeper gene seems to affect the time of onset of metatarsal growth (Cock, 1966) and the eudiplopodia mutation seems to extend the length of time that the limb bud ectoderm responds to mesodermal induction (Fraser and Abbott, 1971). (In the case of creeper, limbs are shortened, and in the case of eudiplopodia, accessory apical ectodermal ridges are induced, resulting in extra digits.) In these examples from nematodes, amphibians, and chickens, single genetic mutations seem primarily to cause alterations in the timing of *events* as opposed to alterations in relative growth rates. The nematode is the only system in which developmental defects can be characterized in sufficient detail to identify the specific cells that are first affected during development. It may be useful to consider the possibility that genetically induced alterations in the timing of developmental events in other systems might also be interpreted as temporal transformations in cell fates.

References

Alberch, P., Gould, S. J., Oster, G. F., and Wake, D. B., 1979, Size and shape of ontogeny and phylogeny, *Paleobiology* **5**:296–317.

Ambros, V., and Fixsen, W., 1987, Cell lineage variation among nematodes, in: *Development as an Evolutionary Process* (R. Raff and E. Raff, eds.), Liss, New York, pp. 139–160.

Ambros, V., and Horvitz, J. R., 1984, Heterochronic mutants of the nematode *Caenorhabditis elegans*, *Science* **226**:409–416.

Ambros, V., and Horvitz, H. R., 1987, The *lin-14* locus of *Caenorhabditis elegans* controls the time of expression of specific postembryonic development events, *Genes Dev.* **1**:398–414.

Brenner, S., 1974, The genetics of *Caenorhabditis elegans*, *Genetics* **77**:71–94.

Chalfie, M., Horvitz, H. R., and Sulston, J. E., 1981, Mutations that lead to reiterations in the cell lineages of *Caenorhabditis elegans*, *Cell* **24**:59–69.

Cock, A. G., 1966, Genetical aspects of metrical growth and form in animals, *Q. Rev. Biol.* **41**:131–190.

Constantini, F., and Lacy, E., 1981, Introduction of rabbit beta-globin into the mouse germ line, *Nature* **294**:92–94.

Cox, G. N., Staphrans, S., and Edgar, R. S., 1981, The cuticle of *Caenorhabditis elegans*. II. Stage-specific changes in ultrastructure and protein composition during postembryonic development, *Dev. Biol.* **86**:456–470.

De Block, M., Herrera-Estrella, L., Van Montagu, M., Schell, J., and Zambryski, R., 1984, Expression of foreign genes in regenerated plants and in their progeny, *EMBO J.* **3**:1681–1690.

Ferguson, E. L., and Horvitz, H. R., 1985, Identification and characterization of 22 genes that affect the vulval cell lineages of the nematode *Caenorhabditis elegans*, *Genetics* **110**:17–73.

Fire, A., 1986, Integrative transformation of *Caenorhabditis elegans*, *EMBO J.* **5**:2673–2680.

Fixsen, W., 1985, *The genetic control of hypodermal cell lineages in the nematode C. elegans*, Ph.D. dissertation, Massachusetts Institute of Technology, Cambridge.

Fixsen, W., Sternberg, P., Ellis, H., and Horvitz, H. R., 1985, Genes that affect cell fate during the development of *Caenorhabditis elegans*, *Cold Spring Harbor Symp. Quantitat. Biol.* **50**:99–128.

Fraser, R. A., and Abbott, U. K., 1971, Studies on limb morphogenesis IV. Experiments with the polydactylous mutant eudiplopedia. *J. Exp. Zool.* **176**:237–248.

Goldschmidt, R., 1981, A preliminary report on some genetic experiments concerning evolution, *Am. Nat.* **52**:28–50.

Goodey, T., 1963, *Soil and Fresh Water Nematodes*, Wiley, New York.

Gottlieb, L. D., 1984, Genetics and morphological evolution in plants, *Am. Nat.* **123**:681–709.

Gould, S. J., 1977, *Ontogeny and Phylogeny*, Harvard University Press, Cambridge.

Greenwald, I. S., Sternberg, P. W., and Horvitz, H. R., 1983, *lin-12* specifies cell fates in *Caenorhabditis elegans*, *Cell* **34**:435–444.

Hechler, H. C., and Taylor, D. P., 1966, The life histories of *Seinura celeris, S. oliveirae, S. oxura* and *S. Steineri* (Nematoda: Aphelenchoididae), *Proc. Helminth. Soc.* **33**(1)71–83.

Hodgkin, J. A., and Brenner, S., 1977, Mutations causing transformation of sexual phenotype in the nematode *Caenorhabditis elegans*, *Genetics* **86**:275–287.

Kallman, K. D., and Schreibman, M. P., 1973, A sex-linked gene controlling gonadotrope differentiation and its significance in determining the age of sexual maturation and size of the platyfish, *Xiphophorus maculatus*, *Gen. Comp. Endocrinol.* **21**:287–304.

Kimble, J., and Hirsch, D., 1979, Post-embryonic cell lineages of the hermaphrodite and male gonads in *Caenorhabditis elegans*, *Dev. Biol.* **70**:396–417.

Morata, G., and Lawrence, P. A., 1977, Homeotic genes, compartments and cell determination in *Drosophila*, *Nature* **265**:211–216.

Rubin, G. M., and Spradling, A. C., 1983, Genetic transformation of *Drosophilia* with transposable element vectors, *Science* **218**:348–353.

Ruvkun, G., Ambros, V., and Horvitz, H. R., 1988, Isolation and characterization of DNA sequences of *lin-14* (in preparation).

Scott, M. P., and Weiner, A. J., 1984, Structural relationships among genes that control development: Sequence homology between the antennapedia, ultrabithorax, and fushi tarazu loci of *Drosophila*, *Proc. Natl. Acad. Sci. USA* **81**:4115–4119.

Sinnott, E. W., 1958, The genetic basis of organic form, *Ann N. Y. Acad. Sci* **71**:1223–1233.

Sternberg, P. W., and Horvitz, H. R., 1981, Gonadal cell lineages of the nematode *Panagrellus redivivus* and implications for evolution by modification of cell lineages, *Dev. Biol.* **88:**147–166.

Sternberg, P. W., and Horvitz, H. R., 1982, Postembryonic nongonadal cell lineages of the nematode *Panagrellus redivivus*: Description and comparison with those of *Caenorhabditis elegans*, *Dev. Biol.* **93:**181–205.

Stinchcomb, D. T., Shaw, J. E., Carr, S. H., and Hirsh, D., 1985, Extrachromosomal DNA transformation of *Caenorhabditis elegans*, *Mol. Cell. Biol.* **5:**3484–3496.

Sulston, J. E., and Horvitz, H. R., 1977, Post-embryonic cell lineages of the nematode *Caenorhabditis elegans*, *Dev. Biol.* **56:**110–156.

Sulston, J. E., and Horvitz, H. R., 1981, Abnormal cell lineages in mutants of the nematode *Caenorhabditis elegans*, *Dev. Biol.* **82:**41–55.

Sulston, J. E., Schierenberg, E., White, J. G., and Thomson, J. N., 1983, The embryonic cell lineage of the nematode *Caenorhabditis elegans*, *Dev. Biol.* **100:**64–119.

Tompkins, R., 1978, Genic control of *Axolotol metamorphosis Am. Zool.* **18:**313–319.

Wigglesworth, V. B., 1940, The determination of characters at metamorphosis in *Rhodnias prolixus* (Hemiptera), *J. Exp. Biol.* **17:**201–222.

Chapter 15

The Abundance of Heterochrony in the Fossil Record

KENNETH J. McNAMARA

1. Introduction

Paleontologists have long recognized that the fossil record seems to indicate a strong relationship between ontogeny and phylogeny. Any attempt to assess the nature of this relationship, however, is plagued by many problems, not the least of which are the dual specters of Ernst Haeckel and Walter Garstang, which have long haunted paleontologists. In the late 19th century the relationship between ontogeny and phylogeny was explained almost entirely in terms of recapitulation. In the same year that Haeckel proposed his biogenetic law, Hyatt (1866), working on fossil cephalopods, formulated a very similar scheme to interpret the evolutionary history of ammonoids. Hyatt's ideas were to have a profound effect on many of his contemporary paleontologist colleagues and some of his students. For example, Jackson (1890, 1912) interpreted many aspects of bivalve and echi-

KENNETH J. McNAMARA • Western Australian Museum, Perth, Western Australia 6000, Australia.

noid evolution in terms of recapitulation, as did Beecher (1893, 1987) on brachiopods and trilobites.

While it is tempting to suggest, as many previous authors have, that these early workers were totally blinkered in their outlook, seeing only recapitulation, and not paedomorphosis, their writings show that they all recognized the occurrence of what they termed "degenerate" forms, which passed through fewer morphological stages during their ontogenies than their ancestors. Most, however, considered such paedomorphic forms, as they later became known, to be relatively uncommon and to have played only a minor role in evolution in the various groups. A few, such as Smith (1914), who has been lumped with the extreme recapitulationists (Donovan 1973, Kennedy 1977), came to realize that the fossil record, at least as far as ammonoids were concerned, seemed to show a dominance of the so-called "degenerate" forms. The biogenetic law was beginning its slide to oblivion.

With its demise paleontologists were quick to seize upon ideas proposed by Walter Garstang on the importance of paedomorphosis with almost as much zealous fervor as followers of Haeckel had pursued recapitulation. Garstang (1922, 1928) sounded the death knell for the biogenetic law by pointing out that ontogeny, rather than just recapitulating phylogeny, creates it. In the last half-century many fossil groups have been said to show evidence of paedomorphosis. Recapitulation had become unfashionable. Even though some workers might have observed the phenomenon, few were prepared to commit themselves to print.

The publication of Gould's (1977) book on ontogeny and phylogeny was an important landmark in the study of heterochrony, as it suggested that neither recapitulation [hereinafter called peramorphosis, following the revised classification of Alberch *et al.* (1979)] nor paedomorphosis was more dominant than the other, but that both can, and do, occur. With this in mind, what does the fossil record show? Does it support the idea of equal opportunity for both peramorphosis and paedomorphosis? Indeed, can it even be used to assess the relative frequencies of these heterochronic phenomena?

In order to examine the evidence for heterochrony in the fossil record, I have largely limited my search to articles published in the last 10 years. I have surveyed all the papers published in *Palaeontology, Journal of Paleontology, Paleobiology*, and *Lethaia* from 1976 to 1985, as well as incorporating as many other references as I could find from other journals published within this period. References to heterochrony *per se* were found in about 100 papers, and in a further 40 the data presented could clearly be interpreted in terms of heterochrony.

I have restricted my review to invertebrates and vertebrates. Not only have I attempted to document the mere existence of heterochrony in the fossil record, but I have tried to analyze the relative frequencies of paedomorphosis and peramorphosis. The fossil record can also be used to examine differences in the frequency of paedomorphosis and peramorphosis between different groups of organisms. Changing frequencies within groups over time can also be assessed. I have also tried to analyze which particular heterochronic processes have occurred and the relative frequencies of these processes in different groups of organisms. Problems with this are discussed below (Section 2).

Descriptions of heterochrony from the fossil record range from intraspecific to the class level. It is therefore possible to try and analyze the importance of heterochrony in the evolution of major morphological novelties, and hence the evolution of higher taxa. However, great care must be taken in such an approach, because of the problems of deciding whether homologous characters are being compared. In recent years research has tended to focus more on analyzing heterochrony at the species level, and, in particular, its role in the development of evolutionary trends by directional speciation. Such heterochronic trends have been described in a number of (mainly invertebrate) groups. The importance of such trends and their ecological significance in different groups are discussed.

2. Identification of Heterochrony from the Fossil Record

2.1. Ancestor–Descendant Relationships

One of the standard criticisms leveled at any attempts to identify heterochrony from the fossil record comes from those who question the ability of paleontologists to identify ancestor–descendant relationships from the fossil record (e.g., Nelson, 1978; Fink, 1982). There seems to be a belief in some quarters that ancestor–descendant relationships can only be deduced by doing cladistic analyses. Paul (1985) has recently written that the sequence of species in the fossil record inevitably reflects the order in which they evolved. Many examples of evolutionary trends produced by heterochrony have been described from the fossil record (see Section 5). These provide a possible test to falsify ancestor–descendant hypotheses. Thus, if with further collecting a morphotype is found that is temporally, but not morphologically, intermediate between an inferred ancestor–descendant pair, or if a presumed descendant is found stratigraphically below its alleged ancestor, then the hypothesis is false. As Levinton (1983) has observed, to dismiss information gathered from the analysis of the temporal sequence of fossils is to reject valuable information in trying to assess phylogenies. While it is never possible to prove ancestor–descendant relationships, biogeographic and stratigraphic criteria allow a more accurate assessment to be made than cladistic techniques, where only morphological criteria based on adult morphologies are used.

In many groups of marine organisms, in particular, ontogenies of fossil species are well known. If it is possible to compare the ontogenies of species pairs that are both morphologically and phylogenetically closely related and have similar, but not entirely overlapping ranges, then it is possible to assess whether the stratigraphically younger species is a paedomorph or a peramorph. Should the last stratigraphic occurrence of a presumed ancestor be separated from the earliest stratigraphic occurrence of its presumed descendant, it is still possible to determine whether the relationship is paedomorphic or peramorphic. This holds even if another taxon is subsequently found in the time gap between the original pair of species.

In assessing heterochrony from the fossil record other factors are also taken into consideration, such as biogeography. This is particularly important if it can be shown that evolutionary relationships are being assessed between endemic species. In addition to all the information that can be gathered from the fossil record, it is also possible to undertake an ancillary cladistic analysis. Such methodology was carried out by Miyazaki and Mickevich (1982). They found a high degree of congruence between the two techniques of cladistic and traditional paleontological analyses.

While the fossil record may not be able to supply us with much information on the underlying genetic causation of heterochrony, it can show patterns of morphological evolution congruent with the hypothesis of heterochrony. In particular, it can demonstrate the frequency of heterochrony in particular groups. It is also possible, as I shall argue in Section 2.2, to assess the frequencies of particular heterochronic processes.

2.2. Problems of Terminology

One problem encountered in this study has been that outmoded or inaccurate terminology has been used in many of the papers. However, it is generally possible to interpret the data and place them within the revised nomenclatural scheme proposed by Alberch *et al.* (1979) (see also McNamara, 1986a; Dommergues *et al.*, 1986). When trying to interpret heterochrony from the fossil record, this scheme has its problems. It suffers from the fact that it combines the elements of size, shape, and time, and not always in a consistent manner. In terms of the hierarchy of heterochrony outlined by McNamara (1986a, Fig. 1), heterochrony itself, as its name implies, involves a change in time; the two possible resultant phenomena, paedomorphosis and peramorphosis, describe shape; the processes that cause these shape changes generally relate to change in shape relative to time, although in some there is a consequent size change. Thus, for example, progenesis implies maturation at an earlier (time) morphological stage of development. Although time-controlled, it has distinctive morphological effects on both size and shape. Similarly with hypermorphosis: although its name suggests a shape difference, it is time-controlled. Like progenesis, change in time affects size and shape. These are generally the only parameters that paleontologists have to deal with.

McKinney (this volume) and Jones (this volume) have both addressed the problem of timing in heterochronic studies involving fossils. The frequent absence of any data on time can be a particular problem when trying to interpret neoteny and acceleration in the fossil record. These processes involve shape and time. The situation is similar with regard to pre- and postdisplacement, though here the displacement can be considered relative to overall body size. In other words, a structure appears at a relatively smaller or larger body size. Size can, in theory, change independently. This is reflected, as I have noted elsewhere (McNamara, 1986a), in some neotenic forms having the same body size as their ancestors, while others are often larger.

The methodology for identifying these processes in the fossil record has been based on the assumption that the rate of change in body size is the same in both ancestor and descendant, with only shape increased or decreased. Thus, if a presumed ancestor–descendant pair shows, for instance, a seemingly neotenic trait in the descendant, size undergoing no relative change, then one can assume true neoteny. However, one could argue that this less morphologically developed trait, which occurs at the same body size in both ancestor and descendant, could conceivably occur by an increase in rate of body size increase, with no change in rate of morphological development at all. The trait will thus appear delayed in development at the same body size. This is used as an argument against identifying processes such as neoteny and acceleration from the fossil record. We can have recourse to McKinney's (this volume) terms "allometric neoteny," "allometric acceleration," and so on. But is it not possible to identify "real" neoteny from the fossil record, even if direct measurement of the time component is not possible? I believe that it may be.

In the scenario outlined above (apparent neoteny resulting from an increase in rate of increase in body size, as it dissociates from shape change), it would be expected that any identified heterochronic lineages would all be paedomorphic, as they had less time to undergo morphological change, if neoteny proceeded at the same rate as in the ancestral form. However, in many instances, such as the heterochronic changes documented in a number of spatangoid echinoids (McNamara, 1987c), there is little change in maximum body size attained by species in lineages, yet some traits are paedomorphic, others peramorphic. Between some species pairs one trait may show no heterochronic change, while another does. The next pair will then show heterochronic changes in both traits, one paedomorphic, the other peramorphic. Such dissociated shape changes, with both paedomorphic and peramorphic traits, could not result simply from speeding up or slowing down the rate of attainment of the same body size. They must reflect true heterochronic changes in rate of shape change.

This can be easily demonstrated between a pair of species of the spatangoid echinoid *Lovenia*. At size x, species 1 has developed, say, 6 adoral and 11 aboral primary tubercles. The descendant species 2 at size x has 8 adoral and 7 aboral tubercles. If it had simply been a case of species 2 attaining size x quicker than species 1, with no change in the rate of tubercle development, then it would have had fewer adoral and aboral tubercles than species 1. Similarly, if it had taken longer to attain the size, both aboral and adoral tubercles would have increased in number. As it is, one set increases, the other decreases: one trait is peramorphic, the other paedomorphic. This can only have occurred if there had been a real change in the rate of tubercle production. These are real examples of, in the one case, acceleration, in the other, neoteny.

It is becoming increasingly clear that such dissociated heterochrony is probably the rule, rather than the exception, in many groups of organisms. The tendency in the past has been to identify a "paedomorphic" or a "peramorphic" species. This can only be done where progenesis or hypermorphosis have occurred. The majority of heterochronic changes, as I shall discuss, are probably neoteny, acceleration, and pre- and postdisplacement. Where more than one of these processes can be seen to have affected specific traits, then, on the basis of the argument I have presented, one can with reasonable confidence assess which heterochronic processes have occurred.

3. Heterochrony in Fossil Invertebrates

3.1. Arthropods

3.1.1. Trilobites

Heterochrony has played an important role in the evolution of trilobites (McNamara, 1978, 1981a,b, 1983a, 1876b). In the Cambrian, 14 of the 17 described

TABLE I. Heterochrony in Trilobites

Name	Morphosis	Process	Taxonomic level	Age	Reference
Ampyxina	Paedo	Progenesis	Genus	Late Ordovician	Brezinski (1986)
"Phacopids"	Paedo	Neoteny	Order	Ordovician	Clarkson (1979)
Missisquoia	Paedo	Neoteny	Genus	Early Ordovician	Fortey and Shergold (1984)
Mucronaspis	Paedo	?Neoteny	Genus	Late Ordovician	Lu and Wu (1983)
Olenellus	Paedo	Progenesis	Species	Early Cambrian	McNamara (1978)
Tretaspis	Paedo	?Neoteny	Subspecies	Late Ordovician	McNamara (1979), Owen (1980)
Galahetes	Paedo	Progenesis	Genus	Middle Cambrian	McNamara (1981a)
Thoracocare	Paedo	Progenesis	Genus	Middle Cambrian	Robison and Campbell (1974)
Acanthopleurella	Paedo	Progenesis	Genus	Early Ordovician	Fortey and Rushton (1980)
Vanuxemella	Paedo	Progenesis	Genus	Middle Cambrian	McNamara (1983a)
Xystridura	Paedo	Neoteny	Species	Middle Cambrian	McNamara (1981a)
Pseudogygites	Paedo	Neoteny	Genus	Middle Ordovician	Ludvigsen (1979)
Daguinaspis	Paedo	Progenesis	Genus	Early Cambrian	Hupé (1953)
Choubertella	Paedo	Progenesis	Genus	Early Cambrian	Hupé (1953)
Tonkinella	Paedo	Progenesis	Genus	Middle Cambrian	Hupé (1953)
Corynexochids	Paedo	—	Order	Middle Cambrian	Robison (1967)

Fallotaspis	Paedo	?Neoteny	Species	Early Cambrian	McNamara (1986b)
Zacanthoidids	Paedo	Progenesis	Genus	Middle Cambrian	McNamara (1986b)
Oryctocephalus	Paedo	Progenesis	Genus	Middle Cambrian	McNamara (1986b)
Metacalymene	Paedo	Neoteny	Genus	Silurian	Siveter (1979)
Calymeninae	Paedo	Neoteny	Subfamily	Silurian	Siveter (1980)
Leonaspis	Paedo/pera	Neoteny/acceleration	Species	Silurian	Chatterton and Perry (1983)
Stelckaspis	Paedo/pera	Neoteny/acceleration	Species	Silurian	Chatterton and Perry (1983)
Ceratocephala	Pera	Acceleration	Species	Silurian	Chatterton and Perry (1983)
Ceratocephalina	Pera	?Acceleration	Species	Silurian/Devonian	Chatterton and Perry (1983)
Redlichiids	Pera	Acceleration	Genus	Early Cambrian	McNamara (1986b)
Olenellids	Pera	Acceleration	Genus	Early Cambrian	McNamara (1986b)
Irvingella	Pera	Acceleration	Species	Late Cambrian	McNamara (1986b)
Amphilichas	Pera	Predisplacement	Genus	Early Ordovician	McNamara (1986b)
Acanthopyge	Pera	Predisplacement	Genus	Early Ordovician	McNamara (1986b)
Remopleuridids	Pera	Predisplacement	Genus	Early Ordovician	McNamara (1986b)
Cheirurids	Pera	Predisplacement and acceleration	Genus	Middle Ordovician	McNamara (1986b)
Cloacaspis	Pera	?Acceleration	Species	Early Ordovician	Fortey (1974)
Poronileus	Pera	—	Species	Early Ordovician	Fortey (1975)
Trinucleids	Pera	Acceleration	Genus	Ordovician	Hughes *et al.* (1975)
Tretaspis	Pera	?Acceleration	Subspecies	Late Ordovician	Ingham (1970)

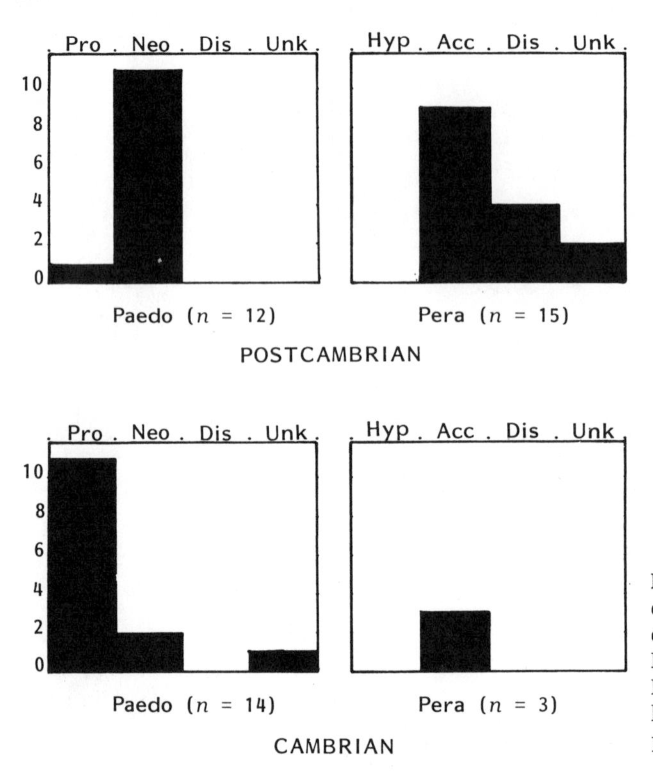

Figure 1. Histograms of frequency of described examples of heterochronic processes in trilobites. PRO, progenesis; NEO, neoteny; DIS, displacement; UNK, unknown process; HYP, hypermorphosis; ACC, acceleration.

examples show paedomorphosis (Table I). Of these, 11 are due to progenesis. Some lineages, such as the Xystridurinae, show both progenesis and neoteny (McNamara, 1981a). The progenesis in this subfamily is not the normal form [termed terminal progenesis by McNamara (1983a)], but is sequential progenesis. This occurred by shortened intermolt periods. These different forms of progenesis probably resulted from changes in the regulation of different hormonal systems: terminal progenesis by cessation of production of juvenilizing hormone in the descendant, at a smaller size; sequential progenesis by earlier production of an ecdysonelike hormone within each molt sequence.

Only three examples of peramorphosis have been documented in Cambrian trilobites: in *Redlichia*, some olenellids, and *Irvingella* (McNamara, 1986b). In post-Cambrian trilobites the frequency of peramorphosis increased greatly. Indeed, with further studies it is likely that peramorphosis will be shown to be more common than paedomorphosis. The so-called "cryptogenetic" appearance of many new trilobite families in the early part of the Ordovician, such as lichids, remopleuridids, phacopids, cheirurids, illaenids, and odontopleurids, occurred, in all likelihood, by an increase in the frequency or selection of forms that had undergone both predisplacement and acceleration (McNamara, 1986b). It is interesting to note that the ratio of progenesis to neoteny (Fig. 1) changed completely, with few examples of progenesis having been documented from post-Cambrian strata.

The change in frequency of heterochronic processes in trilobites through the early Paleozoic was no doubt in part affected by ecological factors, such as the changing nature of biota, in particular the rise of vertebrates. However, it is quite possible that there was also a general "improvement" in the regulation of the developmental system of the organisms. This can be demonstrated, for instance, by the progressive reduction in the degree of variability in segment number at successively higher taxonomic levels through the early Paleozoic. The number varies intraspecifically in many early Cambrian taxa, but is constant at the ordinal level in some post-Cambrian forms.

3.1.2. Crustaceans

Few examples of heterochrony have been described in fossil crustaceans. However, Gramm (1973, 1985) has documented cases of paedomorphosis in cavellinid ostracodes. Gramm has shown how in phylogenetically early forms there are two rows of muscle scars in juveniles. Through ontogeny the number of rows increases. In descendants the biserial condition persists to progressively later in ontogeny. The paedomorphic process appears to have been neoteny. Schram and Rolfe (1982) have also demonstrated paedomorphosis in some Late Pennsylvanian euthycarcinoid arthropods. They consider that the dramatic reduction of limbs noted between sottyxerxids and euthycarcinoids may be a paedomorphic feature, perhaps even an interim stage en route to hexapody.

3.2. Echinoderms

3.2.1. Echinoids

The documentation of heterochrony in echinoids has centered on studies of Cenozoic irregulars. Although there is still a long way to go in our understanding of the patterns and processes of heterochrony in echinoids, because of the paucity of described examples, there is an indication of a difference in the heterochrony between regular and irregular echinoids. Although few examples of heterochrony in regular echinoids have been described, it is possible to recognize a number of paedomorphic forms, such as *Eodiadema, Mesodiadema, Serpianotiaris*, the Saleniidae, and the Tiarechinidae. All are small, progenetic forms. In irregulars, however, progenesis does not appear to have been particularly common, although where it does occur it seems to have resulted in the evolution of major groups of taxa, such as clypeasteroids (Phelan, 1977) and neolampadids (Philip, 1963). There can be little doubt that both paedomorphosis and peramorphosis have been major factors in the evolution and diversification of irregular echinoids. Most examples have been documented (McNamara, 1987b) in spatangoids, holasteroids, oligopygoids, and clypeasteroids (Table II).

Heterochrony is naturally affected by the nature of an organism's ontogeny, and in particular the degree and nature of morphological change. Between many regulars and irregulars there is a difference in the pattern of growth (McNamara, 1987b). The test of regular echinoids grows largely by the addition of new plates, which then increase in size. Differences in allometries between plates are small. Variation in the rate of plate production produces either hypermorphic or progenetic descendants. Rate of plate production also varies among different groups

TABLE II. Heterochrony in Echinoids

Name or feature	Morphosis	Process	Taxonomic level	Age	Reference
Clypeasteroids	Paedo	Progenesis	Order	Early Tertiary	Phelan (1977)
Breynia	Paedo	Neoteny	Species	Miocene–Recent	McNamara (1982b)
Eodiadema	Paedo	Progenesis	Genus	Early Jurassic	Smith (1981)
Irregular teeth	Paedo	Progenesis	Order	Jurassic	Smith (1980)
Mesodiadema	Paedo	Progenesis	Genus	Triassic	This chapter
Serpianotiaris	Paedo	Progenesis	Genus	Triassic	This chapter
Saleniidae	Paedo	Progenesis	Family	Cretaceous–Tertiary	This chapter
Tiarechinidae	Paedo	Progenesis	Family	Late Triassic	This chapter
Neolampadoids	Paedo	Progenesis	Suborder	Tertiary	Philip (1963)
Oligopygus	Paedo/pera	Neoteny, displacement, hypermorphosis	Species	Eocene	McKinney (1984)
Echinolampas	Paedo/pera	Progenesis/acceleration	Species	Tertiary	McNamara and Philip (1980b)
Hemiaster	Paedo/pera	Neoteny/acceleration	Species	Eocene–Miocene	McNamara (1987a)

Psephoaster	Paedo/pera	Neoteny/acceleration	Species	Eocene–Miocene	McNamara (1987a)
Lovenia	Paedo/pera	Neoteny/acceleration	Species	Oligocene–Miocene	McNamara (1987c)
Infulaster/ Hagenowia	Pera	Acceleration	Genus/species	Late Cretaceous	Dommergues *et al.* (1986)
Tube feet	Pera	Acceleration	Genus	Cretaceous–Tertiary	McNamara (1984)
Protenaster	Pera	Acceleration	Species	Eocene–Recent	McNamara (1985a)
Schizaster	Pera	Hypermorphosis/acceleration	Species	Paleocene–Recent	McNamara and Philip (1980a)
Pericosmus	Pera	Acceleration	Species	Oligocene–Miocene	McNamara and Philip (1984)
Mesozoic echinoids	Pera	?Predisplacement	Order	Mesozoic	Raff *et al.* (1984)
Peronella	Pera	Acceleration	Species	Pliocene–Recent	McNamara (1987b)
Rotuloidea/ Heliophora	Pera	Acceleration	Genus/species	Miocene–Recent	McNamara (1987b)

of irregulars. Much of the change in test morphology during ontogeny occurs by allometric growth of some of these plates. There has been pronounced dissociation of plate growth, resulting in pronounced differential plate allometries. This is particularly evident in spatangoids (McNamara, 1987b). Hence, slight changes in these allometries can produce substantial morphological differences in descendants, resulting from either neoteny or acceleration.

3.2.2. Crinoids

While only five examples of heterochrony in crinoids have been described in the last 10 years, this most probably reflects more the interest of researchers than the true frequency of heterochrony. Examples range in age from the Ordovician to the Pennsylvanian and involve four examples of paedomorphosis and one of peramorphosis. In this latter case an unusually rapid rate of development of new branchials and length of arms by acceleration has been described in *Promelocrinus* (Brower, 1976). Change in the length of other food-gathering structures, such as pinnules, has been recorded in disparids (Frest *et al.*, 1979). Here there has been a paedomorphic reduction.

Paedomorphosis has probably played an important role in the evolution of microcrinoids (Lane and Sevastopulo, 1982; Lane *et al.*, 1985). In these crinoids maturity is reached at a very small size, indicating the operation of progenesis. The result is the production of fewer plates. For example, codiacrinoids develop no radials, or if they do appear, it is very late in ontogeny (Lane and Sevastopulo, 1982). A similar paedomorphic loss of plates has been documented in *Bactrocrinites* by McIntosh (1979). He suggests that this indicates that a monocyclic genus could have arisen from a dicyclic inadunate form by the loss of basals or infrabasals. All of the heterochronic changes documented in crinoids affect the rate of plate production, either by acceleration or neoteny, or by progenetic reduction in plate number.

3.2.3. Blastoids

Both paedomorphosis and peramorphosis have been important factors in blastoid evolution. Indeed, Sprinkle (1980) and Brett *et al.* (1983) argued that the blastoids arose from coronoids by paedomorphosis. This occurred by a simplification of the complex brachiolar structures of coronoids. Subsequent evolution in blastoids involved a peramorphic increase in ambulacral development. Such acceleration has been described in the Devonian blastoid *Eleutherocrinus* (Millendorf, 1979). Morphological changes from the typical blastoid plan caused by alteration of the genetic regulatory system governing ontogenetic development occurred in *Eleutherocrinus* by a change in rate of growth on several plate axes. In addition to acceleration of some axes, others show a paedomorphic retardation in growth, such as the development of deltoid axes of the D-ray radial and side plates on the D-ray ambulacrum. The acceleration of growth of zygous basals was an adaptation to low-level suspension feeding.

At the species level both paedomorphosis and peramorphosis have been documented in *Pentremites* lineages (Waters *et al.*, 1985), resulting in the formation of many different cup shapes. Heterochronic change in blastoids and coronoids primarily affected the allometric development of individual plates and not the rate of plate generation, as was the case in crinoids. The consequence of this was the evolution of morphotypes adapted to new feeding strategies.

3.2.4. Edrioasteroids

Although only three cases of heterochrony have been recorded, all three are examples of paedomorphosis produced by progenesis. The cyathocystids are considered by Bockelie and Paul (1983) to have arisen paedomorphically from the Cambrian stromatocystitids by attachment of the ventrodorsal and failure to develop further aboral or ambulacral plates. Such reduction in plate generation, combined with the small adult size, suggests progenesis. Similarly, the dominance of the primordial cover plates in the oral region in the Carneyellidae, along with an extremely simple arrangement of ambulacral cover plates, both features of juvenile lebetodiscinids (Smith, 1985), indicates progenesis. The detailed studies by Bell (1976) and Sprinkle and Bell (1978) of the stages of development of *Timeischytes* compared with other isorophids showed the importance of progenesis in the evolution of this genus.

3.2.5. Carpoids, Asteroids, and Cystoids

Few examples of heterochrony have been described in any of these groups. The only indication in carpoids is the development of an unusually large anal lobe in the early stages of ontogeny in *Iowacystis* (Kolata *et al.*, 1977). In the presumed ancestor, *Scalenocystites*, it is smaller at comparable thecal size. The lobe therefore increased in size by acceleration. Similarly, the scalene outline in *Iowacystis* appears at a larger thecal size. It is possible that variation in plate size in other carpoids occurred by heterochronic changes to plate allometries.

Spencer and Wright (1966) noted that genera within valvatid asteroids are small, with short, wedge-shaped arms and, on the aboral surface, few large ossicles. They consider these features to be paedomorphic. These genera occur in the Ordovician. Later forms show an acceleration in rate of ossicle generation. Heterochronic changes to growth allometries also occur in Late Cretaceous goniasterids. Such changes in other asteroids are probably also attributable to heterochrony.

Reduction in the number of thecal plates in some cystoid genera within the Diploporita (Bockelie, 1984) may also be interpreted as being heterochronic changes. In some genera retention of a small juvenile number of plates into the adult stage is accompanied by an increase in the relative size of these plates. In *Haplosphaeronis* the increase in number of grooves per ambulacrum in stratigraphically younger species indicates peramorphosis. Variation in thecal growth rates in *Haplosphaeronis* resulted in important structural changes, such as an increase in the periporal floor, causing an improvement in respiratory capability.

3.2.6. Synthesis

Allowing for the limited data available, it is possible to infer that the different frequencies of paedomorphosis and peramorphosis and of processes in all echi-

Table III. Frequency of Heterochrony in Echinoderm Classes and Its Relationship to Mode of Skeletal Growth

Group	Paedomorphosis	Peramorphosis	Processes	Plate production	Plate growth
Crinoids	4	1	Neoteny, progenesis, acceleration	×	
Carpoids	0	1	Acceleration		×
Echinoids					
Regular	5	0	Progenesis	×[a]	×[b]
Irregular	9	13	All, but progenesis rare	×[b]	×[a]
Cystoids	1	1	Neoteny, acceleration	×	
Edrioasteroids	3	0	Progenesis	×	
Blastoids	4	3	Neoteny, acceleration		×
Asteroids	1	0	Progenesis	×	

[a] Major.
[b] Minor.

noderm classes are related to the nature of their ontogenetic development (Table III). Thus, progenesis seems to have occurred more often in those echinoderm classes that grow fundamentally by the addition of many new plates, such as crinoids, edrioasteroids, and perhaps asteroids. On the other hand, neoteny and acceleration may have occurred with greater relative frequency in carpoids and blastoids, as in some irregular echinoids, where plate allometries played a more significant role in growth.

In many of these echinoderm classes heterochronic changes have resulted in pronounced and rapid morphological changes, often resulting in changes in feeding strategies. Examples include burrowing in irregular echinoids, development of a greater variety of food-gathering apparatus in blastoids, and changes in arm length and pinnule number in crinoids. Perhaps one of the more interesting aspects of recent work on heterochrony in echinoids is the realization that dissociated heterochrony is a common occurrence, some traits of a species being paedomorphic, others peramorphic (McKinney, 1984; McNamara, 1987a–c).

3.3. Molluscs

3.3.1. Ammonoids

Studies of heterochrony in molluscs have been predominantly on ammonites. The dominance of paedomorphosis in ammonites that the data from studies made in the last 10 years seems to indicate, with 16 of the 25 examples (Table IV) of heterochrony being paedomorphosis, is probably more apparent than real, being to a large degree influenced by the reaction to the all-pervasive influence of recapitulation in ammonite studies made at the turn of the century. Recent work, such as that by Dommergues (1986), Dommergues *et al.* (1986), and Landman (this volume), seems to indicate similar frequencies of both peramorphosis and paedomorphosis. Some of these studies have revealed the existence of dissociated heterochrony.

The structures in ammonoids that are affected by heterochronic changes are mainly the suture pattern and ornamentation of the shell. Landman (this volume) has noted how sutural changes are mainly reflected at high taxonomic levels, while heterochronic changes in the ornamentation of the shell, as well as shell shape, occur principally at lower taxonomic levels, that is, generic and below. Apart from the occurrence of a number of progenetic forms (McNamara, 1985b) (Table IV), there is a general trend in lineages of nonheteromorph ammonites for suture lines to show peramorphosis by acceleration and predisplacement. However, in heteromorphs, paedomorphosis is more common (Wiedmann and Kullmann, 1980). The relatively large number of examples of progenesis (Table IV) is perhaps not a true reflection of the ubiquity of this process, but is partly a function of the relative ease of identifying these small forms. In all likelihood, hypermorphosis was probably equally as common.

3.3.2. Bivalves

Three-quarters of the 12 examples of heterochrony in fossil bivalves documented in the last decade are of paedomorphosis. The majority of these are progenetic (Table V). It is possible that, as with the ammonoids, this is partly a reflection of the relative ease of identifying progenetic forms. The counterpart of progenesis, hypermorphosis, can be recognized in bivalve lineages. In fact, hypermorphosis has operated in all of the three examples of peramorphosis, in *Myalina* and bakevilliid genera (Stanley, 1972) and *Chesapecten* (Miyazaki and Mickevich, 1982). Heterochrony in fossil bivalves has only been documented at the generic level and below. However, it is likely that it played a significant role in the evolution of morphological novelties at higher taxonomic levels (Waller, 1986).

3.3.3. Gastropods

Relatively few examples of heterochrony in fossil gastropods have been described. The few that have suggest that it operated at a number of taxonomic levels, from subspecific (Gould, 1969) to the family level, with the Fissurellidae (McLean, 1984), Scissurellidae, Eotomariidae, and Temnotropidae (Batten, 1975) all possibly have arisen by neotony. One of the few examples of heterochrony in gastropods at the species level to have been documented in recent years, in *Glossaulax*, shows dissociated heterochrony (Majima, 1985).

3.4. Brachiopods

Recent descriptions of heterochrony in brachiopods have been overwhelmingly dominated by paedomorphosis, suggesting that, in this instance, it is a real phenomenon. Of 30 examples of heterochrony, 25 are of paedomorphosis (Table VI), and of these, 16 were the result of progenesis. Not only has progenesis been

TABLE IV. Heterochrony in Ammonoids

Name	Morphosis	Process	Taxonomic level	Age	Reference
Reynesocoeloceratinae	Paedo	Neoteny	Genus	Early Jurassic	Dommergues (1986)
Gaudryceras	Paedo	Neoteny	Species	Late Cretaceous	Henderson and McNamara (1985)
Brahmaites	Paedo	Neoteny	Genus	Late Cretaceous	Henderson and McNamara (1985)
Falloticeras	Paedo	Progenesis	Genus	Late Cretaceous	Kennedy and Cooper (1977)
Algericeras (Sakondryella)	Paedo	?Neoteny	Genus	Late Cretaceous	Kennedy and Wright (1981)
Scaphites	Paedo	Progenesis	Species	Late Cretaceous	Mancini (1978a)
Adkinsia	Paedo	Progenesis	Species	Late Cretaceous	Mancini (1978a)
Naramoceras	Paedo	Progenesis	Genus	Early Cretaceous	McNamara (1985b)
Flickia	Paedo	Progenesis	Genus	Late Cretacous	Wright and Kennedy (1979)
Pseudoxybeloceras/ Solenoceras	Paedo	Progenesis	Genus/species	Late Cretaceous	Ward and Mallory (1977)

Salaziceras	Paedo	Progenesis	Genus	Early Cretaceous	Wright and Kennedy (1979)
Protacanthoceras	Paedo	Progenesis/ neoteny	Genus/species	Late Cretaceous	Wright and Kennedy (1980)
Euhystrichoceras	Paedo	Progenesis	Genus	Early Cretaceous	Kennedy (1977)
Eoscaphites	Paedo	Progenesis	Genus	Early Cretaceous	Kennedy (1977)
Baculitids	Paedo	Progenesis	Family	Late Cretaceous	Kennedy (1977)
Liparoceras/ Beanbiceras/ Aegoceras	Paedo	Neoteny	Genus/species	Jurassic	Dommergues *et al.* (1986)
Pavlovia	Pera	?Predisplacement	Genus	Late Jurassic	Cope (1978)
Reynesocoeloceratinae	Pera	Acceleration	Genus	Early Jurassic	Dommergues (1986)
Pachydiscus	Pera	Acceleration	Species	Late Cretaceous	Henderson and McNamara (1985)
Subperrinites/ Perrinites	Pera	Acceleration	Species	Early Permian	Tharalson (1984)
Aegoceras	Pera	Acceleration/ hypermorphosis	Species	Jurassic	Dommergues *et al.* (1986)
Amaltheus	Pera	Acceleration	Genus/species	Jurassic	Dommergues *et al.* (1986)

TABLE V. Heterochrony in Bivalves

Name	Morphosis	Process	Taxonomic level	Age	Reference
Gryphaea	Paedo	Progenesis	Species	Jurassic	Hallam (1978)
Gryphaea	Paedo	Neoteny	Species	—	Hallam (1982)
"Bivalves"	Paedo	Neoteny	Genus	Late Cretaceous	Mancini (1978a)
Nuculites	Paedo	Progenesis	Species	Late Ordovician	Snyder and Bretsky (1971)
Brachiodontes	Paedo	—	Genus	—	Stanley (1972)
Arcoids	Paedo	Progenesis	Genus	—	Stanley (1972)
Inoceramus (Mytiloides)	Paedo	?Progenesis	Subgenus	Cretaceous	Stanley (1972)
Crassinella	Paedo	Progenesis	Genus	Cenozoic	Stanley (1972)
Carditids	Paedo	Progenesis	Genus	Post-Paleocene	Stanley (1972)
Myalina	Pera	Hypermorphosis	Species	Early Pennsylvanian	Stanley (1972)
Bakevilliidae	Pera	?Hypermorphosis	Genus	Permian–Cretaceous	Stanley (1972)
Chesapecten	Pera	Acceleration/hypermorphosis	Species	Miocene–Pliocene	Miyazaki and Mickevich (1982)

recorded at the species and genus levels, particularly in the evolution of many small brachiopod genera, but it has been an important factor in the evolution of some suborders, such as the Thecideidinea (Backhaus, 1959; Rudwick, 1968; Pajaud, 1970; Baker, 1983), and superfamilies, such as the Orthacea and Strophomenacea (Williams and Hurst, 1977). Of the three superfamilies and five suborders thought to have arisen by heterochrony (Table VI), all show paedomorphosis. Heterochronic changes in brachiopods affect all characters of the shell, its shape and ornamentation as well as internal structures such as lophophore supports and shell microstructure. Examples have been documented from the Ordovician to the Recent (Table VI).

3.5. Microfossils

3.5.1. Foraminifers

One of the few papers to exploit the potential for heterochronic studies in foraminifers is that by Scott (1982) on the evolution of *Globorotalia puncticulata*. He observed that ancestral forms developed a keel late in ontogeny. Through a series of intermediate forms the ancestral *G. mizoea* gives rise to a paedomorphic

TABLE VI. Heterochrony in Brachiopods

Name	Morphosis	Process	Taxonomic level	Age	Reference
Enallothecida	Paedo	Progenesis	Species	Mid Jurassic	Baker (1983)
Thecideidines	Paedo	Progenesis	Genus	Jurassic	Baker (1983, 1984), Pajaud (1970)
Thecideidines	Paedo	Progenesis	Genus	Cretaceous	Backhaus (1959)
Thecideidines	Paedo	Progenesis	Genus	Permian	Rudwick (1968)
Mimikonstantia	Paedo	Progenesis	Genus	Mid Jurassic	Baker and Elston (1984)
Thecidiopsis	Paedo	Progenesis	Genus	Late Cretaceous	Williams (1973)
Neothecidella	Paedo	Progenesis	Genus	Jurassic/ Cretaceous	Baker and Laurie (1978)
Tegulorhynchia	Paedo	Progenesis	Species	Tertiary	McNamara (1983b)
Amphithyris	Paedo	Progenesis	Genus	Recent	Williams and Hurst (1977)
Gwynia	Paedo	Progenesis	Genus	Recent	Williams and McKay (1979)
Pumilus	Paedo	Progenesis	Genus	Recent	Williams and Hurst (1977)
Thaumotosia	Paedo	Progenesis	Genus	Recent	Williams and Hurst (1977)
Thecospira	Paedo	Progenesis	Genus	Triassic	Williams and Hurst (1977)
Craniaceans	Paedo	?Progenesis	Superfamily	Ordovician–Recent	Williams and Hurst (1977)
Various	Paedo	Progenesis	Genus	Cretaceous	Surlyk (1972)
Lacazella	Paedo	Progenesis	Genus	Tertiary	Pajaud (1970)
Thecidellina	Paedo	Progenesis	Genus	Tertiary	Pajaud (1970)
Dalmanella	Paedo	?Neoteny	Species	Mid Ordovician	Lockley (1983)
Tegulorhynchia	Paedo	Neoteny	Species	Tertiary	McNamara (1983b)
Notosaria	Paedo	Neoteny	Genus	Tertiary	McNamara (1983b)
Strophomenaceans	Paedo	Neoteny	Superfamily	Ordovician	Williams and Hurst (1977)
Productines	Paedo	Neoteny	Suborder	Ordovician	Williams and Hurst (1977)

(continued)

TABLE VI. (*Continued*)

Name	Morphosis	Process	Taxonomic level	Age	Reference
Atrypidines	Paedo	Neoteny	Suborder	Ordovician	Williams and Hurst (1977)
Orthaceans	Paedo	—	Superfamily	Ordovician	Williams and Hurst (1977)
Chonetidines	Paedo	—	Suborder	Ordovician	Williams and Hurst (1977)
Oldhaminidines	Paedo	—	Suborder	Permian/ Carboniferous	Williams and Hurst (1977)
Eospirigerina	Pera	Hypermorphosis/ ?acceleration	Genus	Early Silurian	Copper (1982)
Protatrypa	Pera	?Acceleration	Genus	Early Silurian	Copper (1982)
Atrypoidea	Pera	Predisplacement	Species	Late Silurian	Jones (1979)

form, *G. puncticulata,* which lacks the keel. As this species is smaller, Scott suggests that the process was progenesis. Interestingly, Scott calculated that the change from keeled to unkeeled forms occurred in only about 90,000 years. The descendant, unkeeled *G. puncticulata,* then persisted for 2.6 million years. Scott considered that the paedomophic evolution occurred at a time of climatic and biogeographic change.

Insufficient research has been carried out to determine the relative frequencies of paedomorphosis and peramorphosis. Some indication of the ubiquity of heterochrony in foraminiferal evolution is indicated by comments of Brasier (1982). He noted that in the Textulariina, biserial growth appeared later than uniserial, having a possible secondary origin from "retardation" of the juvenile stage and acceleration of the ephebic stage in multiform tests. He also suggested that *Haplophragmoides* may have evolved by the retardation in onset of development of the uniserial stage in *Ammobaculites,* or by reduced growth translocation in *Trochammina.* Clearly, the potential for studies of heterochrony in foraminifers is great.

3.5.2. Conodonts

One of the problems in assessing the role of heterochrony in this group is, as Broadhead and McComb (1983) pointed out, that little research has been carried out on the analysis of evolutionary trends and processes. The few studies that have been carried out indicate that both peramorphosis and paedomorphosis occur in conodonts. For instance, Broadhead and McComb (1983) described the paedomorphic evolution of *Icriodus,* probably from *Pedavis.* They suggested that over just a few generations there was a paedomorphic reduction in the two anteriorly directed lateral processes. As conodonts add denticles through ontogeny, it is possible that phylogenetic increase or decrease in denticle number occurred by heterochrony (Merrill and Powell, 1980). Paedomorphosis occurs in epigondolellid conodonts, with denticle reduction (Mosher, 1973), and in *Neognathodus* (Merrill, 1972), where a paedomorphocline of simplification of ornamentation has been described. In *Histiodella*, on the other hand, a peramorphocline exists among a number of species that show a progressive increase in denticulation, from smooth to minutely serrate to prominently denticulate (McHargue, 1982). This change probably occurred by acceleration.

3.6. Colonial Organisms

3.6.1. Graptolites

Because of the effect of variation in both individual ontogenies and in the astogenetic development of the colony as a whole, heterochrony is less easy to establish in colonial organisms. Pandolfi (this volume) has argued that astogenetic heterochrony occurs more often than ontogenetic heterochrony. Most examples of heterochrony have been described in graptolites. Elles (1923) observed that "progressive development is first indicated in the proximal, therefore youthful, region of the colony." This results in peramorphosis. Where innovations first appeared in the distal region, then spread to the proximal region, this showed paedomorphosis. While a number of examples of paedomorphosis in graptolites have been described, principally by Rickards (1977) and Finney (1986), Urbanek (1960) earlier stressed the problems of interpreting such colonial morphological changes as paedomorphosis. Colony shape can be affected by the time of budding, suggesting that pre- and postdisplacement may have been important factors in graptolite evolution. Urbanek interpreted these changes in timing as being under the influence of changing hormonal regimes. In this and other groups of organisms, changes in the regulation of hormonal secretion are likely to have been important factors in producing heterochronic changes.

Both peramorphosis and paedomorphosis have been described in graptolites. Urbanek (1973) described both effects in monograptid colonies. More recently, Finney (1986) documented peramorphosis by acceleration in the evolution of *Apoglossograptus* from *Glossograptus*, involving changes in the curvature of stipes. Progenesis appears to have occurred a number of times in graptolite evolution (Finney, 1986; Rickards, 1977). It is possible that heterochronic changes in the shape of graptolite colonies resulted in ecological separation into different parts of the water column. The cryptic appearance of many taxa was perhaps due to the dominant influence of heterochrony in graptolite evolution (Finney, 1986). In his review of heterochrony in colonial organisms, Pandolfi (this volume, Table II) documents six cases of peramorphosis and 12 of paedomorphosis. Peramorphosis affects thecal development primarily, while paedomorphosis was involved more with stipe reduction and early ontogenetic development.

3.6.2. Bryozoans

Although astogenetic peramorphosis and paedomorphosis in cheilostomatous and cyclostomatous bryozoans had been described (Dzik, 1975, 1981), there had been no specific research on the heterochronic processes that have occurred in bryozoan evolution until the work of Anstey (1987) and Pandolfi (this volume). While Schopf (1977) considered that heterochrony had "not yet been recognized to be significant for bryozoans," there is every indication that changes in the developmental history of colonies through time had a significant impact on bryozoan evolution. For example, Podell and Anstey (1979) observed in some colonies a developmental change from positive allometry in the protoecial cone to negative allometry in the ancestral disc. Heterochronic changes in colony growth can be affected by external factors. For instance, Podell and Anstey (1979) described how removal by borers of the monarchozooid, which releases a morphogenetic substance to regulate subcolony growth, can have a substantial effect on the whole growth pattern of the colony. Variation in the timing of release of such substances by changes in the genetic regulation of their production may thus have significant heterochronic effects.

Bryozoans show an enormous degree of developmental flexibility and thus the potential for changes in growth patterns by heterochronic changes. As Lidgard (1985) commented, bryozoan evolution involves changes in developmental patterns of both zooids and colonies, although to what extent these two are interdependent is not known. In his review of the impact of heterochrony in bryozoan evolution, Anstey documented 18 cases of peramorphosis and 22 of paedomorphosis. Pandolfi (this volume) showed seven and three cases, respectively.

While paedomorphosis and peramorphosis have been found to occur with similar frequencies in bryozoans, there is a tendency for each to act on different structural aspects of the organism, and at different times in the geological history of the group. Anstey (1987) has shown that paedomorphosis in Paleozoic bryozoans seems to have affected the colony more than individual zooids. Conversely, peramorphic features occur more often in zooidal structures than in the colony as a whole. Different structures are not equally affected by the two heterochronic expressions. In shallow water bryozoans peramorphosis mainly affects monticule and subcolony characters, whereas paedomorphosis occurs in characters related particularly to branch astogeny and zooidal density. It is possible that there is an interdependence between the two effects: peramorphosis in some features may occur only when other characters are paedomorphic. Anstey (1987) has also documented a predominance of peramorphosis in early Paleozoic bryozoans, while paedomorphosis became more frequent in later Paleozoic forms.

3.6.3. Corals

Few examples of heterochrony in fossil corals have been documented. Whether this is as a result of factors such as the irregular manner of septal insertion in some genera (Jull, 1976), making the identification of heterochronic descendants difficult, is not clear. The only detailed example of heterochrony in corals is the early study by Carruthers (1910), who documented a peramorphocline, formed by predisplacement in septal insertion, in *Amplexizaphrentis*.

4. Heterochrony in Fossil Vertebrates

4.1. Fish

The importance of changes in developmental regulation in vertebrate evolution has long been recognized. Recent studies on fish have shown both paedomorphosis and peramorphosis, particularly involving changing skull allometries and the timing of fusion of different skull plates. Variation in the timing of fusion of skull elements has been documented in the Pteraspididae (White, 1958; Elliott, 1984). Progressive delay in the fusion of plates occurred as the family evolved. Elliott demonstrated peramorphosis in the family between the Cyathaspidinae, where shield growth occurs at maturity, to the Anchipteraspidinae, where initiation of shield growth occurred in more than one center and before the onset of maturity, to the Pteraspidinae, where the growth was initiated early in ontogeny at small body size.

Heterochronic changes in the allometric growth pattern in the skull have been described in bothriolepid placoderms (Werdelin and Long, 1986), and osteolepiforms (Schultze, 1984). The typical ontogenetic postorbital growth in osteolepiforms is reduced in panderichthyids and in labyrinthodonts (Schultze and Arsenault, 1985). In *Osteolepis* the orbit size is constant during growth. There is a change in orbit size during ontogeny in *Eusthenopteron*. The relative size of the orbit in tetrapods is similar to that in juvenile osteolepiforms.

The evolution of postcranial skeletal elements was also affected, to a large degree, by the ontogenetic modification of existing morphological plans, resulting in the evolution of new adaptive patterns in rhipidistians and teleosts (Laerm, 1979, 1982). Likewise, the evolution of enamel from enameloid in the dermal skeleton of osteolepiform rhipidistian fishes occurred by regulatory changes (Meinke and Thomson, 1983). Individual structures, such as the timing of scale and fin spine development, have been affected by heterochrony. Scale development is significantly delayed in *Tarrasius* relative to *Paratarrasius* (Lund and Melton, 1982). The evolution of the growth pattern of hybodont fin spines also shows paedomorphosis, the transition from the juvenile costate condition to an adult tuberculate pattern occurring at a later stage of development in later genera, such as *Hybodus* and *Aerodon*, compared with the earlier *Asteracanthus* (Maisey, 1978).

Paedomorphosis appears to have played a particularly important role in the evolution of dipnoans. While Berman (1976a) has shown how the onset of growth acceleration in different regions of the skull occurs at different stages of growth in some Early Permian lungfish, Bemis (1984) has documented, in some detail, the paedomorphic changes that have taken place in dipnoan evolution. These mainly involve the loss of the heterocercal tail; fusion of median fins; reduction of fin rays; loss of cosmine; change in scale shape from rhombic to round; reduction in ossification; and an increase in cell size. Bemis suggested that accompanying the increase in cell size there was a corresponding increase in DNA content. A similar correlation of high DNA content and large cell size has also been noted in some paedomorphic salamanders (Morescalchi and Serra, 1974; Morescalchi, 1979). As cell size influences the length of the cell cycle, there is a possible mechanistic link between cell size and developmental rates in both lungfish and salamanders. It will be interesting to see if trends in increasing cell size in other groups of organisms show any such correlation with a paedomorphic morphology.

4.2. Amphibians

Most observations of heterochrony in fossil vertebrates have been made on amphibians. The overwhelming number of cases are of paedomorphosis, with 17

TABLE VII. Heterochrony in Amphibians

Name or feature	Morphosis	Process	Taxonomic level	Age	Reference
Doleserpeton	Paedo	Progenesis	Genus	Early Permian	Bolt (1977)
Tersomius	Paedo	Progenesis	Genus	Early Permian	Bolt (1977)
Branchiosaurus	Paedo	Progenesis	Genus	Permian	Credner (1882–1893), Boy (1972, 1978)
Microbrachis	Paedo	Progenesis	Genus	Pennsylvanian	Milner (1980b)
Branchiosaurids	Paedo	Progenesis	Family	Mid Pennsylvanian	Milner (1982), Boy (1971, 1972)
Temnospondyls	Paedo	Progenesis	Genus	Carboniferous	Milner *et al.* (1986)
Trimerorachis	Paedo	Progenesis	Genus	Early Permian	Olson (1979)
"Salamanders"	Paedo	?Progenesis/ neoteny	Genus	Paleozoic	Bolt (1977)
Ptyonius	Paedo	?Neoteny	Genus	Late Carboniferous	Boyd (1984)
Urocordylus	Paedo	?Neoteny	Genus	Late Carboniferous	Boyd (1984)
Gerrothorax	Paedo	Neoteny	Genus	Mid Pennsylvanian	Milner (1982)
Diplocaulus	Paedo	Neoteny	Species	Early Permian	Olson (1951)
Greererpeton	Paedo	?Neoteny	Genus	Late Mississippian	Holmes (1980)
Discosaurids	Paedo	—	Family	Early–Mid Permian	Carroll (1977)
Eugyrinurus	Paedo	—	Genus	Late Carboniferous	Milner (1980a)
Amphibamus	Paedo	—	Species	Mid Pennsylvanian	Milner (1982)
Dvinosaurus	Paedo	—	Genus	Late Permian	Watson (1926)
Pteroplax	Pera	Postdisplacement	Genus	Carboniferous	Boyd (1980)
Amphiuma	Pera	Acceleration	Genus	Paleocene	Estes (1969)
Tetrapod skull	Pera	Acceleration	Class	Carboniferous	Westoll (1980)

of the 20 recorded heterochronic examples (Table VII). Many families of Paleozoic and Mesozoic salamanders have been shown to be paedomorphic, in particular branchiosaurids (Boy, 1971, 1972; Milner, 1982). Size–shape relationships suggest that both progenesis and neoteny have caused the paedomorphosis. Genera such as *Doleserpeton, Tersomius* (Bolt, 1977). *Microbrachis* (Milner, 1980b), and *Branchiosaurus* (Boy, 1972, 1978) are probably progenetic. As with fish, structures affected by heterochrony are cranial elements and limbs in particular. The most common paedomorphic event in amphibians has been the retention of gills, expressed in the fossils as the presence of gill ossicles. Thus, the paedomorphic microsaur *Microbrachis* never exceeds 300 mm in length, has lateral line sulci, gill ossicles, unossified carpels and tarsels, small limbs, and elongate trunk (Milner, 1980b). Other such paedomorphic forms, including *Branchiosaurus*, are characterized by their wide mouth and simple, peglike teeth. They are thought to be secondarily aquatic forms (Boy, 1972, 1978).

Changes in skull allometries have been documented in *Diplocaulus* (Olson, 1951) and *Trimerorhacis* (Olson, 1979). Indeed, Westoll (1980) commented on how the fusion of the intercranial joint at progressively earlier stages of ontogeny across the dermal gap was one of the features of the evolution of tetrapods from osteolepidids. Within tetrapods, differences in gross limb morphology have been considered (Rackoff, 1980) to be attributable to variant systems of genetic regulation and developmental control. Such heterochronic differences in limb structure have been shown in the interclavicles of the temnospondyls *Ptyonius* and *Urocordylus,* comparable in morphology with the interclavicle of juvenile *Batrachiderpeton* (Boyd, 1984). Other postcranial heterochronic changes have affected features of the axial skeleton. Thus, in *Pteroplax,* peramorphosis has been recognized (Boyd, 1980), there having been precocious development of the lateral boss in juvenile intercentra.

There can be little doubt that there was great adaptive advantage to paedomorphic amphibians. For instance, Milner (1982) commented on how branchiosaurids, which were facultatively or permanently paedomorphic relatives of small terrestrial temnosponyls, exploited plankton-feeding niches in Autnian intermontane lakes. The ontogenetic changes associated with terrestriality had been suppressed. Milner (1982) considers that paedomorphosis in the Branchiosauridae may have occurred a number of times. The question needs to be asked, however, as to whether the dominance of paedomorphosis over peramorphosis in amphibians is a true reflection of the greater frequency of occurrence of paedomorphic processes. While research on living salamanders (e.g., Alberch and Alberch, 1981) has shown the common occurrence of paedomorphosis, it is possible that there has been some bias in the fossil record. Paedomorphic amphibians that have adopted a permanent aquatic mode of life are much more likely to have been fossilized than forms that were terrestrial.

4.3. Reptiles and Birds

Although few studies have been carried out on heterochrony in fossil reptiles, its effect can be recognized in many different groups. As in amphibians and fishes, skull, teeth development, and, to a lesser extent, postcranial elements were affected by heterochrony. While examples of paedomorphosis appear to be more common than peramorphosis, much more analysis is required to verify this.

Heterochronic changes in the development of cranial skeletal elements can be seen in the Oligo-Miocene amphisbaenian *Dyticonastis,* where there is a paedomorphic delay in the closure of sutures of the occipitootic region of the braincase. In most other rhineurids these sutures closed early in ontogeny (Berman, 1976b). Retardation in onset of fusion of sutures also occurred in the Paleocene anguid *Glyptosaurus* (Sullivan, 1979). Other paedomorphic features include unfused hexagonal dermal armour and tooth shape. The relatively large size of these lizards suggests that neoteny was the paedomorphic process. As Sullivan stated, neoteny has resulted in the evolution of a highly specialized lizard, particularly in terms of skull morphology.

Variations in allometries of many cranial and postcranial elements were an important aspect of dinosaur evolution (Hopson, 1977). Both peramorphosis and paedomorphosis can be seen in a single ichthyosaurian lineage (McGowan, 1986). Following a peramorphic acceleration in rostral growth rate in *Excalibosaurus* compared with *Ichthyosaurus*, the descendant *Eurhinosaurus* shows a decrease in mandibular growth rate, while retaining the high rostral growth rate. McGowan (1986, p. 456) noted how relatively small genetic changes had profound morphological and biological consequences.

Changes in dentition were also affected by heterochrony. For instance, in the Triassic rhyncosaurs, such as *Captorhinus*, *Ladibosaurikos*, and *Moradisaurus*, the presence of multiple row dentition has been interpreted by Benton (1984) as being the retention in adults of an ancestral embryonic type of dentition. Paedomorphic tooth reduction has also been observed in rhineurids (Berman, 1976b).

In his detailed analysis of the phylogeny of the Late Jurassic plesiosaurs, Brown (1981) documented many changes in allometries of postcranial elements. Most involve acceleration. For instance, the pectoral and pelvic bars in Early and Middle Jurassic forms failed to develop. However, by the Late Jurassic they occurred ossified in adults. Furthermore, fusion of neural arches occurred in adults only in the Late Jurassic. Tooth and phalange numbers increase through the Jurassic. Paedomorphosis in limbs has been recorded in other fossil reptiles, such as in the small limbs of the Triassic lizardlike reptile *Lacertulus* (Carroll and Thomson, 1982).

Progenesis does not appear to have occurred as frequently in reptiles as in amphibians, but some Mesozoic "dwarf" crocodiles, such as atoposaurids (Joffe, 1967), retained juvenile characters (Langston and Rose, 1978) and may well have been progenetic.

Paedomorphosis may have played a significant role in the evolution of birds. The very large orbits, inflated braincase, retarded dental development, and overall limb proportions of *Archaeopteryx* led Thulborn (1985) to conclude that there is a case for early birds being interpreted as paedomorphic theropod dinosaurs. He suggested that feathers were present on juvenile theropods. Their retention into adult descendants was clearly of tremendous adaptive significance. Flightless birds are probably paedomorphic. One example is the Hawaiian Pleistocene goose *Thambetochen*. This lacks a keel, and has reduced wings and an open angle between the coracoid and scapula, all characteristics present in juveniles of flying geese, but lost in their adults (James and Olson, 1983).

4.4. Mammals

Although many of the allometric changes in fossil mammal lineages, such as horses (Radinsky, 1984) and rhinoceroses, could, with further study, be analyzed in terms of heterochrony, documented examples of heterochrony in fossil mammals are few and far between. Hypermorphosis has for some time been recognized as having played a major role in the evolution of the "Irish Elk," *Megaloceros* (Gould, 1974), and, combined with predisplacement and acceleration, titanotheres (McKinney and Schoch, 1985).

Peramorphosis may also be invoked for the evolution of molars in elephants. During ontogeny, teeth form by the addition of plates. Thus, in an early, Middle Pleistocene form only seven plates were formed in M3. The three descendant lineages show varying degrees of peramorphic increase in number of plates in each molar, reaching a maximum of 30 in some species of *Mammuthus* (Maglio, 1972). The rate of increase in plate number was least in the *Loxodonta* lineage. Accompanying the increased plate number was a reduction in thickness of enamel. These heterochronic changes probably had significant effects on the diets of descendants. The peramorphic changes led to evolution into "new adaptive zones" (Maglio, 1972), with improved shearing capability. Progenesis occurred in another lineage of *Elephas*. In *E. falconeri*, molar plate number was reduced and the plates became thicker and less folded. Other paedomorphic traits were lack of a frontal crest, small tusks and tusk sockets, and rounded cranium (Maglio, 1973).

Peramorphosis appears to have been significant in the evolution of some Oligocene rabbits. Both interspecific and intergeneric morphological changes in *Palaeolagus* would seem to be related to developmental changes in growth rate (Gawne, 1978). For example, relative to *Palaeolagus*, *Chadrolagus* possessed a number of precociously advanced characters. Detailed analysis of changing allometries in fossil mammal lineages is likely to reveal the impact of heterochrony in mammal evolution.

4.5. Synthesis

The pattern that emerges in fossil vertebrates is: first, heterochrony has played a major part in evolution at many taxonomic levels; second, heterochrony has affected all skeletal structures, in particular cranial elements; and third, there is a preponderance of paedomorphosis. However, it is possible that a more detailed analysis of vertebrate groups will reveal more examples of peramorphosis.

One significant pattern to emerge is that at low taxonomic levels heterochronic changes have mainly involved changes in allometry of particular skeletal elements. However, heterochronic changes involving timing of onset of development of particular elements, such as teeth or bones, have resulted in more fundamental structural changes, which are expressed at a higher taxonomic level. Thus, as in echinoderms and arthropods, vertebrates show a two-tiered structure of growth: first, the production of individual skeletal elements, whose appearance relative to body size may be altered phylogenetically by changes in the timing of onset of development; then, second, the subsequent growth of these elements, which can evolve by changes to their allometries—the finer tuning of evolution.

5. Discussion

This survey of heterochrony in invertebrates and vertebrates from the fossil record has revealed an overall preponderance of paedomorphosis. Of a total of 272 examples documented, 179 are of paedomorphosis and 93 of peramorphosis. What is the significance of this distribution? What differences are there between groups? An important factor to try to remove at the outset is the impact of historical preconditioning. For instance, whereas ammonites were the prime model for peramorphosis in the late 19th century and then for paedomorphosis in the mid-20th century, they are now considered to show both paedomorphosis and peramorphosis.

The most detailed studies of heterochrony carried out during the last 5 years have been in trilobites, echinoids, ammonites, bryozoans, and graptolites. These all show both paedomorphosis and peramorphosis as important factors in their evolution, although paedomorphosis is still predominant, with 90 examples to 65. Of the remainder of groups, where less detailed analysis has been carried out in recent times, paedomorphosis also predominates. However, it is significant that in mammals few examples of paedomorphosis have been described, peramorphosis seeming to be more significant. One possible reason for paedomorphosis generally occurring more often than peramorphosis could be that existing developmental programs are utilized. Moreover, morphological structures have been tried and tested. Peramorphosis requires the extension of preexisting developmental pathways and the production of entirely novel baseplans.

In these groups that have been better studied in recent years, a number of significant patterns have emerged. These highlight the reasons there is a variation in frequency of particular heterochronic processes between different groups and also during the history of particular groups. These differences arise from variations in ontogenetic growth patterns between groups and in the extent of ecological change during individual ontogenies. Analysis of these patterns in other groups may provide a means of predicting the likely prevalence of different heterochronic processes.

One pattern seen in trilobites, but which is not readily evident in other groups, is a shift in emphasis from paedomorphosis to peramorphosis, during the phylogenetic history of the group. This largely reflects a change from a dominance of progenesis in Cambrian forms compared with post-Cambrian forms and an increase in the frequency of acceleration and predisplacement in post-Cambrian taxa. In order for progenesis to be effective as a means of generating new taxa, two criteria need to be satisfied. First, there must be appreciable morphological change during ontogeny. In other words, allometries, either positive or negative, must be some distance from isometry, or individual structures such as thoracic segments must be generated at a high rate. Precocious maturation will thus produce a mature individual morphologically markedly different from the ancestral form. Second, ecological separation between juveniles and adults must be great. This is particularly so in trilobites, where juveniles are thought to have been pelagic and adults benthic. Progenesis in such situations will have produced immediate niche partitioning between adults and consequent ecological and perhaps genetic isolation. The decline in progenesis in post-Cambrian trilobites may reflect a breakdown in the ecological separation between juveniles and adults, but, as I argued earlier, it more likely reflects an improvement in development control.

Progenesis would also appear to have been an important process in amphibians and brachiopods. While the morphological difference between juvenile and adult amphibians is not great in many characters, in a few critical features it was of paramount importance. However, ecological separation between juveniles and adults is great, from aquatic to terrestrial. Retention of juvenile gill filaments due to progenesis allowed retention of a juvenile mode of life. Again the morphological difference between many juvenile and adult brachiopods is often not great, but a critical ecological difference is often the mechanism of attachment to the substrate. Thus, where there is a marked ecological distinction between juveniles and ancestors, progenesis may be a significant process. It is interesting to note that in reptiles progenesis is much rarer than in amphibians. Perhaps this is because ecological separation of juveniles and adults is much less in this group.

In echinoids and in bryozoans, paedomorphosis and peramorphosis occur with relatively equal frequency. In echinoids, much of the research has been carried out on spatangoids. This group shows two important patterns: the rarity of progenesis and the common occurrence of dissociated heterochrony. Again the rarity of progenesis probably reflects minimal ecological separation of juveniles and adults. The frequent occurrence of peramorphosis in this group is probably a function of the nature of growth of the echinoid test. Details of this are more fully documented elsewhere (McNamara, 1987a,b). In essence, there is a two-tiered pattern of growth, combined with pronounced plate allometries. Small changes in allometries early in development will have pronounced morphological effects on descendant adults, allowing niche partitioning of adults. I have suggested that earlier in the history of the echinoids, changes in the rate of plate production may have been more significant. This may be seen from the greater number of examples of progenesis in regular echinoids than in irregular echinoids, where plate allometries are more pronounced.

This same two-tiered pattern of growth occurs in vertebrates and in arthropods such as trilobites. Thus, structures, such as individual skeletal elements in vertebrates, coronal plates in echinoids, thoracic segments in trilobites, are produced at a certain rate. Variation in the rate of production of these elements will have immediate, pronounced structural significance. The phylogenetic effect of such meristic heterochrony, as it may be termed, may be abrupt and may appear in the fossil record as a punctuated event. The second tier is the modification of the growth allometries of those structural elements that have been generated. Heterochronic changes in these allometries may produce what can be termed allometric heterochrony. These changes are more likely to appear in the fossil record as gradual changes.

There is an interesting similarity between this two-tiered model and the growth pattern of colonial organisms. In recent years both graptolite colonies (Finney, 1986) and bryozoan colonies (Taylor and Furness, 1978; Highsmith, 1982) have been treated as individuals. Pandolfi (this volume) has highlighted the two-tiered nature of the growth of such colonies, involving astogeny (the growth of the colony) and ontogeny (the growth of individual elements in the colony). Could we perhaps equate the production of meristic elements in vertebrates, echinoids, and arthropods with the ontogenetic production of individuals in the colonies, and the sum of the allometric changes with the astogenetic development of colonial organisms? If so, astogenetic heterochrony would equate with allometric heterochrony and ontogenetic with meristic heterochrony. As meristic heterochrony can have profound, and relatively rapid, morphological effects, so do large morphological differences ensue when highly integrated colonies display ontogenetic heterochrony (Pandolfi, this volume).

I have mentioned the probable importance of dissociated heterochrony in echinoids. This phenomenon has also been recorded in ammonites, gastropods, and reptiles. Examples are also to be found in trilobites (Edgecombe and Chatterton, 1987). Future research will, I suspect, show dissociated heterochrony to be a common event in many groups. Its occurrence has important bearings on the interpretation of the target of selection. While, historically, morphological changes produced by heterochrony have always been viewed as the targets of selection, a view supported by the occurrence of dissociated heterochrony, McKinney (1986) has recently argued that size may often be the target. Any consequent morphological changes will be incidental.

McKinney (1986) found in his analysis of the ontogenetic development of fossil echinoids from the Tertiary of southeastern United States that 15 of 17 related pairs of species showed that the larger form, which had evolved by slower, neotenic and/or extended, hypermorphic growth, occurred predominantly in stable environments. The target of selection, therefore, appeared to be on reproductive timing and/or body size. Changes in shape that accompanied the size changes were merely allometric by-products: the effects, not the cause. McKinney further suggested that Cope's rule may to a large degree be explained by *K*-selective pressure favoring neotonic and hypermorphic forms. Does this mean that all previous workers in their adaptationist approaches were barking up the wrong (phylogenetic) trees?

The data set accumulated in this study can be used to test McKinney's hypothesis. This can be done in three ways. A comparable set of Tertiary echinoids from southern Australia can be analyzed to see if it conforms to the same pattern as the set from southeastern United States; documented examples of progenesis can be analyzed to see if they conform to a pattern of *r*-selection, favoring selection of small individuals; and heterochronic trends in various groups (paedomorphoclines and peramorphoclines) can be analyzed in conjunction with ecological data to assess whether there are indications of selection favoring size rather than shape.

Analysis of changes in the maximum sizes attained by echinoid species in lineages of *Echinolampas* (three lineages), *Lovenia, Hemiaster, Psephoaster, Brissopsis, Eupatagus, Pericosmus, Schizaster,* and *Protenaster* from the Tertiary of southern Australia shows that between species pairs, 16 show a trend toward increased size, 11 show a decrease, and 5 show no change (on the basis of the change having to exceed 5%). All lineages show evolution into finer grained sediments and can be assumed to be evolving into more stable environments on the basis of criteria used by McKinney (1986). I have elsewhere argued [see McNamara (1987c) for a review] that selection favored the evolution of morphologies adapted to the finer grained sediments. So, while 50% of species pairs show an increase in size, the other 50% show no change or a decline in maximum size. All, however, show evolution of morphological traits.

Progenetic species, it has been suggested (Gould, 1977; McKinney, 1986), may be *r*-selected. In other words, they are characterized by early maturation, small size, occupation of unstable environments, and high fecundity. While the fossil records supports this view in general, what is the target of selection? Mancini (1978a,b) considered that progenetic species may result from fluctuating, unstable environments. Likewise, Snyder and Bretsky (1971) have suggested that "dwarfed" (i.e., progenetic) bivalve faunas are *r*-stragegists, living in unstable, immature environments. The small body size attained by progenesis in oysters and ammonites in the Late Cretaceous Grayson micromorph fauna led Mancini (1978b) to conclude that it was an adaptation to the soft, unstable substrates. The same is probably true for many progenetic brachiopods (Surlyk, 1972). High fecundity is also a feature of some progenetic species, in particular, edrioasteroids, which often occur in large concentrations (Sprinkle and Bell, 1978). They were also adapted to shallow, high-energy environments. Progenetic trilobites also tend to occur in relatively large numbers (McNamara, 1983a).

Thus, in general, it seems likely that in most cases, as McKinney (1986) suggested, small size and precocious maturation were the principal targets of selection. However, that is not to say that the consequent morphological features were of no adaptive significance. If the progenetic form evolved a shape that was entirely unsuited to the unstable environment, the selection of that form as a new species would not have occurred. The target of selection in many cases of heterochrony was probably a combination of *both* size and shape. In some instances, however, as in *r*-selective environments, size or reproductive timing was more important. It is perhaps significant that none of the echinoid lineages from southern Australia show progenesis to any degree, yet the only example of significant hypermorphosis, in the *Schizaster* lineage, shows a strong degree of size increase accompanying the shape changes. Progenesis and hypermorphosis are size-related. When these processes have occurred, selection may in part have been targeted on size. However, there need be no size change involved when neoteny, acceleration, or pre- or postdisplacement occur, suggesting that when these processes occur the target of selection is shape, not size.

I have discussed in detail elsewhere (McNamara, 1982a) the possible mechanism for the selection of heterochronic morphotypes along environmental gradients, resulting in the formation of either paedomorphoclines or peramorphoclines. These can be seen in many diverse groups of organisms, such as trilobites, ammonoids, brachiopods, corals, graptolites, echinoids, bivalves, gastropods, and mammals (Table VIII). Such heterochronic trends are generally adaptively significant along an environmental gradient such as deep to shallow water (or vice versa), coarse to fine-grained sediments, or infaunal to epifaunal. They are therefore also developing along environmental stability gradients. While the functional adaptations of the morphologies in many of the heterochronic morphoclines have been interpreted, no attempt has been made to assess the significance of size changes. If McKinney's hypothesis is to be supported, these paedomorphoclines and peramorphoclines should show size increases from unstable to more stable environments.

Whereas the echinoid morphoclines show increasing, decreasing, and no size change as the stability gradient increases, examples from trilobites, ammonites, and brachiopods show a preponderance of decreasing size as the stability gradient decreases. Thus, while there is no overall trend of increasing size with stability, the converse has been documented from a number of groups: decreasing size with decreasing stability.

There can be little doubt that heterochrony has played a significant role in the evolution of the metazoa. Variations in the complex interplay of morphological change and behavioral differences during ontogeny have led to the evolution of a wide range of heterochronic morphotypes in many groups of organisms. The seeming predominance of paedomorphosis in many groups may signify that per-

Table VIII. Heterochrony and Evolutionary Trends

	Paedomorphocline	Environmental gradient	Stability gradient	Size change
Trilobites	*Olenellus*	Deep/shallow water	Decreasing	Decrease
	Tretaspis			None
	Zacanthoidids	Deep/shallow water	Decreasing	Decrease
	Oryctocephalids			Decrease
Ammonoids	*Protacanthoceras*			Decrease
	Pseudoxybeloceras			Decrease
Brachiopod	*Tegulorhynchia*	Deep/shallow water	Decreasing	None
Graptolite	*Petalograptus*			Decrease
Echinoids	*Hemiaster*	Coarse/fine sediments	Increasing	None
	Psephoaster	Coarse/fine sediments	Increasing	None
	Lovenia	Coarse/fine sediments	Increasing	Decrease
Bivalve	*Gryphaea*			Increase
Gastropod	*Glossaulax*			None
	Peramorphocline	**Environmental gradient**	**Stability gradient**	**Size change**
Trilobites	*Leonaspis*	Deep/shallow water	Decreasing	None
	Redlichiids			Decrease
Ammonoids	Perrinitids			
Coral	*Amplexizaphrentis*			None
Echinoids	*Hagenowia*	Shallow/deep burial		Decrease
	Protenaster	Coarse/fine sediments	Increasing	Increase
	Hemiaster	Coarse/fine sediments	Increasing	None
	Psephoaster	Coarse/fine sediments	Increasing	None
	Lovenia	Coarse/fine sediments	Increasing	Decrease
	Schizaster	Coarse/fine sediments	Increasing	Increase
	Pericosmus	Coarse/fine sediments	Increasing	Increase
Bivalves	*Chesapecten*	Shallow/deep water	Increasing	Increase
	Myalina	Infaunal/epifaunal	Decreasing	Increase

turbations of existing developmental programs are more common than extensions to the developmental system. However, the correlation of a decline in paedomorphosis with a refinement of the developmental system in trilobites, and a possible similar pattern in echinoids, may indicate that in those few groups, such as mammals, where peramorphosis may be more common, regulation of the developmental system is more rigidly controlled.

References

Alberch, P., and Alberch, J., 1981. Heterochronic mechanisms of morphological diversification and evolutionary change in the neotropical Salamander *Bolitoglossa occidentalis* (Amphibia: Plethodontidae), *J. Morphol.* **167:**249–264.

Alberch, P., Gould, S. J., Oster, G. F., and Wake, D. B., 1979, Size and shape in ontogeny and phylogeny, *Paleobiology* **5**:296–317.

Anstey, R. L., 1987, Astogeny and phylogeny: evolutionary heterochrony in Paleozoic bryozoans, *Paleobiology* **13**:20–43.

Backhaus, E., 1959, Monographie der cretacischen Thecideidae (Brachiopoden), *Mitt. Geol. Staatsinst. Hamburg* **28**:5–90.

Baker, P. G., 1983, The diminutive thecideidine brachiopod *Enallothecida pygmaea* (Moore) from the middle Jurassic of England, *Palaeontology* **26**:663–669.

Baker, P. G., 1984, New evidence of a spiriferid ancestor for the Thecideidina (Brachiopoda), *Palaeontology* **27**:857–866.

Baker, P. G., and Elston, D. G., 1984, A new polyseptate thecideacean brachiopod from the Middle Jurassic of the Cotswolds, England, *Palaeontology* **27**:777–791.

Baker, P. G., and Laurie, K., 1978, Revision of Aptian thecideidine brachiopods of the Faringdon Sponge Gravels, *Palaeontology* **21**:555–570.

Batten, R. L., 1975, The Scissurellidae—are they neotenously derived fissurellids? (Archaeogastropoda), *Am. Mus. Novit.* **2567**:1–29.

Beecher, C. E., 1893, Some correlations of ontogeny and phylogeny in the Brachiopoda, *Am. Nat.* **27**:599–604.

Beecher, C. E., 1987, Outline of a natural classification of the trilobites, *Am. J. Sci. Ser. 4* **3**:89–106, 181–207.

Bell, B. M., 1976, Phylogenetic implications of ontogenetic development in the Class Edrioasteroidea (Echinodermata), *J. Paleontol.* **50**:1001–1019.

Bemis, W. E., 1984, Paedomorphosis and the evolution of the Dipnoi, *Paleobiology* **10**:293–307.

Benton, M. J., 1984, Tooth form, growth, and function in Triassic rhynchosaurs (Reptilia, Rapsida), *Palaeontology* **27**:737–776.

Berman, D. S., 1976a, Cranial morphology of the Lower Permian lungfish *Gnathorhiza* (Osteichthyes: Dipnoi), *J. Paleontol.* **50**:1020–1033.

Berman, D. S., 1976b, A new amphisbaenian (Reptilia: Amphisbaenia) from the Oligocene–Miocene John Day Formation, Oregon, *J. Paleontol.* **50**:165–174.

Bockelie, J. F., 1984, The Diploporita of the Oslo region, Norway, *Palaeontology* **27**:1–68.

Bockelie, J. F., and Paul, C. R. C., 1983, *Cyathotheca suecica* and its bearing on the evolution of the Edrioasteroidea, *Lethaia* **16**:257–264.

Bolt, J. R., 1977, Dissorophoid relationships and ontogeny, and the origin of the Lissamphibia, *J. Paleontol.* **51**:235–249.

Boy, J. A., 1971, Zur Problematik der Branchiosaurier (Amphibia, Karbon-Perm), *Paläontol. Z.* **45**:107–119.

Boy, J. A., 1972, Die Branchiosaurier (Amphibia) des saarpfälzischen Rotliegenden (Perm, SW-Deutschland), *Abh. Hess. Landesamt. Bodenforsch.* **65**:1–137.

Boy, J. A., 1978, Die tetrapodenfauna (Amphibia, Reptilia) des saarpfälzischen Rotliegenden (Unter-Perm; SW-Deutschland). I. *Branchiosaurus, Mainzer Geowiss. Mitt.* **7**:27–76.

Boyd, M. J., 1980, The axial skeleton of the Carboniferous amphibian *Pteroplax cornutus*, *Palaeontology* **23**:273–285.

Boyd, M. J., 1984, The Upper Carboniferous tetrapod assemblage from Newsham, Northumberland, *Palaeontology* **27**:367–392.

Brasier, M. D., 1982, Architecture and evolution of the foraminiferid test—A theoretical approach, in: *Aspects of Micropalaeontology* (P. T. Bauner and A. R. Lord, eds.), pp. 1–41, Allen and Unwin, London.

Brett, C. E., Frest, T. J., Sprinkle, J., and Clement, C. R., 1983, Coronoidea: A new class of blastozoan echinoderms based on taxonomic reevaluations of *Strephanocrinus*, *J. Paleontol.* **57**:627–651.

Brezinski, D. K., 1986, An opportunistic Upper Ordovician trilobite assemblage from Missouri, *Lethaia* **19**:315–325.

Broadhead, T. W., and McComb, R., 1983, Paedomorphosis in the conodont family Icriodontidae and the evolution of *Icriodus*, *Fossils Strata* **15**:149–154.

Brower, J. C., 1976, *Promelocrinus* from the Wenlock at Dudley, *Palaeontology* **19**:651–680.

Brown, D. S., 1981, The English Upper Jurassic Plesiosauridea (Reptilia) and a review of the phylogeny and classification of the Plesiosauria, *Bull. Br. Mus. Nat. Hist. (Geol.)* **35**:253–347.

Carroll, R. L., 1977, Patterns of amphibian evolution: an extended example of the incompleteness of the fossil record, in: *Patterns of Evolution, As Illustrated by the Fossil Record* (A. Hallam, ed.), pp. 405–437, Elsevier, Amsterdam.

Carroll, R. L., and Thompson, P., 1982, A bipedal lizardlike reptile from the Karroo, *J. Paleontol.* **56:**1–10.

Carruthers, R. G., 1910, On the evolution of *Zaphrentis delanouei* in Lower Carboniferous times, *Q. J. Geol. Soc. Lond.* **264:**523–536.

Chatterton, B. D. E., and Perry, D. G., 1983, Silicified Silurian odontopleurid trilobites from the MacKenzie Mountains, *Palaeontogr. Can.* **1:**1–126.

Clarkson, E. N. K., 1979, The visual system of trilobites, *Palaeontology* **22:**1–22.

Cope, J. C. W., 1978, The ammonite faunas and stratigraphy of the upper part of the Upper Kimmeridge Clay of Dorset, *Palaeontology* **21:**469–533.

Copper, P., 1982, Early Silurian atrypoids from Manitoulin Island and Bruce Peninsula, Ontario, *J. Paleontol.* **56:**680–702.

Credner, H., 1882–1893, Die Stegocephalen und Saurier aus dem Rothliegenden des Plauenschen Grundes bei Dresden, *Z. Deutsch. Geol. Ges.* III (34), IV (35), V, (37), VI (38), IX (42), X (35).

Dommergues, J.-L., 1986, Les Dactylioceratidae du Carixien et du Domerian basal, un groupe monophyletique. Les Reynesocoeloceratinae nov. subfam. *Bull. Sci. Bourg.* **39:**1–26.

Dommergues, J. L., David, B., and Marchand, D., 1986, Les relations ontogenèse–phylogenèse: applications paléontologiques, *Geobios* **19:**335–356.

Donnovan, D. T., 1973, The influence of theoretical ideas on ammonite classification from Hyatt to Trueman, *Paleontol. Contrib. Univ. Kansas* **62:**1–16.

Dzik, J., 1975, The origin and early phylogeny of the cheilostomatous bryozoa, *Acta. Paleontol. Pol.* **20:**395–423.

Dzik, J., 1981, Evolutionary relationships of the early Palaeozoic 'cyclostomatous' bryozoa, *Palaeontology* **24:**827–861.

Edgecombe, G. D., and Chatterton, B. D. E., 1987, Heterochrony in the Silurian radiation of encrinurine trilobites, *Lethaia* **20:**337–351.

Elles, G. L., 1923, Evolutionary palaeontology in relation to the Lower Palaeozoic rocks, *Rep. Br. Assoc. Adv. Sci.* **91:**83–107.

Elliott, D. K., 1984, A new subfamily of the Pteraspididae (Agnatha, Heterostraci) from the Upper Silurian and Lower Devonian of Arctic Canada, *Palaeontology* **27:**169–197.

Estes, R., 1969, The fossil record of amphiurid salamanders, *Breviora* **322:**1–11.

Fink, W. L., 1982, The conceptual relationship between ontogeny and phylogeny, *Paleobiology* **8:**254–264.

Finney, S. C., 1986, Heterochrony, punctuated equilibrium, and graptolite zonal boundaries. in: *Palaeoecology and Biostratigraphy of Graptolites* (C. P. Hughes and R. B. Rickards, eds.) pp. 103–113, Geological Society Special Publication No. 20.

Fortey, R. A., 1974, The Ordovician trilobites of Spitsbergen I Olenidae, *Norsk Polarinstitutt* **160:** 1–81.

Fortey, R. A., 1975, The Ordovician trilobites of Spitsbergen II Asaphidae, Nileidae, Raphiophoridae and Telephinidae of the Valhallfonna Formation, *Norsk Polarinstitutt* **162:**1–124.

Fortey, R. A., and Rushton, A. W. A., 1980, *Acanthopleurella* Groom 1902: Origin and life-habits of a miniature trilobite, *Bull. Br. Mus. Nat Hist. (Geol.)* **33:**79–89.

Fortey, R. A., and Shergold, J. H., 1984, Early Ordovician trilobites, Nova Formation, central Australia, *Palaeontology* **27:**315–366.

Frest, T. J., Strimple, H. L., and McGinnis, M. R., 1979, Two new crinoids from the Ordovician of Virginia and Oklahoma, with notes on pinnulation in the Disparida, *J. Paleontol.* **53:**399–415.

Garstang, W., 1922, The theory of recapitulation: A critical restatement of the biogenetic law, *J. Linn. Soc. Zool.* **35:**81–101.

Garstang, W., 1928, The morphology of the Tunicata, and its bearing on the phylogeny of the Chordata. *Q. J. Microscop. Sci.* **72:**51–187.

Gawne, C. E., 1978, Leporids (Lagomorpha, Mammalia) from the Chadronian (Oligocene) deposits of Flagstaff Rim, Wyoming, *J. Paleontol.* **52:**1103–1118.

Gould, S. J., 1969, An evolutionary microcosm: Pleistocene and Recent history of the land snail P. (*Poecilozonites*) in Bermuda, *Bull. Mus. Comp. Zool.* **138:**407–532.

Gould, S. J., 1974, The evolutionary significance of "bizarre" structures: antler size and skull size in the "Irish Elk," *Megaloceros giganteus*, *Evolution* **28:**191–220.

Gould, S. J., 1977, *Ontogeny and Phylogeny*, Harvard University Press, Cambridge.

Gramm, M. N., 1973, Cases of neoteny in fossil ostracodes, *Paleontol. J.* **1:**3–12.

Gramm, M. N., 1985, The muscle scar in cavellinids and its importance for the phylogeny of platycope ostracodes, *Lethaia* **18:**39–52.

Hallam, A., 1978, How rare is phyletic gradualism and what is its evolutionary significance? *Paleobiology* **4:**16–25.

Hallam, A., 1982, Patterns of speciation in Jurassic *Gryphaea, Paleobiology* **8:**354–366.

Henderson, R. A., and McNamara, K. J., 1985, Maastrichtian non-heteromorph ammonites from the Miria Formation, Western Australia, *Palaeontology* **28:**35–88.

Highsmith, R. C., 1982, Reproduction by fragmentation in corals, *Mar. Ecol. Progr. Ser.* **7:**207–226.

Holmes, R., 1980, *Proterogyrinus scheeli* and the early evolution of the labyrinthodont pectoral limb, in: *The Terrestrial Environment and the Origin of Land Vertebrates* (A. L. Panchen, ed.), pp. 351–376, Systematics Association Special Volume 15.

Hopson, J. A. 1977, Relative brain size and behaviour in archrosaurian reptiles, *Annu. Rev. Ecol. Syst.* **8:**429–448.

Hughes, C. P., Ingham, J. K., and Addison, R., 1975, The morphology, classification and evolution of the Trinucleidae (Trilobita), *Phil. Trans. R. Soc. Lond. B* **272:**537–607.

Hupé, P., 1953, Contribution à l'etûde du Cambrien Inférieur et du Précambrien III de l'Anti-Atlas Marocain, *Notes Mem. Serv. Min. Carte Géol. Maroc.* **103:**1–402.

Hyatt, A., 1866, On the agreement between the different periods in the life of the individual shell and the collective life of the tetrabranchiate cephalopods, *Proc. Boston Soc. Nat. Hist.* **10:**302–303.

Ingham, J. K., 1970, A monograph of the upper Ordovician trilobites from the Cautley and Dent districts of Westmorland and Yorkshire, *Palaeontol. Soc. [Monogr.]* **7:**1–58.

Jackson, R. T., 1890, Phylogeny of the Pelecypoda: the Aviculidae and their allies, *Mem. Boston Soc. Nat. Hist.* **4**(8)**:**277–400.

Jackson, R. T., 1912, Phylogeny of the Echini, with a review of Palaeozoic species, *Mem. Boston Soc. Nat. Hist.* **7:**1–443.

James, H. F., and Olson, S. L. 1983, Flightless birds, *Nat. Hist.* **92**(9):30–40.

Joffe, J., 1967, The 'dwarf' crocodiles of the Purbeck Formation, Dorset: a reappraisal, *Palaeontology* **10:**629–639.

Jones, B., 1979, *Atrypoidea erebus* n. sp. from the Late Silurian of Arctic Canada, *J. Paleontol.* **53:**187–196.

Jull, R. K., 1976, Septal development during hystero-ontogeny in the Ordovician tabulate coral *Foerstephyllum, J. Paleontol.* **50:**380–391.

Kennedy, W. J., 1977, Ammonite evolution, in *Patterns of Evolution, As Illustrated by the Fossil Record* (A. Hallam, ed.), pp. 251–304, Elsevier, Amsterdam.

Kennedy, W. J., and Cooper, M. R., 1977, The micromorph Albian ammonite *Falloticeras* Parona and Bonarelli, *Palaeontology* **20:**793–804.

Kennedy, W. J., and Wright, C. W., 1981, *Euhystrichoceras* and *Algericeras*, the last mortoniceratine ammonites, *Palaeontology* **24:**417–435.

Kolata, D. R., Strimple, H. L., and Levorson, C. O., 1977, Revision of the Ordovician carpoid family Iowacystidae, *Palaeontology* **20:**529–557.

Laerm, J., 1979, On the origin of rhipidistian vertebrae, *J. Paleontol.* **53:**175–186.

Laerm, J., 1982, The origin and homology of the neopterygian vertebral column, *J. Paleontol.* **56:**191–202.

Lane, N. G., and Sevastopulo, G. D., 1982, Microcrinoids from the Middle Pennsylvanian of Indiana, *J. Paleontol.* **56:**103–115.

Lane, N. G., Sevastopulo, G. D., and Strimple, H. L., 1985, *Amphipsalidocrinus:* A monocyclic camerate microcrinoid, *J. Paleontol.* **59:**79–84.

Langston, W., and Rose, H., 1978, A yearling crocodilian from the Middle Eocene Green River Formation of Colorado, *J. Paleontol.* **52:**122–125.

Levinton, J. S., 1983, Stasis in progress: the empirical basis of macroevolution, *Annu. Rev. Ecol. Syst.* **14:**103–137.

Lidgard, S., 1985, Zooid and colony growth in encrusting cheilostome bryozoans, *Palaeontology* **28:**255–291.

Lockley, M. G., 1983, A review of brachiopod dominated palaeocommunities from the type Ordovician, *Palaeontology* **26:**111–145.

Lu Yan-Hao and Wu Hong-Ji, 1983, Ontogeny of the trilobite *Dalmanitina (Dalmanitina) nanchengensis* Lu, *Palaeontologia Cathayana* **1:**123–144.

Ludvigsen, R., 1979, The Ordovician trilobite *Pseudogygites* Kobayashi in eastern and arctic North America, *Life Sci. Contrib. R. Ont. Mus.* **120:**1–41.

Lund, R., and Melton, W. G., 1982, A new actinopterygian fish from the Mississippian Bear Gulch Limestone of Montana, *Palaeontology* **25:**485–498.

Maglio, V. J., 1972, Evolution of mastication in the Elephantidae, *Evolution* **26:**638–658.

Maglio, V. J., 1973, Origin and evolution of the Elephantidae, *Trans. Am. Philos. Soc.* **63**(3):1–142.

Maisey, J. G., 1978, Growth and form of finspines in hybodont spines, *Palaeontology* **21:**657–666.

Majima, R., 1985, Intraspecific variation in three species of *Glossaulax* (Gastropoda: Naticidae) from the Late Caenozoic strata in central and southwest Japan, *Trans. Proc. Palaeontol. Soc. Japan* N.S. **138:**111–137.

Mancini, E. A., 1978a, Origin of micromorph faunas in the geologic record, *J. Paleontol.* **52:**311–322.

Mancini, E. A., 1978b, Origin of the Grayson micromorph fauna (Upper Cretaceous) of north-central Texas, *J. Paleontol.* **52:**1294–1314.

McGowan, C., 1986, A putative ancestor for the sword-like ichthyosaur *Eurhinosaurus*, *Nature* **322:**454–456.

McHargue, T. R., 1982, Ontogeny, phylogeny, and apparatus reconstruction of the conodont genus *Histiodella*, Joins Fm., Arbuckle Mountains, Oklahoma, *J. Paleontol.* **56:**1410–1433.

McIntosh, G. C., 1979, Abnormal specimens of the Middle Devonian crinoid *Bactrocrinites* and their effect on the taxonomy of the genus, *J. Paleontol.* **53:**18–28.

McKinney, M. L., 1984, Allometry and heterochrony in an Eocene echinoid lineage: morphological change as a by-product of size selection, *Paleobiology* **10:**407–419.

McKinney, M. L., 1986, Ecological causation of heterochrony: test and implications for evolutionary theory, *Paleobiology* **12:**282–289.

McKinney, M. L., and Schoch, R. M., 1985, Titanothere allometry, heterochrony, and biomechanics: revising an evolutionary classic, *Evolution* **39:**1352–1363.

McLean, J. H., 1984, A case for derivation of the Fissurellidae from the Bellerophontacea, *Malacologia* **25:**3–20.

McNamara, K. J., 1978, Paedomorphosis in Scottish olenellid trilobites (Early Cambrian), *Palaeontology* **21:**635–655.

McNamara, K. J., 1979, Trilobites from the Coniston Limestone Group (Ashgill Series) of the Lake District, England, *Palaeontology* **22:**53–92.

McNamara, K. J., 1981a, Paedomorphosis in Middle Cambrian xystridurine trilobites from northern Australia, *Alcheringa* **5:**209–224.

McNamara, K. J., 1981b, The Role of Paedomorphosis in the Evolution of Cambrian Trilobites, Open-File Report of the U.S. Geological Survey No. 81-743, pp. 126–129.

McNamara, K. J., 1982a, Heterochrony and phylogenetic trends, *Paleobiology* **8:**130–142.

McNamara, K. J., 1982b, Taxonomy and evolution of living species of *Breynia* (Echinoidea: Spatangoida) from Australia, *Rec. West. Aust. Mus* **10:**167–197.

McNamara, K. J., 1983a, Progenesis in trilobites, in: *Trilobites and Other Early Arthropoda: Papers in Honour of Professor H. B. Whittington, FRS* (D. E. G. Biggs and P. D. Lane, eds.), pp. 59–68, Special Papers in Palaeontology 30.

McNamara, K. J., 1983b, The earliest *Tegulorhynchia* (Brachiopoda: Rhynchonellida) and its evolutionary significance, *J. Paleontol.* **57:**461–473.

McNamara, K. J., 1984, Observations on the light-sensitive tube feet of the burrowing echinoid *Protenaster australis* (Gray, 1851), *Rec. West. Aust. Mus.* **11:**411–420.

McNamara, K. J., 1985a, Taxonomy and evolution of the Cainozoic spatangoid echinoid *Protenaster*, *Palaeontology* **28:**311–330.

McNamara, K. J., 1985b, A new micromorph ammonite genus from the Albian of South Australia, *Spec. Publ. S. Aust. Dep. Mines Energy* **5:**263–268.

McNamara, K. J., 1986a, A guide to the nomenclature of heterochrony, *J. Paleontol.* **60:**4–13.

McNamara, K. J., 1986b, The role of heterochrony in the evolution of Cambrian trilobites, *Biol. Rev.* **61:**121–156.

McNamara, K. J., 1987a, Taxonomy, evolution, and functional morphology of southern Australian Tertiary hemiasterid echinoids, *Palaeontology* **30:**319–352.

McNamara, K. J., 1987b, Heterochrony and the evolution of echinoids, in: *Phylogeny and Evolutionary Biology of Echinoderms* (C. R. C. Paul and A. B. Smith, eds.), Oxford University Press, Oxford.

McNamara, K. J., 1987c, The role of heterochrony in the evolution of spatangoid echinoids, in: *Ontogenèse et Evolution* (J. Chaline *et al.*, eds.), CNRS, Paris.

McNamara, K. J., and Philip, G. M., 1980a, Australian Tertiary schizasterid echinoids. *Alcheringa* **4**:47–65.

McNamara, K. J., and Philip, G. M., 1980b, Tertiary species of *Echinolampas* (Echinoidea) from southern Australia, *Mem. Nat. Mus. Vict.* **41**:1–14.

McNamara, K. J., and Philip, G. M., 1984, A revision of the spatangoid echinoid *Pericosmus* from the Tertiary of Australia, *Rec. West. Aust. Mus.* **11**:319–356.

Meinke, D. K., and Thomson, K. S., 1983, The distribution and significance of enamel and enameloid in the dermal skeleton of osteolepiform rhipidistian fishes, *Paleobiology* **9**:138–149.

Merrill, G. K., 1972, Taxonomy, phylogeny and biostratigraphy of *Neognathodus* in Appalachian Pennsylvanian rocks, *J. Paleontol.* **46**:817–829.

Merrill, G. K., and Powell, R. J., 1980, Paleobiology of juvenile (nepionic?) conodonts from the Drum Limestone (Pennsylvanian, Missourian-Kansas City area) and its bearing on apparatus ontogeny, *J. Paleontol.* **54**:1058–1074.

Millendorf, S. A., 1979, The functional morphology and life habits of the Devonian blastoid *Eleutherocrinus cassedayi* Shumard and Yandell, *J. Paleontol.* **53**:553–561.

Milner, A. R., 1980a, The temnospondyl amphibian *Dendrerpeton* from the Upper Carboniferous of Ireland, *Palaeontology* **23**:125–141.

Milner, A. R., 1980b, The tetrapod assemblage from Nyrany, Czechoslovakia, in: *The Terrestrial Environment and the Origin of Land Vertebrates* (A. L. Panchen, ed.), pp. 439–496, Systematics Association Special Volume 15.

Milner, A. R., 1982, Small temnospondyl amphibians from the Middle Pennsylvanian of Illinois, *Palaeontology* **25**:635–664.

Milner, A. R., Smithson, T. R., Milner, A. C., Coates, M. I., and Rolfe, W. D. I., 1986, The search for early tetrapods, *Modern Geology* **10**:1–28.

Miyazaki, J. M., and Mickevich, M. F., 1982, Evolution of *Chesapecten* (Mollusca: Bivalvia, Miocene–Pliocene) and the biogenetic law, in: *Evolutionary Biology*, Vol. 15 (M. K. Hecht, B. Wallace, and R. Prance, eds.), pp. 368–409, Plenum Press, New York.

Morescalchi, A., 1979, New developments in vertebrate cytotaxonomy. I. Cytotaxonomy of the amphibians, *Genetica* **50**:179–193.

Morescalchi, A., and Serra, V., 1974, DNA renaturation kinetics in some paedogenetic urodeles, *Experientia* **30**:487–489.

Mosher, L. C., 1973, Evolutionary, ecologic and geographic observations on conodonts during their decline and extinction, *Geol. Soc. Am. Spec. Pap.* **141**:143–152.

Nelson, G. J., 1978, Ontogeny, phylogeny, paleontology, and the biogenetic law, *Syst. Zool.* **27**:324–345.

Olson, E. C., 1951, *Diplocaulus*, a study in growth and variation, *Fieldiana: Geol.* **11**(2):57–154.

Olson, E. C., 1979, Aspects of the biology of *Trimerorhacis* (Amphibia: Temnospondyli), *J. Paleontol.* **53**:1017.

Owen, A. W., 1980, The trilobite *Tretaspis* from the Upper Ordovician of the Oslo region, Norway, *Palaeontology* **23**:715–747.

Pajaud, D., 1970, Monographies des Thecidees (Brachiopodes), *Mem. Soc. Geol. Fr. (N.S.)* **49**(112):1–349.

Paul, C. R. C., 1985, The adequacy of the fossil record reconsidered, *Spec. Pap. Palaeontol.* **33**:7–15.

Phelan, T. F., 1977, Comments on the water vascular system, food grooves, and ancestry of the clypeasteroid echinoids, *Bull. Mar. Sci.* **27**:400–422.

Philip, G. M., Two Australian Tertiary neolampadids, and the classification of cassiduloid echinoids, *Palaeontology* **6**:718–726.

Podell, M. E., and Anstey, R. L., 1979, The interrelationship of early colony development, monticules and branches in Paleozoic bryozoans. *Palaeontology* **22**:965–982.

Rackoff, J. S., 1980, The origin of the tetrapod limb and the ancestry of tetrapods, in: *The Terrestrial Environment and the Origin of Land Vertebrates* (A. L. Panchen ed.), pp. 255–292, Systematics Association Special Volume 15.

Radinsky, L. B., 1984, Ontogeny and phylogeny in horse skull evolution, *Evolution* **38**:1–15.

Raff, R. A., Anstrom, J. A., Huffman, C. J., Leaf, D. S., Loo, J.-H., Showman, R. M., and Wells, D. E., 1984, Origin of a gene regulatory mechanism in the evolution of echinoderms, *Nature* **310**:312–314.

Rickards, R. B., 1977, Patterns of evolution in the graptolites, in: *Patterns of Evolution, As Illustrated by the Fossil Record* (A. Hallam, ed.), pp. 333–358, Elsevier, Amsterdam.
Robison, R. A., 1967, Ontogeny of *Bathyuriscus fimbriatus* and its bearing on affinities of corynexochid trilobites, *J. Paleontol.* **41**:213–221.
Robison, R. A., and Campbell, D. P., 1974, A Cambrian corynexochoid trilobite with only two thoracic segments, *Lethaia* **7**:173–282.
Rudwick, M. J. S., 1968, The feeding mechanisms and affinities of the Triassic brachiopods *Thecospira* Zugmayer and *Bactrynium* Emmrich, *Palaeontology* **11**:329–360.
Schopf, T. J. M., 1977, Patterns and themes of evolution among Bryozoa, in: *Patterns of Evolution, As Illustrated by the Fossil Record* (A. Hallam, ed.), pp. 159–207, Elsevier, Amsterdam.
Schram, F. R., and Rolfe, W. D. I, 1982, New euthycarcinoid arthropods from the Upper Pennsylvanian of France and Illinois, *J. Paleontol.* **56**:1434–1450.
Schultze, H.-P., 1984, Juvenile specimens of *Eusthenopteron foordi* Whiteaves, 1881 (Osteolepiform Rhipidistian, Pisces) from the late Devonian of Miguasha, Quebec, Canada, *J. Vert. Paleontol.* **4**:1–16.
Schultze, H.-P. and Arsenault, M., 1985, The panderichthyid fish *Elpistostege*; a close relative of tetrapods?, *Palaeontology* **28**:293–309.
Scott, G. H., 1982, Tempo and stratigraphic record of speciation in *Globorotalia puncticulata, J Foram. Res.* **12**:1–12.
Siveter, D. J., 1979, *Metacalymene* Kegel, 1927, a calymenid trilobite from the Kopanina Formation (Silurian) of Bohemia, *J. Paleontol.* **53**:367–379.
Siveter, D. J., 1980, Evolution of the Silurian trilobite *Tapinocalymene* from the Wenlock of the Welsh Borderland, *Palaeontology* **23**:783–802.
Smith, A. B., 1981, Implications of lantern morphology for the phylogeny of post-Palaeozoic echinoids, *Palaeontology* **24**:779–801.
Smith, A. B., 1985, Cambrian eleutherozoan echinoderms and the early diversification of edrioasteroids, *Palaeontology* **28**:715–756.
Smith, J. P., 1914, Acceleration of Development in Fossil Cephalopoda, Leland Stanford Junior University Publications University Series, pp. 1–30, Stanford University.
Snyder, J., and Bretsky, P. W., 1971, Life habits of diminutive bivalve molluscs in the Maquoketa Formation (Upper Ordovician), *Am. J. Sci.* **271**:227–251.
Spencer, W. K., and Wright, C. W., 1966, Asterozoans, in: *Treatise on Invertebrate Paleontology*, Part V, *Echinodermata 3* (R. C. Moore, ed.), pp. U5–U107, Geological Society of America and University of Kansas Press, Lawrence.
Sprinkle, J., 1980, Origin of blastoids: new look at an old problem, in: *Geological Society of America, Abstracts with Programs*, Vol. 12, No. 7, p. 528.
Sprinkle, J., and Bell, B. M., 1978, Paedomorphosis in edrioasteroid echinoderms, *Paleobiology* **4**:82–88.
Stanley, S. M., 1972, Functional morphology and evolution of byssally attached bivalve mollusks, *J. Paleontol.* **46**:165–212.
Sullivan, R. M., 1979, Revision of the Paleogene genus *Glyptosaurus* (Reptilia, Anguidae), *Bull. Am. Mus. Nat. Hist.* **163**:1–72.
Surlyk, F., 1972, Morphological adaptations and population structures of the Danish chalk brachiopods (Maastrichtian, Upper Cretaceous), *Biol. Skr. K. Dan. Vidensk. Selsk.* **19**:1–57.
Taylor, P. D., and Furness, R. W., 1978, Astogenetic and environmental variation in zooid size within colonies of Jurassic *Stomatopora* (Bryozoa, Cyclostomata), *J. Paleontol.* **52**:1093–1102.
Tharalson, D. B., 1984, Revision of the Early Permian ammonoid family Perrinitidae, *J. Paleontol.* **58**:804–833.
Thulborn, R. A., 1985, Birds as neotenous dinosaurs, *Rec. N. Z. Geol. Surv.* **9**:90–92.
Urbanek, A., 1960. An attempt at biological interpretation of evolutionary changes in graptolite colonies, *Acta Palentol. Pol.* **5**:127–233.
Urbanek, A., 1973, Organisation and evolution of graptolite colonies, in: *Animal Colonies* (R. S. Boardman, A. H. Cheetham, and W. A. Oliver, eds.), pp. 441–514, Dowden, Hutchinson and Ross, Stroudsburg, Pennsylvania.
Waller, T. R., 1986, The evolution of ligament systems in the Bivalvia, *Am. Malacol. Bull.* (4) **1986**:111–112.
Ward, P. D., and Mallory, V. S., 1977, Taxonomy and evolution of the lytoceratid genus *Pseudoxybeloceras* and relationship to the genus *Solenoceras*, *J. Paleontol.* **51**:606–618.

Waters, J. A., Horowitz, A. S., and Macurda, D. B., 1985, Ontogeny and phylogeny of the Carboniferous blastoid *Pentremites*, *J. Paleontol.* **59:**701–712.

Watson, D. M. S., 1926, The evolution and origin of the Amphibia, *Philos. Trans. R. Soc. Lond. B* **214:**189–257.

Werdelin, L., and Long, J. A., 1986, Allometry in the placoderm *Bothriolepis canadensis* and its significance to antiarch evolution, *Lethaia* **19:**161–169.

Westoll, T. S., 1980, Prologue: problems of tetrapod origin, in: *The Terrestrial Environment and the Origin of Land Vertebrates* (A. L. Panchen, ed.) pp. 1–10, Systematics Association Special Volume 15.

White, E. I., 1958, Original environment of the craniates, in: *Studies on Fossil Vertebrates* (T. S. Westoll, ed.), pp. 212–234, Athlone Press, London.

Wiedmann, J., and Kullmann, J., 1980, Ammonoid sutures in ontogeny and phylogeny, in: *The Ammonoidea* (M. R. House and J. R. Senior, eds.), pp. 215–255, Systematics Association Special Volume 18.

Williams, A., 1973, The secretion and structural evolution of the shell of thecideidine brachiopods, *Philos. Trans. R. Soc. Lond. B.* **264:**439–478.

Williams, A., and Hurst, J. M., 1977, Brachiopod evolution, in: *Patterns of Evolutions, As Illustrated by the Fossil Record* (A. Hallam, ed.), pp. 79–121, Elsevier, Amsterdam.

Williams, A., and McKay, S., 1979, Differentiation of the brachiopod periostracum, *Palaeontology* **22:**721–736.

Wright, C. W., and Kennedy, W. J., 1979, Origin and evolution of the Cretaceous micromorph ammonite family Flickiidae, *Palaeontology* **22:**685–704.

Wright, C. W., and Kennedy, W. J., 1980, Origin, evolution and systematics of the dwarf acanthoceratid *Protacanthoceras* Spath, 1923 (Cretaceous Ammonoidea), *Bull. Br. Mus. Nat. Hist. (Geol.)* **34:**65–107.

Chapter 16

Heterochrony in Evolution

An Overview

MICHAEL L. McKINNEY

1. Introduction

The chapters in this book essentially form a collective argument for the abundance of heterochrony in evolution. In a sense, this is hardly surprising, since changing the rate or time at or during which a trait grows is certainly the simplest way to alter form (McKinney, this volume, Chapter 2). Why then all the ruckus? As Gould (this volume) has noted, there are a number of reasons, not the least of which is that a growing knowledge of developmental mechanisms is allowing us to define more precisely the link between ontogeny and evolution. But how far will heterochrony take us? At minimum, it may have little impact on the Darwinian view. These abundant cases could amount to little more than hanging polysyllabic names on the obvious mechanics of growth (larger traits have grown faster or for a longer time, so what?). However, at maximum, it may turn out that there are

MICHAEL L. McKINNEY • Department of Geological Sciences, and Graduate Program in Ecology, University of Tennessee, Knoxville, Tennessee 37996-1410.

important ramifications not assimilated into the traditional view of evolution. Among the most important of these are previously unrealized heterochronic effects on rate and direction of evolution.

In the textbooks at least, the Darwinian (or Modern Synthetic) view that evolution is generally slow and externally steered is often counterposed against the saltationist view that it is rapid and internally steered. But is this polarized view of effects on rate and direction justified? In this chapter, I say no, based upon two aspects of heterochrony that have not been fully considered in past formulations of this dichotomy: (1) heterochrony can indeed be a process of rapid, drastic change, but it can also act only locally and incrementally, and (2) even where saltations occur, the products are not "random monsters," as they are often depicted; even profound heterochronic changes occur along an axis of trait covariation such that traits are already highly integrated internally and highly functional externally. Most importantly, this axis of traits is often already aligned with an environmental axis: heterochronic extrapolations of ancestral ontogeny are already "aimed" at adaptive peaks extrapolated from ancestral environments.

In sum, I argue that, by the first aspect of heterochrony just mentioned, gradations of rate can occur, from fast to slow; and by the second aspect, even where internal forces do all of the creating, all (or, in many cases, most—again a gradation) of the morphology is usually attributable to previously effected environmental demands and not developmental by-products. Thus, heterochrony may often (perhaps nearly always) operate in the traditional Darwinian mode of gradual, externally directed change. However, the key point is that, in terms of developmental capability, the door is open to rapid, internally driven change. Whether or not this latter occurs often or at all must await further empirical data.

These points are developed in the sections that follow. First is a brief discussion to clarify the meaning of external vs. internal direction, followed by a section each on the effects of heterochrony on the rate of evolution, the direction of evolution, and macroevolution. This last briefly discusses such issues as species selection and origin of higher taxa. There is nothing seminal in these sections. My goal is simply to digest and synthesize, to provide a theoretical framework for further work.

2. A Hierarchy of Depth and Breadth

2.1. Introduction

In an ideal universe, the adaptive changes required by organisms could be created *ex nihilo*, shaping whatever was needed with no dependence on time or space. Life being the contingent series it is, however, evolution is instead limited in its change by how much time has passed and what existed at the start. [In Jacob's (1977) analogy, this use of preexisting parts is "tinkering"; Frazzetta (1975) pointed out that there is yet a further limitation: the tinkering must be carried out while the "machine is running," i.e., the organism must be viable at all times.]

To minimize these constraints, an organism must be able to change some of its components as rapidly as possible while, inevitably, leaving some unaltered, especially those that work best as they are. Furthermore, these changes must permit the components to function as an integrated unit when finished. To carry this out, a hierarchy of breadth (number of parts affected) and depth (degree of effect) is required. As noted in Chapter 2, heterochrony operates in a hierarchy of breadth and depth, permitting biological change to minimize the constraints of contingency (and Frazzetta's constraint as well; by changing only the rate and timing, minimum disruption of function and integration occurs). It is at the extreme ends of the hierarchy that we encounter the distinction between internal and external creativity.

Consider the analogy of a builder (or more familiarly, a watchmaker), who, like natural selection, constructs a functional unit from preexisting parts. If he is supplied with many components of unlimited size and shape from which he can choose (i.e., small breadth of change possible), the resultant construct will be finely tuned and every part will be of optimal form. This is essentially the neo-Darwinian view, whereby selection, unconstrained by development, which supplies most everything needed, is the creator. However, this is a slow and inefficient process, since each time a part is altered, cascading effects necessitate changes elsewhere, which often limits depth of change (on the original part). In evolution, selection would thus not only have to wait for a mutation to alter one part, but one for each affected part as well.

As noted by Waddington (1966), this "piecemeal" problem is overcome by changes that alter whole integrated systems instead of each part. In the extreme case of this, the builder (selection) would be supplied with only a few "prefabricated" units wherein the components are already positioned and integrated (i.e., large breadth and depth of change possible). This greatly reduces the creative input of selection, which simply deploys these larger units, and increases the influence of the supplier (development). However, it is also much faster and more efficient *if* the integrated units already meet specifications required by the construct to function well. The main drawback is that ever-occurring environmental changes would mean that specifications could often change. How long will it take the supplier to alter the components? Until such changes are made, the "prefabricated" units will contain some suboptimal aspects. To the evolutionist, such "locked-in" covariation of parts is a major form of developmental "constraint" on evolution.

Thus, the influence of development on evolution varies depending on how extensive (breadth) and intensive (depth) change can be. The above-mentioned heterochronic hierarchy being as flexible as it is (Chapter 2), the input of development on evolution will vary. Where environmental change is gradual and only adaptive "fine-tuning" occurs, then of course evolution is externally directed. However, the ease and effectiveness of rapid, global changes lead to the possibility of significant internal influence. While most of the resultant morphology would likely be environmentally adaptive (see below), there is a much greater chance here for allometric extrapolations and other types of "unintended" covariations to lead to suboptimal traits, or, with adaptively neutral traits, to show purely developmental effects.

More important than "who's in control" is the lability inherent in hierarchies. Rapid, large-scale change is possible where it is advantageous (to reach adaptive peaks otherwise unattainable or just more quickly), but more often what is required are simple adjustments to adaptive peaks.

2.2. Hierarchy As By-Product?

The importance of integrated change to evolution (e.g., in solving the "incipient organ" problem) emphasized by Waddington (1966) and others (e.g., Riedl, 1978) often seems to imply that such a hierarchy has been "selected for." However, I would argue that it exists primarily as a by-product (as opposed to cause) of the hierarchical nature of life itself. Any assembly of smaller units into larger ones necessarily results in a series of contingent events with progressively less general effects, since a perturbation can occur at many points in the series and with many magnitudes.

This in turn leads to a hierarchy of selection along the spectrum, from parts to individuals. Internal selection acts to keep the changes interwoven with developmental processes, while external selection acts to keep them functional in the environment (Cheverud, 1984; Riska, 1986). These selection pressures are what drives the continual forming (covariation), breaking up (dissociation), and re-forming of trait associations.

3. Heterochrony and Rates of Evolution

3.1. Slow Rates

The genetic bases of timing and rate changes in ontogeny are obviously variable, though largely unknown (Raff and Kaufman, 1983). As noted in Chapter 2, the changes may be either polygenic or monogenic.

Where rate and/or timing is polygenic, it is fairly easy to apply the mechanics of orthodox Mendelian cumulative change to heterochrony (Slatkin, 1987). Mild mutations in such genes can be selected for in "bean bag" fashion and gradually spread throughout the population, resulting either in anagenetic or cladogenetic change, depending on the circumstances. Indeed, this point has been made by evolutionists of neo-Darwinian bent (Mayr, 1982; Grant, 1985). For instance, Mayr in his strong advocacy of speciation via gradual population processes, noted (Mayr, 1982, p. 618) that "undoubtedly regulatory genes are participating in these changes or are largely responsible for them." Quite possibly, most heterochronically induced species divergence occurs in this way. The dissociations of timing and rate changes that separate species would not happen at once, but accumulate through ecological time scales (ostensibly via peripheral isolation), appearing to be abrupt only in the record. Thus, in Fig. 1, the ontogenetic changes viewed between ancestral and descendant species would be derived through gradual environmental selection, with intermediate forms present.

3.2. Rapid Rates

Rapid morphological changes (Fig. 1) are more likely to result from relatively simple changes involving mutations in one or a few genes (McKinney, this volume, Chapter 2; Hall, 1984; Ambros, this volume). Because of trait dissociation,

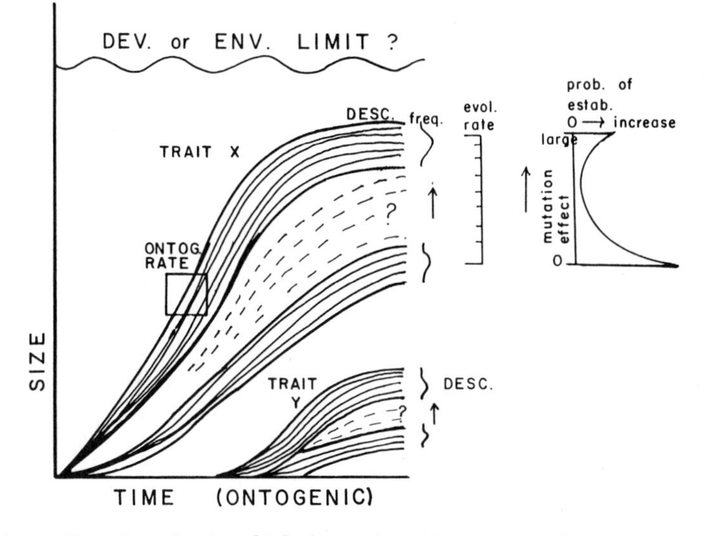

Figure 1. Ontogenetic trajectories in which descendant (DESC) is accelerated (see McKinney, this volume, Chapter 2) for both traits relative to ancestor. Traits are assumed to be normally distributed in both species; dashed line = possible intermediates. Probability of establishment may increase after a certain amount of acceleration due to niche separation, as discussed in the text.

the rapid change may affect only one or a few traits. However, if the change is extreme, problems of developmental and functional integration with the unaffected parts of the organism can result (Riedl, 1978; Cheverud, 1984). For this reason, high rates of evolution (large depth) are usually correlated among a number of traits (large breadth).

4. Heterochrony and the Direction of Evolution

For cumulative processes of incremental heterochronies, it is clear that environmental effects both move and steer evolution. However, in the case of rapid heterochrony involving one or a few genes, the case is less clear. More importantly, while the discussion above implies that radical ontogenetic change can occur in one generation, the real problem with saltation has always been in establishing such mutations in a population, not in having them. This in turn has much bearing on the direction of evolution.

4.1. Establishing Monsters

The establishment of a mutation is largely a function of three factors: whether the resulting organism is viable (physiologically functional), is competitively equal or superior, and can transmit the gene(s) to future generations. The chances of viability are greatly enhanced by the use of preexisting pathways. Indeed, as noted, changes that involve many traits are often better for their developmental and functional integration.

Regarding competition, larger mutations may again have an advantage over smaller ones, as argued perhaps most persuasively by Arthur (1984). As shown in Fig. 1, larger mutations could sometimes move the affected organism into a sufficiently different lifestyle, eliminating or minimizing competition (see also Pimm, 1986). While body size changes are the best known for causing niche separation [for review see Werner and Gilliam (1984)], drastic changes in restricted growth fields, e.g., food-acquiring organs, could have similar effects.

Finally, there is the question of "who does the monster mate with"? Again, Arthur (1984) discusses the possible answers. Among the most credible is that the mutation(s) could be sex-linked or recessive, effectively hiding gene expression until it has been passed along to a number of individuals. Also possible is the nonrandom occurrence of many such mutations, leading to its independent appearance in more than one individual (Alberch, 1982).

4.2. Internal Motor but Not Steering

For these reasons, saltations may occur and, in that sense, could "drive" evolution. However, few evolutionists doubt that organisms are well adapted to their environment and the fact remains that even if internal factors are pushing, to be successful they must push in the direction that external factors demand if environmental controls are strong. In Fig. 2 we see a case where a trait (e.g., body size) is changing in size, following a series of adaptive environmental peaks. Whether the increase is "pulled" by cumulative external shaping or "pushed" by internal forces makes little difference.

These peaks are often arrayed along environmental gradients, as documented so well by McNamara (1982, 1986, and this volume). For body size, I have compiled a number of environmental factors that may control such gradients (Table I). Obviously many traits will often covary along this gradient. Figure 3A illustrates a simple cline of "allometric hypermorphosis" resulting from the simultaneous acceleration of traits in Fig. 1. McNamara (1982) has termed these heterochronic-environmental gradients "paedo- and peramorphoclines." The finely tuned functional nature of the heterochrony in these clines (McNamara, 1982, 1986, and this volume) provides clear empirical evidence that the directionality is environmentally determined. Traits that covary along these gradients often, but not always, include body size (McNamara, this volume). A well-documented morphocline is seen in Fig. 4. Note that in this case intermediates are found in at least one transition. These developmental and functional suites are not limited to morphology, but may include life history (fecundity, longevity, etc.) and behavioral traits (e.g., Gould, 1977; Balon, 1981; McKinney, 1986; Breven, 1987).

Suggestions that such trends represent mainly developmentally directed evolution (Jablonski *et al.*, 1986) seem improbable in the face of such close morphological–environmental associations. These certainly are migrations up and down ontogenetic trajectories, but they are tracking environmental trajectories to which they are coadapted to begin with.

Is there then anything qualitatively different about such internal "pushes"? For one thing, it gets the affected organisms to the adaptive peak faster. For an-

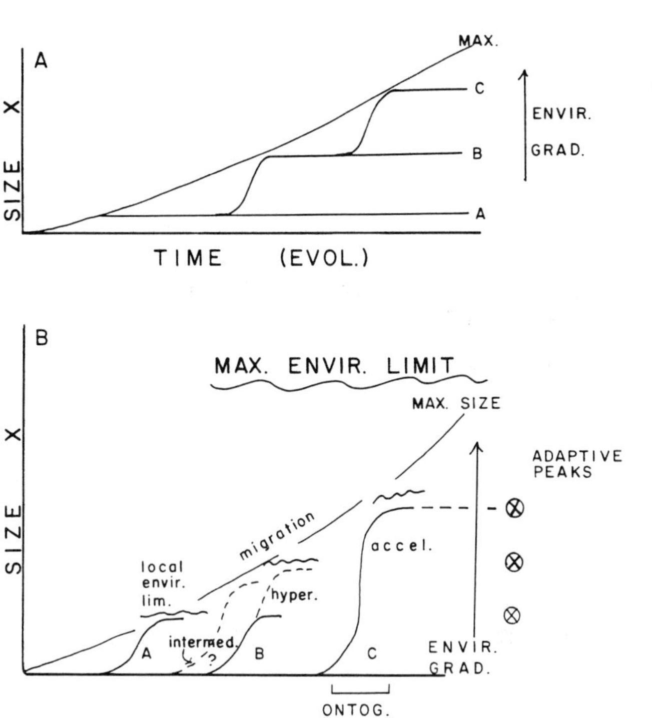

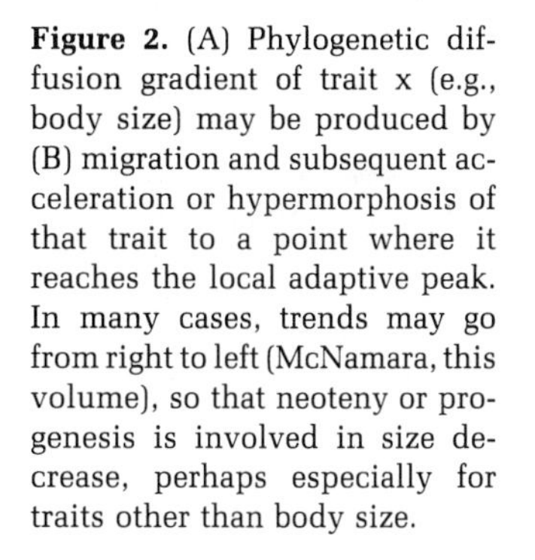

Figure 2. (A) Phylogenetic diffusion gradient of trait x (e.g., body size) may be produced by (B) migration and subsequent acceleration or hypermorphosis of that trait to a point where it reaches the local adaptive peak. In many cases, trends may go from right to left (McNamara, this volume), so that neoteny or progenesis is involved in size decrease, perhaps especially for traits other than body size.

other, it may be that in some cases, external selection would not get them there at all, since they could not get across the adaptive valley between the peaks, to continue Wright's (1932) landscape analogy. As the ontogenetic–environmental axis is already pointed in that direction, the saltation essentially propels the organism across the valley.

Table I. Environmental/Adaptive Correlates Fostering Larger Size, in Crudely Ascending Order of Importance[a]

Correlate	Group
Regularly abundant food	All
Cooler temperatures	Many vertebrates
Warmer temperatures	Reptiles
Prey size	Predators
Seasonality	All
Predation	Marine organisms
Brooding	Marine organisms
Postarboreality	Vertebrates
Water density	Planktonic foraminifera
Water temperature	Benthic foraminifera

[a] Based on a compilation of sources (McKinney, in preparation).

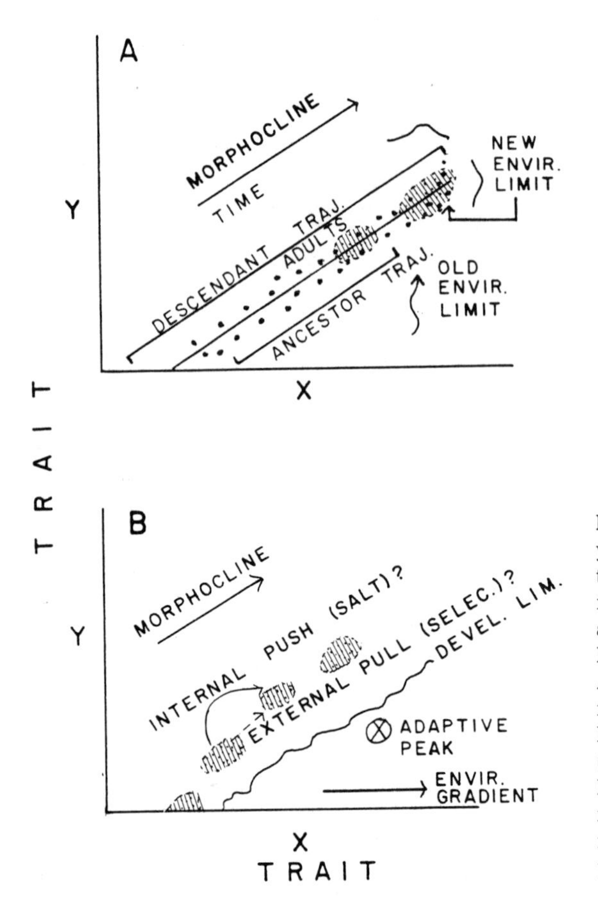

Figure 3. (A) Morphocline of two covarying traits. Ontogenetic trajectories show that the descendant species is "allometrically hypermorphic," being produced by acceleration of x and y (Fig. 1) (see McKinney, this volume, Chapter 2). Adult specimens (ovals) are bivariate normal in distribution of two traits. (B) Depiction of a series of related species (adults only) along the same trajectory. How often do such "locked-in" trajectories prevent a species from reaching the morphology most useful for an adaptive peak?

4.3. Developmental Effects

Due to this dominance of environmental control, development is usually mentioned as a constraint on selection, as opposed to the reverse (Alberch, 1982, 1985). Thus, although development may not actively direct evolution, wherever it does not produce the raw material necessary for optimal design, it has an important effect. As noted above, a major cause of this disparity between what is usable and what is producible is the necessity of integration among parts. While dissociation can eliminate this disparity, we have no idea how "easy" or common dissociation is in many trait suites.

Where dissociations are "difficult" or slow in coming, traits may be evolutionarily carried along solely by virtue of their developmental associations. In functional traits, this may result in suboptimal states; in neutral traits, they are simply there [e.g., human hairlessness? (Gould, 1977)]. The classic evolutionary examples of heterochrony, such as the "Irish Elk" (Gould, 1974) and titanotheres (McKinney and Schoch, 1985), are thought to involve covarying traits tagging along with body size. Whether the developmental "byproducts" of antler and horn size are fully adaptive or suboptimal is unknown.

This matter of developmental constraint is illustrated in Fig. 3B, where a

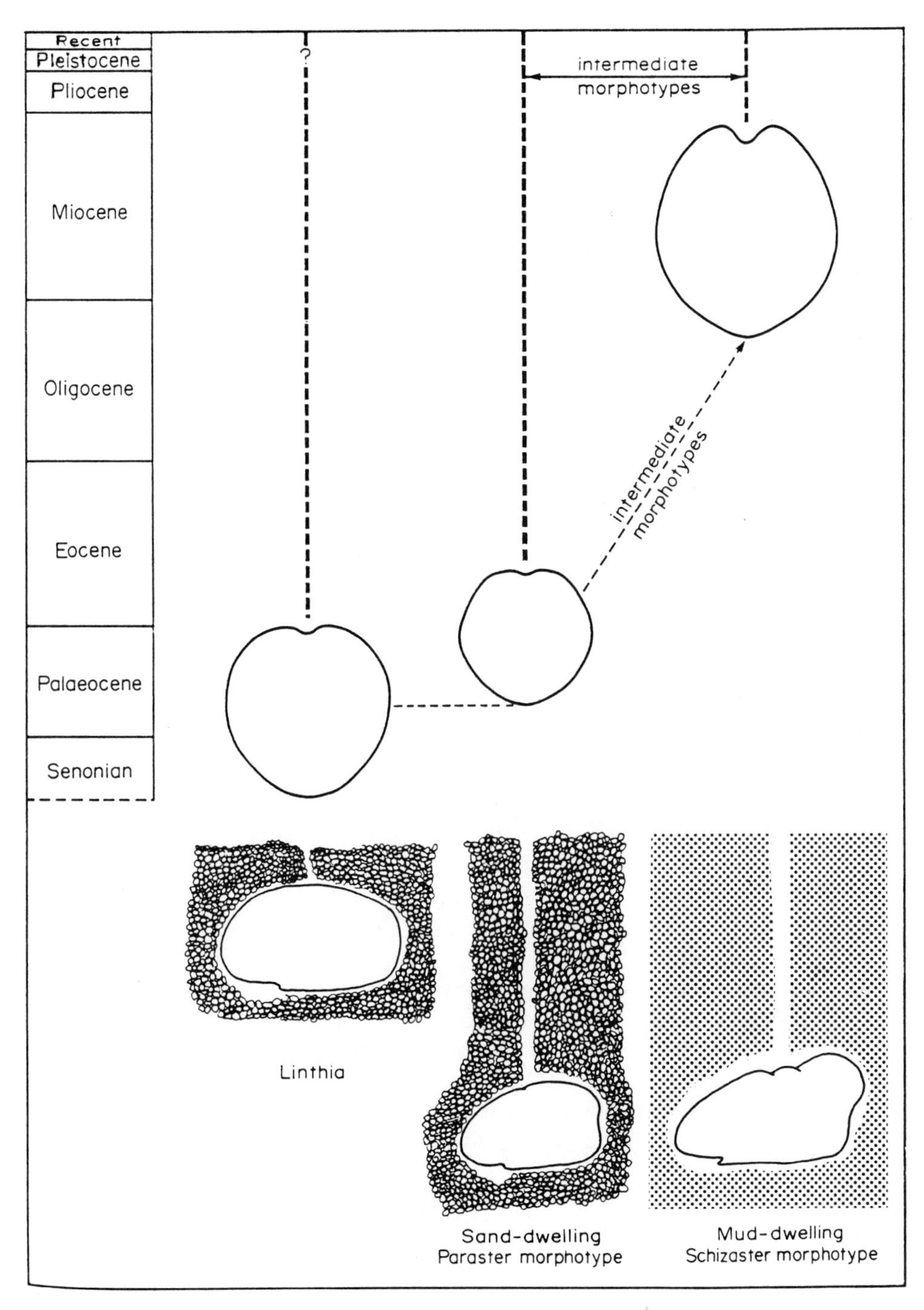

Figure 4. A well-documented morphocline. Modified from McNamara and Philip (1980).

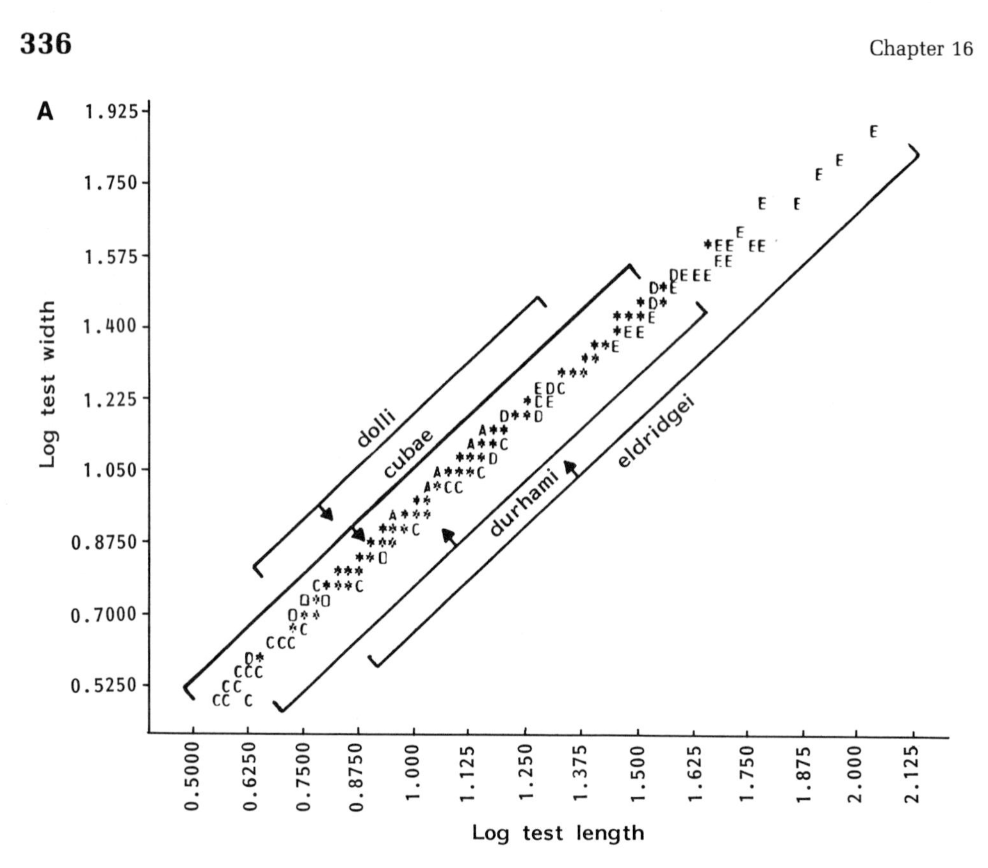

Figure 5. (A) Plot of log test length vs. log test width for 231 individuals from four species of the extinct echinoid family Neolaganidae. Arrows = points of sexual maturation as shown by genital pore development. Letters denote species: C, *cubae*; A, *dalli*; D, *durhami*; E, *eldridgei*.*, overlap. (B) Plot of log test length vs. log test height, showing much greater variation.

tightly constrained trajectory has prevented the morphological optimum most efficient for the adaptive peak. But how often does this occur? As already mentioned, such constrained trajectories are generally "aimed" at the adaptive dimensions to begin with: shape demands with size increase can be fairly regular and predictable from past ontogenetic needs. In addition, there is heterochronic dissociation: even where the trajectory is inappropriate, the morphology can be "fine tuned".

For example, in Fig. 5A length–width dimensions for species of extinct sand dollars are shown. Even though length increases by a factor of 20, there is no significant change in the shape. Yet, for length–height (Fig. 5B) the proportions can vary much more. This is what would be expected from an adaptationist interpretation: test height is more sensitive to environmental changes in sediment and oxygen parameters than is test width (McKinney, in press).

While the non-Darwinian effect of internal processes on evolutionary direction has long-standing appeal (Mayr, 1982), and, we see here, some theoretical justification, data collection on its presence or abundance is greatly impeded by our inability to clearly define and measure a trait's "adaptedness." In cases of simple functional morphology, it may be possible to define and measure an op-

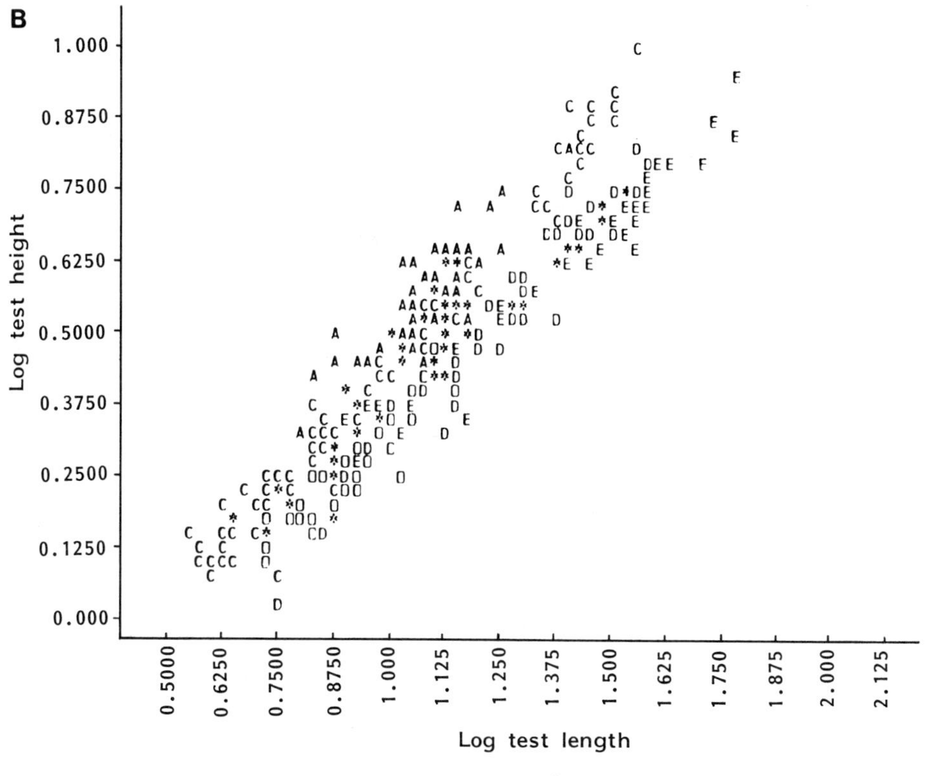

Figure 5. (*continued*)

timum by subtracting an allometric deviation such as in Fig. 3B (McKinney, M. L., in preparation). But who can say if a titanothere horn was optimal for its use at a given body size?

4.4. Summary: Environment–Development Interplay

While we often speak of environment or development as driving or steering evolution, it is of course the interplay between the two that ultimately does the job. Figure 6 summarizes this schematically, pointing out that the developmental limits (what is currently producible) within the morphospace of all conceivable forms sometimes exceeds and sometimes falls short of the environmental limits. The latter limits represent what morphologies could survive in the available environments. Mutations can change the developmental limits. External selection and internal "pushing" allow occupation of adaptive peaks, while environmental limits undergo flux (see caption). As discussed above, the environment is shown to vary predictably along a gradient, as does the corresponding "realizable" morphospace that has evolved along with it or into it. Mutations that expand the developmental envelope into environmentally favorable morphospace are selected for, while those that exceed it are selected out and the envelope therefore shrinks.

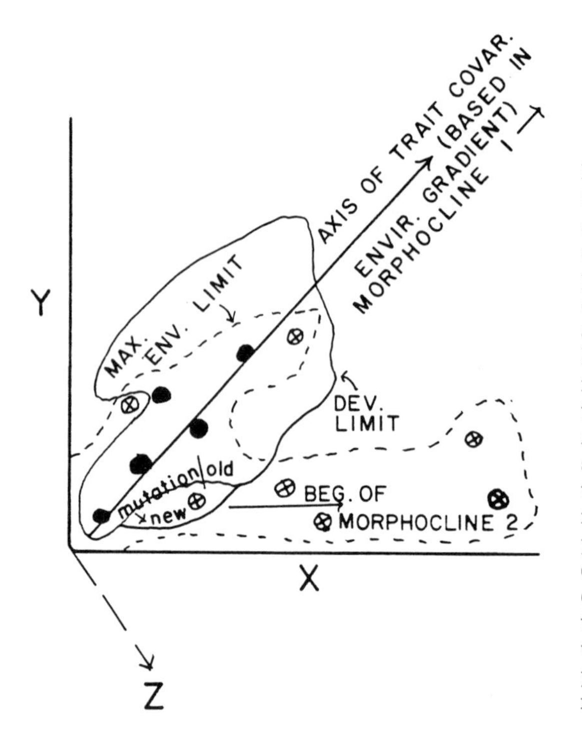

Figure 6. Morphological covariation gradient (here traits x, y, and z) aligned with an environmental gradient. Circles with crosses represent unoccupied adaptive peaks, dark ovals represent adults of species (as in Figs. 3A and 3B) occupying adaptive peaks. Maximum environmental limit (dashed line) encloses that part of morphospace that is viable and competitive in at least some part of the biosphere. As this shrinks or expands through time, species will go extinct or have new opportunities to expand, respectively. Developmental limit (solid line) represents morphologies currently producible by the clade genome. Mutations leading to global or dissociated heterochronies can expand this envelope and can become fixed if within environmental limits. Whether expansion is gradual or rapid is of course a major unknown.

5. Evolutionary Patterns: Conclusions and Prospects

5.1. Trends

From the above discussion, it seems that the existence of trends in evolution can seemingly be well explained by the heterochronic–environmental patterns discussed here. As a kind of biogenic Walther's law, spatial patterns of ontogeny are recast as time-transgressive phylogenetic ones. While the story is not as simple as the traditional environmental orthoselection of the past, the additional developmental information now included does not seriously detract from the externally directed explanation. Rather, it rightly adds the notion of evolution as a change in ontogenies rather than adults.

More importantly, it provides a testable theoretical framework for determining precisely the effects of development on specific phylogenetic patterns, a long-neglected area. For instance, as Alberch (1982) pointed out, the nonrandom morphologies produced by these patterned ontogenetic processes probably account for the common convergent, parallel, and iterative forms of evolution that we see in the record.

5.2. Species Selection

Also of potential import is the impact of development as a species-level property. The most often cited example of species selection is the possession or non-possession of planktotrophic larvae (Jablonski, 1986). Yet, as Freeman (1982) has pointed out, the derivation of nonplanktotrophic larvae via heterochrony is quite simple and probably common. Thus, any tendency toward those heterochronies that cause nonplanktrophy (and not the possession of that mode itself) could be a major factor in clade proliferation.

As another case, the development of canids is much more labile than that of felids (Wayne, 1986). Could this clade-level property not significantly affect rates of species origination independent of individual selection?

5.3. Origin of Higher Taxa

Much of this chapter has addressed the question of saltations. Perhaps the classical accumulation of local heterochronic changes can account for most higher taxa; or perhaps advantages of niche separation, adaptive "aiming," and other factors noted above make small saltations common.

But what about large leaps producing truly monstrous hopefuls? If such saltations ever were successful, it seems likely that it was in the late Precambrian and early Paleozoic. Environmental constraints would have been greatly reduced (*sensu* Fig. 6, an expanded environmental envelope), since most of the ecospace was "empty" (Valentine, 1986). Further, the developmental limits would have been similarly less confining (*sensu* Fig. 6, mutational expansions of the developmental envelope could have been rapid and large), because simpler organisms have fewer interdependent developmental pathways, making radical changes easier and more likely to produce viable organisms (Valentine, 1986; Langridge, 1987). With both constraints lessened, evolution would have been much freer to "experiment," as it seems to have done. Thus, as Arthur (1984) has argued, such large saltations may have led to the origin of major body plans. As the biosphere "barrel" filled up and *Bauplan* complexity increased during the rest of the Phanerozoic, the chances of successful monsters would have progressively diminished.

Only further empirical evidence will tell. The main point here is that on the basis of much new information on heterochrony and developmental mechanisms for change, no possibility can be eliminated *a priori*. The flexibility intrinsic to the hierarchical processes means that both cumulative change through small dissociations and integrated (and hence physiologically viable) large-scale changes are readily accomplished. The empirical data must focus on the other side; how readily are these producible, viable, and "allometrically preadapted" saltations established?

References

Alberch, P., 1982, Developmental constraints in evolutionary processes, in: *Evolution and Development* (J. T. Bonner, ed.) pp. 313–332, Springer-Verlag, Berlin.

Alberch, P., 1985, Developmental constraints: Why St. Bernards often have an extra digit and Poodles never do, *Am. Nat.* **126**:430–433.

Arthur, W., 1984, *Mechanisms of Morphological Evolution*, Wiley, New York.

Balon, E. K., 1981, Saltatory processes and altricial to precocial forms in the ontogeny of fishes, *Am. Zool.* **21**:573–596.

Berven, K. A., 1987, The heritable basis of variation in larval developmental patterns within populations of the wood frog (*Rana sylvatica*), *Evolution* **41**:1088–1097.

Cheverud, J. M., 1984, Quantitative genetics and developmental constraints on evolution by selection, *J. Theor. Biol.* **110**:155–171.

Frazzetta, T., 1975, *Complex Adaptations in Evolving Populations*, Sinauer, Sunderland, Massachusetts.

Freeman, G. L., 1982, What does the comparative study of development tell us about evolution?, in: *Evolution and Development* (J. T. Bonner, ed.), pp. 155–168, Springer-Verlag, Berlin.

Gould, S. J., 1974, The evolutionary significances of 'bizarre' structures: Antler size and skull size in the 'Irish Elk', *Megaloceras gigantans*, *Evolution* **28**:191–220.

Gould, S. J., 1977, *Ontogeny and Phylogeny*, Harvard University Press, Cambridge.

Grant, V., 1985, *The Evolutionary Process*, Columbia University Press, New York.

Hall, B. K., 1984, Developmental processes underlying heterochrony as an evolutionary mechanism, *Can. J. Zool.* **62**:1–7.

Jablonski, D., 1986, Causes and consequences of mass extinctions, in: *Dynamics of Extinction* (D. K. Elliot, ed.), pp. 183–229, Wiley, New York.

Jablonski, D., Gould, S. J., and Raup, D. M., 1986, The nature of the fossil record: A biological perspective, in: *Patterns and Processes in the History of Life* (D. M. Raup and D. Jablonski, eds.), pp. 7–22, Springer-Verlag, Berlin.

Jacob, F., 1977, Evolution and tinkering, *Science* **196**:1161–1166.

Langridge, J., 1987, Old and new theories of evolution, in: *Rates of Evolution* (K. S. Campbell and M. F. Day, eds.), pp. 248–262, Allen & Unwin, Boston.

Mayr, E., 1982, *The Growth of Biological Thought*, Harvard University Press, Cambridge.

McKinney, M. L., 1986, Ecological causation of heterochrony: A test and implications for evolutionary theory, *Paleobiology* **12**:282–289.

McKinney, M. L., in press, Roles of allometry and ecology in echinoid evolution, in: *Echinoderm Phylogeny and Evolutionary Biology* (A. B. Smith and C. R. C. Paul, eds.), Wiley, New York.

McKinney, M. L., and Schoch, R. M., 1985, Titanothere allometry, heterochrony, and biomechanics: Revising an evolutionary classic, *Evolution* **39**:1352–1363.

McNamara, K. J., Philip, G., 1980, Australian Tertiary schizasterid echinoids, *Alcheringa* **4**:47–65.

McNamara, K. J., 1982, Heterochrony and phylogenetic trends, *Paleobiology* **8**:130–142.

McNamara, K. J., 1986, A guide to the nomenclature of heterochrony, *J. Paleontol.* **60**:4–13.

Pimm, S. L., 1986, Filling niches carefully, *Trends Ecol. Evol.* **1**:86–87.

Raff, R. A., and Kaufman, T. C., 1983, *Embryos, Genes, and Evolution*, Macmillan, New York.

Riedl, R., 1978, *Order in Living Systems*, Wiley, New York.

Riska, B., 1986, Some models for development, growth, and morphometric correlation, *Evolution* **40**:1303–1311.

Slatkin, M., 1987, Quantitative genetics of heterochrony, *Evolution* **41**:799–811.

Valentine, J. W., 1986, Fossil record of the origin of bauplane and its implications, in: *Patterns and Processes in the History of Life* (D. M. Raup and D. Jablonski, eds.), pp. 209–231, Springer-Verlag, Berlin.

Waddington, C. H., 1986, Fields and gradients, in: *Major Problems in Developmental Biology* (M. Locke, ed.) pp. 105–124, Academic Press, New York.

Wayne, R. K., 1986, Cranial morphology of domestic and wild canids: The influence of development on morphological change, *Evolution* **40**:243–261.

Werner, E. E., and Gilliam, J. F., 1984, The ontogenetic niche and species interactions in size-structured populations, *Annu. Rev. Ecol. Syst.* **15**:393–425.

Wright, S., 1932, The roles of mutation, inbreeding, cross-breeding, and selection in evolution, *Proc. XI Int. Cong. Genet.* **1**:356–366.

Index